UNIVERSAL FOOD SECURITY

UNIVERSAL FOOD SECURITY

How to End Hunger While Protecting the Planet

GLENN DENNING

Columbia University Press

New York

Columbia University Press
Publishers Since 1893
New York Chichester, West Sussex
cup.columbia.edu

Library of Congress Cataloging-in-Publication Data
Names: Denning, G. L., author.
Title: Universal food security : how to end hunger while protecting the planet / Glenn Denning.
Description: New York : Columbia University Press, [2022] |
Includes bibliographical references and index.
Identifiers: LCCN 2022019327 (print) | LCCN 2022019328 (ebook) |
ISBN 9780231197601 (hardback) | ISBN 9780231197618 (trade paperback) |
ISBN 9780231552257 (ebook)
Subjects: LCSH: Food security. | Food supply—Forecasting. |
Food industry and trade. | Food consumption.
Classification: LCC HD9000.5.D4368 2022 (print) | LCC HD9000.5 (ebook) |
DDC 338.1/9—dc23/eng/20220524
LC record available at https://lccn.loc.gov/2022019327
LC ebook record available at https://lccn.loc.gov/2022019328

Columbia University Press books are printed
on permanent and durable acid-free paper.
Printed and bound by CPI Group (UK) Ltd, Croydon, CR0 4YY

COVER DESIGN: Milenda Nan Ok Lee
COVER IMAGE: Tuul & Bruno Morandi/Getty Images

To Pam

To Paolo, Kinlay, and Camille

And to Silverio

CONTENTS

PART III: STRATEGY

PART IV: IMPLEMENTATION

PREFACE

Imagine a world where every person enjoys a healthy diet derived from sustainable food systems. Seriously. Take a moment to visualize a genuinely food-secure world—where everyone gets to consume the right quantity and quality of food to live a healthy, active, and productive life. No hunger. No obesity. No malnutrition. Or in the words of Donella Meadows and her colleagues from their ground-breaking report *The Limits to Growth* (1972), imagine a world where "each person has an equal opportunity to realize [their] individual human potential."[1] Because of what they eat!

Now, take a further step to imagine a world where the food needed for those healthy diets is produced, managed, and distributed in ways that do not harm the environment. From production to consumption, let us contemplate the possibility of food systems that halt environmental degradation and even begin to repair the damage that has been done in our desperate efforts to feed humanity.

Healthy diets for all, from sustainable food systems.[2] I call this *universal food security.*

I have chosen the term *universal* over the more commonly used *global* to establish the aspiration to reach every person on Earth. After all, food security is a human right, embodied in the Universal Declaration of Human Rights in 1948 and elaborated in 1966 in the International Covenant on Economic, Social, and Cultural Rights.[3] In arriving at this concept of universal food security, I am indebted to the visionaries who crafted the Rome Declaration on World Food Security that emerged from the World Food Summit of 1996. They established the ambition of universal food security in stressing access to food for "all people, at all times."[4]

Nine years earlier the World Commission on Environment and Development, chaired by Norwegian prime minister Gro Harlem Brundtland, highlighted the issue of intergenerational equity by defining sustainable development as "development that meets the needs of the present without compromising the ability of future generations to meet their own needs."[5] These ambitious global declarations provided the moral foundation for the 2030 Agenda for Sustainable Development, a collective pledge made by the world's governments in September 2015 that "no one will be left behind."[6] Six years later, in synthesizing the outcomes of the Food Systems Summit of 2021, United Nations (UN) secretary-general António Guterres reiterated the commitments embodied in the 2030 Agenda and the Sustainable Development Goals, declaring, "We do not need new goals; we need to move boldly—now—to implement the transformative actions needed to achieve the goals we have."[7]

This book shows that we have the *know-how* and the *do-how* to achieve universal food security. My objective is to introduce readers to the science, policies, and practices that will need to be understood, contextualized, adapted, and—most important—*applied* across our hungry and damaged planet. It is a shameful fact that about half of the world's more than eight billion people are malnourished. At least two billion people are experiencing undernutrition—deficiencies of one kind or another that adversely affect the human body's growth and development. There are also more than two and a half billion people who consume low-quality diets or too much food, leading to excessive weight gain and a raft of chronic ailments like diabetes and heart disease.[8] The United Nations food agencies estimated that three billion people could not afford a healthy diet in 2019.[9] And that was prior to the COVID-19 pandemic and its impacts. Unhealthy diets come at great cost to society by curtailing the realization of human potential.

By 2050 the world population will be close to ten billion. With current trajectories of prevalence in malnutrition—a combination of slow progress in lowering undernutrition and essentially no progress in reversing overweight and obesity trends—coupled with land degradation, water resource depletion, biodiversity loss, and climate change, we are headed toward a food systems apocalypse. With that grim outlook, why would I have the audacity to suggest that an alternative future is possible?

First, to quote Isaac Newton, "If I have seen further it is by standing on the shoulders of Giants."[10] I have had the privilege to access literature, expertise, and experience from food systems "giants" of the past and present. This book is my interpretation of the science and practice from hundreds of leading authorities from diverse disciplines in the related fields of agriculture, food, and nutrition.

Second, I am now and always have been *a practitioner*. In all my professional roles over four decades, I have focused on putting good ideas into practice. With the support of colleagues and perhaps more than my fair share of serendipity, I have had enough positive experiences (and learned sufficiently from my disappointments) to believe that the future is, to a large extent, what we decide to create. I consider myself a pragmatic optimist.

Third, over the past thirteen years or so, I have also been a teacher: a professor of practice. As founding director of Columbia University's Master of Public Administration in Development Practice (MPA-DP), I have been given the opportunity to teach, mentor, and learn from more than five hundred young professionals from eighty countries; 70 percent of my students are women. Much more than hope, this experience has given me confidence—a genuine belief that transformation of our food systems is urgently needed, that universal food security is a moral imperative, and that both are attainable by 2050.

With these three factors working to my advantage, I decided to write this book. In exploring the rather messy and sometimes divisive worlds of science and practice, I have sought the truth. On matters of science, I have relied heavily on the peer-reviewed literature in agriculture and food systems. In recent years, there has been a plethora of high-level reports and scientific journal publications emerging from prominent commissions, task forces, and institutions. On matters of both science and practice, I have also sought expert opinion, drawing on colleagues, mentors, and other professionals with real-world experience in global, national, and local food systems. From these sources and hundreds of peer-reviewed research papers, I have attempted to interpret and synthesize the state of science and practice in most areas of relevance to food systems transformation and food security.

In assembling the case for universal food security, I have also drawn extensively on my personal experiences. I started out as an inexperienced twenty-two-year-old undertaking field research on the island of Bali for my master's degree with the University of Queensland. From there, I spent three years in conflict-affected Mindanao in the southern Philippines, working as an agronomist on a large rural development project. Then followed eighteen years with the International Rice Research Institute (IRRI), also in the Philippines, where the Asian Green Revolution had been launched more than a decade earlier. While at IRRI, I carved out time for a PhD degree from the University of Reading and an MPA degree from the Harvard Kennedy School. After IRRI, I moved to Africa, serving for six years with the World Agroforestry Centre in Kenya, followed by a further five years in Kenya, establishing and leading the Millennium Development Goals

Centre for East and Southern Africa. Finally (perhaps), in 2009 I joined Columbia University in New York City. It has been an exciting, often challenging, and mostly rewarding journey in food systems transformation, with a strong focus on agriculture and rural development.

Throughout the book, I share experience and perspectives gained from the many programs and projects in which I played a significant role or that I observed at close range. I have met with many thousands of farmers and hundreds of scientists and practitioners from governments, businesses, and international organizations in more than fifty countries across all regions of the world, though mainly in Asia and Africa. You will find anecdotes from my travels, events, and conversations. I use these vignettes to help introduce, contextualize, personalize, and illustrate the challenges and opportunities for achieving food systems transformation.

My opinions and biases will occasionally surface. I make every effort to ensure a balanced, transparent, evidence-based presentation of the issues. I am not one for the false dichotomies that emerge from being either for or against a particular policy or practice. In most cases, I have been drawn to a middle ground or a blended solution. Science, context, needs, and rights—not ideology—must shape our decisions on policies and practices. But ultimately, I accept that this book is *my* take on what is required to achieve a food-secure world.

The book is organized into four sections: context, knowledge, strategy, and implementation.

In part I, I present the historical context of today's global food system. From Malthus's essay on population in 1798 to the Food Systems Summit of 2021, there have been regular calls to anticipate and confront the challenges of food insecurity, hunger, and malnutrition. *Prophets of doom* (chapter 1), prompted by recurring food crises, have served an important purpose: they have challenged us to take action on population, food supply, and the environment. In response, there have been periods of success and subsequent optimism, like the Asian Green Revolution (chapter 2). The agricultural productivity and food security gains in Asia during the second half of the twentieth century provided a forceful rebuttal to the Malthusian hypothesis. Poverty, hunger, and land conversion to agriculture were all reduced. But a balanced assessment of the Asian Green Revolution reveals some important limitations and caveats and has spawned efforts to learn and adapt from our experience. In chapter 2 I also present the largely unrecognized progress of the more recent *African* Green Revolution where, contrary to popular narratives, cereal production growth modestly outperformed that in Asia during the fifteen years following their respective launch dates.

To bring meaningful improvements to our food systems, it is necessary to grasp the underlying science, policies, and practices of agriculture, food, and nutrition.

For readers without formal training in these areas, it can be a challenge to navigate the language of food systems. Personally, as an agronomist, I had to work especially hard to build basic competency in the field of human nutrition, and it remains work in progress. I may have understood the nutritional needs of rice plants, but I knew little of what constituted a healthy, nutritious diet. We all need to fill gaps in our knowledge to be effective *food systems* practitioners.

To that end, part II introduces the fundamental components of most terrestrial food systems: soil and land (chapter 3), water (chapter 4), seeds (chapter 5), climate (chapter 6), and human nutrition (chapter 7). In chapter 8 I provide a holistic framework for integrating these diverse but interrelated topics into a food systems transformation strategy, including an overview of structural transformation—the ubiquitous reallocation of economic activity across agriculture, industry, and services. Inevitably, there is no escaping jargon and acronyms as you venture into these areas of technical knowledge and field practice. I try to keep these insider terms to the minimum necessary to communicate effectively in a diversity of professional settings. Becoming familiar with the language and concepts of food systems will be greatly beneficial when it comes to designing and executing policies and programs.

Part III lays out my strategy for transforming food systems to achieve the goal of universal food security. I have been refining this strategy in earnest since early 2014, when I was invited to make a presentation on sustainable agriculture, food security, and nutrition to the Open Working Group of the United Nations General Assembly in the lead-up to the Sustainable Development Summit of 2015.[11] The strategy I present here, incorporating feedback gained from numerous public lectures and panels, consists of five interrelated, mutually reinforcing themes of food systems investment: sustainable intensification (chapter 9), market infrastructure (chapter 10), postharvest stewardship (chapter 11), healthy diets (chapter 12), and social protection (chapter 13). In each of these themes, I describe policies and programs to incentivize and support positive change. I call these investment areas the *Big Five*.

These five investment clusters are not meant to exclude other important and complementary areas for food systems transformation. I know many colleagues who will bristle at the absence of water, sanitation, and hygiene (known by the acronym WASH) from my Big Five, considering the importance of these elements in effective food utilization and overall health. Others would like to see women's empowerment or good governance or urban development take their place among the highest-priority investment areas.

There is no question that these additional themes (and others) are important, even critical, components of a comprehensive strategy to attain universal food security. I have addressed these complementary issues at different points throughout the book. After much deliberation, however, I concluded that, if I was forced to name just five priority areas of investment, my Big Five are a sound, strategic, and practical place to start.

In addition to discussing the Big Five individually, I have included—in part III—a chapter on COVID-19 and food security (chapter 14). The fragilities and inequalities of our food systems have been exposed by the pandemic and most recently by the Russian invasion of Ukraine. We must all learn from these experiences to be better prepared to withstand the next pandemic and other shocks that can disrupt our strategies for achieving universal food security.

In part IV I ask the all-important question: So what? How will we implement the Big Five strategy? In chapter 15 I reiterate the value of beginning with a clear vision of success, namely, universal food security. I reinforce the importance of a context-determined strategy of transformative investments: the Big Five. And I highlight the role of *agents of transformation* and call for a *whole-of-society approach*, emphasizing the importance of coherent contributions from the public, private, and third (not-for-profit) sectors, as well as from universities and the public at large. I propose institutional innovations and reforms in the way we address food security at the international, national, and local levels.

In the final chapter I focus on the pivotal role of transformative leadership (chapter 16). Inspired by what I have seen at Columbia's MPA-DP program over more than a decade, I suggest that universities and other higher education institutions can and should step up and help ignite food systems transformations. They can do so by investing in the education and development of *practitioner-leaders*: a cadre of professionals who will directly and indirectly influence the actions and impacts of the Big Five investment areas.

To write a book invites scrutiny and criticism. I welcome the feedback and advice of readers, especially those agents of transformation on the front lines of science, policy, and practice. We can all make a commitment to learn, reflect, share, collaborate, act, and move constructively toward a common goal of universal food security.

ABBREVIATIONS

ADB	Asian Development Bank
AE	agronomic efficiency
AfDB	African Development Bank
AFOLU	Agriculture, Forestry, and Other Land Use
AGRA	Alliance for a Green Revolution in Africa
ATP	adenosine triphosphate
BMGF	Bill and Melinda Gates Foundation
BMI	body mass index
BRI	Belt and Road Initiative
Bt	*Bacillus thuringiensis*
CA	Comprehensive Assessment
CAADP	Comprehensive Africa Agriculture Development Programme
CALA	Centre for African Leaders in Agriculture
CARDI	Cambodian Agricultural Research and Development Institute
CCAFS	Climate Change, Agriculture, and Food Security
CCT	conditional cash transfer
CEC	cation exchange capacity
CFS	Committee on World Food Security
CGIAR	Consultative Group on International Agricultural Research
CIAP	Cambodia–IRRI–Australia Project
CIMMYT	International Maize and Wheat Improvement Center (Centro Internacional de Mejoramiento de Maíz y Trigo)
CSA	climate-smart agriculture
DHS	Demographic and Health Survey
ENSO	El Niño–Southern Oscillation

EPA	Environmental Protection Agency
FAO	Food and Agriculture Organization
FBS	Food Balance Sheet
FCC	Functional Capability Classification
FDPIR	Food Distribution Program on Indian Reservations
FEWS NET	Famine Early Warning Systems Network
FFA	Food for Assets
FFPI	FAO Food Price Index
FIES	Food Insecurity Experience Scale
FISP	Farm Input Subsidy Program
FNG	Fill the Nutrient Gap
FSIN	Food Security Information Network
GAFSP	Global Agriculture and Food Security Program
GAIN	Global Alliance for Improved Nutrition
GDP	gross domestic product
GE	genetic engineering
GHG	greenhouse gas
GMO	genetically modified organism
GNP	gross national product
GWP	global warming potential
HDDS	Household Dietary Diversity Score
HLPE	High Level Panel of Experts on Food Security and Nutrition
HLTF	High-Level Task Force on Global Food and Nutrition Security
HYV	high-yielding variety
ICARDA	International Center for Agricultural Research in the Dry Areas
ICRISAT	International Crops Research Institute for the Semi-Arid Tropics
ICT	information and communication technology
IFAD	International Fund for Agricultural Development
IFI	international financial institution
IFPRI	International Food Policy Research Institute
IFSTAL	Interdisciplinary Food Systems Teaching and Learning program
IPC	Integrated Food Security Phase Classification
IPCC	Intergovernmental Panel on Climate Change
IRRI	International Rice Research Institute
ITC	International *Musa* Germplasm Transit Centre
MAM	moderate acute malnutrition
MDER	Minimum Dietary Energy Requirement

MDG	Millennium Development Goal
MENA	Middle East and North Africa
MPA-DP	Master of Public Administration in Development Practice
MVP	Millennium Villages Project
NARS	national agricultural research systems
NCD	noncommunicable disease
NDC	nationally determined contribution
NGO	nongovernmental organization
NRCS	Natural Resources Conservation Service
NUE	nitrogen use efficiency
OPV	open-pollinated variety
PADAP	Philippines–Australia Development Assistance Programme
PDR	People's Democratic Republic
PNAE	Brazil's National School Feeding Program (Programa Nacional de Alimentação Escolar)
PNB	Philippine National Bank
PoU	prevalence of undernourishment
PSNP	Productive Safety Net Programme
RBA	Rome-based agency
RF	Rockefeller Foundation
RUTF	ready-to-use therapeutic food
SAM	severe acute malnutrition
SDG	Sustainable Development Goal
SDP	Strategic Development Plan
SDSN	Sustainable Development Solutions Network
SI	sustainable intensification
SIPA	School of International and Public Affairs
SMEC	Snowy Mountains Engineering Corporation
SNAP	Supplemental Nutrition Assistance Program
SOC	soil organic carbon
SOM	soil organic matter
SSSA	Soil Science Society of America
t/ha	metric tons per hectare
UN	United Nations
UNDP	United Nations Development Programme
UNEP	United Nations Environment Programme
UNHCR	United Nations Refugee Agency
UNICEF	United Nations Children's Fund
UPLB	University of the Philippines Los Baños

USAID	United States Agency for International Development
USDA	United States Department of Agriculture
VAD	vitamin A deficiency
VC	Viet Cong
VNR	voluntary national review
WASH	water, sanitation, and hygiene
WEF	World Economic Forum
WF	water footprint
WFP	World Food Programme
WHO	World Health Organization
WRAP	Waste and Resources Action Programme
WRB	World Reference Base
WRI	World Resources Institute
WRR	World Resources Report

UNIVERSAL
FOOD SECURITY

PART I

Context

1

PROPHETS OF DOOM

On an appropriately bleak and chilly October afternoon in 2019, I wandered through a tree-lined cemetery in the town of Armagh, Northern Ireland. For the past 170 years this solemn, stone-walled graveyard has quietly accommodated the mortal remains of former residents of the nearby workhouse, where hundreds of men, women, and children perished through malnutrition and disease—all casualties of the Great Irish Famine (1845–1852). Accompanied by my good friend, the Irish American scholar of the famine Dr. Frank Costello, I was imbibing the historical context of a lecture I was about to deliver at the nearby Armagh Robinson Library, founded in 1771 by Archbishop Richard Robinson.

Ambitiously entitled "Achieving a Food-Secure World by 2050," my talk would argue that a world without hunger and malnutrition is entirely possible; we need only act—holistically, coherently, and with a sense of urgency—on the vast body of knowledge and experience at hand. But before the event, I had a few moments to reflect on a national tragedy of historic proportions, one that, even in the mid-nineteenth century, could have been avoided if political leaders had deployed timely policy interventions and complementary humanitarian acts.

The Great Irish Famine graphically illustrated the vulnerability of subsistence farmers to production shocks, a grim reality I have observed repeatedly across Asia and Africa.[1] Potatoes were introduced to Ireland from South America via Spain in the late sixteenth century. The crop rapidly spread and emerged as the preferred staple food of the poor, most of whom were tenant farmers, growing potatoes on small plots of land while also working as laborers on large commercial farms that grew wheat and other cash crops for the market.

By the 1840s most smallholder farmers grew a single favored variety called Lumper, a high-yielding type that performed well on poor land. Beginning in

1845, however, fields of Lumper were devastated by potato blight, caused by a fungus-like pathogen. Potato blight could ravage a crop, especially in damp and mild weather conditions. Across Ireland, production in 1845 was halved; the crop was almost entirely wiped out in 1846 and 1848, and the disease persisted into 1851 and 1852 in some areas.[2]

While earlier weather-related crop failures had occurred in Ireland, the Great Irish Famine is remembered as the most devastating in scale. Although potato blight spread to much of northern Europe, the most severe impact was in Ireland, where potatoes dominated the rural landscape and were prominent on the dinner plates of the poor. The impact of the famine in Ireland was unprecedented: more than one million people died from starvation and malnutrition-induced diseases. More than a million others fled Ireland in search of a better life. The impact of the potato blight was intensified by the narrow genetic base of the staple food, the poverty of the subsistence farmers, and the failure of central government policy to alleviate the crisis, notably at a time when food was being exported from Ireland.[3]

Britain's culpability for the Great Irish Famine has long been debated. Few would argue that London's response was adequate, as it allowed the export of grain and cattle while Ireland's poor starved. In a letter read at an event in Millstreet, County Cork, in 1997 commemorating the famine, then British prime minister Tony Blair acknowledged that those actions and inactions by politicians, whether through neglect or intent, had turned a widespread crop failure into a catastrophic human tragedy: "That one million people died in what was then part of the richest and most powerful nation in the world is something that still causes pain as we reflect on it today."[4]

Of course, famines did not begin or end with the Great Irish Famine. In the twentieth century, famines in the Soviet Union (1921–1922), India (Bengal, 1943), China (1958–1962), Ethiopia (1983–1985), and North Korea (1995–1999), among others, have highlighted the vulnerability of small-scale, predominantly subsistence farmers to the effects of weather, and the failure of governments to act effectively to meet food shortfalls. Then, as now, hunger and famine were not and should not be regarded as inevitable.

1798 AND 1898

Following my talk to a standing-room-only crowd of around thirty staunch supporters of the Armagh Robinson Library, I was invited by the dean of Armagh and keeper of the library, the Very Reverend Gregory Dunstan, to observe the institution's most prized possession: Jonathan Swift's personal copy of *Gulliver's Travels*, complete with notes in Swift's handwriting, apparently correcting his publisher's misguided editing. Fascinating. But I was interested in something else and asked the good Reverend if the library held any works of a fellow clergyman, the influential philosopher and economist Thomas Robert Malthus.

Dunstan hurried away and returned a few minutes later with a pristine fourth edition (volume 1) of *An Essay on the Principle of Population; or a View on Its Past and Present Effects on Human Happiness*. First published anonymously in 1798, these writings of Malthus were to fuel generations of doomsayers through his thesis on what he described as the "natural inequality" of the forces of population and food production.[5] Without constraints, Malthus argued, the human population increases geometrically (exponentially), while food production advances arithmetically (linearly). He warned that "the power of population is so superior to the power of the Earth to produce subsistence for man, that premature death must in some shape or other visit the human race."[6] To Malthus, unchecked exponential population growth would have disastrous consequences and therefore must be the focus of intervention.

Malthus calculated that food production in England would need to double every twenty-five years to meet the demand of a growing population. He rejected as infeasible an alternative proposition whereby farmers could simply expand their cultivated land area by working harder. Such an outcome, Malthus argued, could only be achieved by "ploughing up all the grazing countries and putting an end almost entirely to the use of animal food."[7] If successful, such a scheme would lead the population to adopt a largely vegetarian diet! Malthus further noted that manure from grazing animals was the main source of dressing (fertilizer). Without animals and their excrement, land productivity would further decline. He then asked what happens in the twenty-five years after that, when the population again doubles. At some point there must be no further room for the expansion of croplands.

Malthus maintained that population increase was limited by the means of subsistence, and that unless contained by "extermination, sickly seasons, epidemics, pestilence, and plague," there would be a "gigantic inevitable famine" that "with one mighty blow levels the population with the food of the world."[8] In other words, relentless population growth was a threat to food security, and a catastrophic outcome was a real prospect. Malthus argued for a demand-side response: lowering population growth through birth control, postponement of marriage, and celibacy. He was perhaps the first of the apocalyptic environmentalists. And it is worth noting that the world population in 1798 was around one billion.[9]

One hundred years later, Sir William Crookes, in his presidential address to the British Association for the Advancement of Science, examined the same "inequality," but he believed this dilemma could be best addressed by increasing food supply. By 1898 the world population had grown to 1.6 billion, largely ignoring, it seems, the stern moral advice of Malthus. Crookes, a noted chemist and physicist, urged greater attention to the problem of soil fertility depletion and the resulting decline in wheat productivity, warning his audience of the prospect of starvation by the 1930s. While his motivations were unashamedly racist—with references to "the great Caucasian race" who run the risk of being "squeezed out of existence by races to whom wheaten bread is not the staff of life"—Crookes drew scientific attention to the same dilemma that troubled Malthus, though with an alternative supply-side solution. Without ever using the term *sustainability*, Crookes captured the challenge as follows: "I am constrained to show that our wheat-producing soil is totally unequal to the strain put upon it."[10] More concisely, in today's language: we are exhausting our soils.

In contrast to Malthus, Crookes called for correcting the "natural inequality" through increasing food supply rather than curbing demand. He argued for the intensification of production or, at least, a reversal of nutrient depletion that accompanied continuous cropping on England's once-fertile soils. He identified artificial fixation of nitrogen gas (N_2), which constitutes 78 percent of Earth's atmosphere, as a highly promising solution—"one of the great discoveries awaiting the ingenuity of chemists."[11] The challenge was taken up by Fritz Haber and Carl Bosch, who in subsequent decades developed and commercialized the Haber–Bosch process for producing nitrogen fertilizer, a topic I return to in chapter 3.

MIDCENTURY WARNINGS

The first half of the twentieth century saw sharp increases in food production associated with the expansion of agricultural land, mechanization of farm operations, use of newly bred crop varieties, and increased use of synthetic fertilizers and other chemicals. In 1900 agriculture in the United States was labor intensive, employing about 40 percent of the nation's workforce and deploying twenty-two million horses and mules for draft power. By 1960, 4.7 million tractors were in use, and the number of work animals had dropped sharply to three million.[12] In the early 1930s hybrid maize (corn) was introduced and rapidly adopted by farmers. Between 1933 and 1950 maize yields rose from 1.5 to 2.4 metric tons per hectare (t/ha), as the use of hybrid seed increased from virtually nil to 78 percent of the maize crop. Yields in the U.S. state of Iowa rose from 2.0 to 3.5 t/ha between 1933 and 1943, as hybrid seed adoption increased from 1 percent to 99 percent. These increases were mainly the result of genetic improvements. The additional impacts of synthetic fertilizer and other improved agronomic practices were yet to be fully realized.[13]

In *The State of Food and Agriculture*, a report of the Rome-based United Nations Food and Agriculture Organization (FAO) in 1960, B. R. Sen, the organization's director-general, noted the impressive advances made in agricultural science during the first sixty years of the twentieth century. He reminded readers of FAO's flagship report, however, that these advances were largely confined to North America, Europe, Australia, and New Zealand. Sen observed: "In these regions agricultural yields and productivity have risen rapidly since the [Second World] war, in some countries so fast as to have led to the accumulation of surplus stocks. In the less-developed regions, on the other hand, the increased production so far achieved has come primarily from an enlargement of the cultivated area, and for most products the potential contribution from higher yields and productivity has as yet scarcely been tapped." He concluded: "Today, with an accelerating growth of population and an increasingly insistent demand for higher living standards, no country can afford any longer to be content with such slow progress."[14]

FAO took the position that improvements in agriculture would be possible through national "programming," early UN-speak that meant *government-led* efforts to increase investment and support for agriculture though infrastructure, credit, input supply, extension services, marketing, and land tenure reforms. In a similar vein, around this time, the Rockefeller Foundation and the Ford Foundation were well advanced in mobilizing and deploying resources to improve research on agriculture in the developing world. The Rockefeller Foundation had worked in Mexico since the early 1940s to improve agricultural productivity—efforts that would ultimately lead to the Green Revolution, which I discuss in chapter 2.

Despite some early indications of success emerging from the Green Revolution in the 1960s, a growing sense of crisis and impending doom led the more pessimistic observers to return to the writings of Malthus. Doubts were growing that the modernization of agriculture in the United States and other industrialized countries would be sufficient to feed a world population that, by 1960, had passed three billion.

This gloomy outlook was most dramatically expressed by William and Paul Paddock in their best seller *Famine—1975! America's Decision: Who Will Survive?* (1967). The Paddock brothers concluded that the combination of exploding populations and static agriculture in the developing world was beyond the capacity of food aid from surplus-producing countries, at least in the short term. Noting that the United States was the only feasible source of significant food aid, they argued that "even if it fully cultivates all its land, even if it opens every spigot of charity, [the United States] will not have enough wheat and other foodstuffs to keep alive all the starving."[15] They proposed a system of triage, classifying countries according to the likelihood they could be rescued by the generosity of the United States. Three categories of potential beneficiaries were characterized starkly as "can't-be-saved," "walking wounded," and "should receive food." According to the Paddock brothers, India could not be saved, but Pakistan should indeed receive food aid, suggesting that Cold War politics had perhaps crept into the assessment methodology.

The roots of the midcentury Malthusian revival, exemplified if not caricatured by the writing of William and Paul Paddock, can be traced to two books published in 1948: Fairfield Osborn's *Our Plundered Planet* and William Vogt's *Road to Survival*.[16] Osborn and Vogt influenced a generation of environmental activists who viewed Earth's population growth as unsustainable and unstoppable without radical action. Twenty years later Paul Ehrlich published *The Population Bomb*, again highlighting the imbalance of population and food supply.[17] With its opening sentence—"The battle to feed all of humanity is over"—Ehrlich controversially called for radical actions by governments to control population growth and endorsed the Paddock brothers' triage approach by coupling food aid with population control and agricultural development.[18]

More narrowly focused on the environment, but no less influential in the debates around agriculture and food supply, was Rachel Carson's *Silent Spring*, published in 1962.[19] One of the most important books of the twentieth century, *Silent Spring* documented the environmental impact of synthetic pesticides on human health and natural ecosystems. Carson called out chemical companies and the U.S. government for misinformation and lack of transparency. Her writing challenged the use of pesticides, a major element of the twentieth-century effort to modernize agriculture, and resulted in important new streams of research and policy in the United States and around the world.

Ten years after the publication of *Silent Spring*, a team of researchers at Massachusetts Institute of Technology published *The Limits to Growth*. The report, commissioned by the Club of Rome, modeled five factors that they argued would limit economic growth: population, agricultural production, natural resources, industrial production, and pollution. Climate change was not yet an issue of wide concern, even in the scientific community. The conclusions of *The Limits to Growth* are worth repeating verbatim with my emphasis (in italics) on the second sentence of the second point:

> 1. If the present growth trends in world population, industrialization, pollution, food production, and resource depletion continue unchanged, the limits to growth on this planet will be reached sometime within the next one hundred years. The most probable result will be a rather sudden and uncontrollable decline in both population and industrial capacity.
>
> 2. It is possible to alter these growth trends and to establish a condition of ecological and economic stability that is sustainable far into the future. *The state of global equilibrium could be designed so that the basic material needs of each person on Earth are satisfied and each person has an equal opportunity to realize [their] individual human potential.*
>
> 3. If the world's people decide to strive for this second outcome rather than the first, the sooner they begin working to attain it, the greater will be their chances of success.[20]

The Limits to Growth remains a profound and compelling report. The definition of a new "state of global equilibrium" is relevant to our understanding of sustainable development and, more specifically, the goal of a food-secure world. Food shortages, widespread hunger, and even famines were clearly undermining "individual human potential" in the less-developed world.

MALTHUS REEMERGES IN THE TWENTY-FIRST CENTURY

Malthus, Crookes, the Paddock brothers, and the Club of Rome all pointed to future catastrophic outcomes when food supply falls behind demand. Even Dr. Norman Borlaug, father of the Green Revolution and recipient of the Nobel Peace Prize in 1970, recognized the futility of production increases in the face of what he called the "Population Monster." In his Nobel lecture in Oslo on December 11, 1970, Borlaug warned, "The green revolution has won a temporary success in man's war against hunger and deprivation; it has given man a breathing space. If fully implemented, the revolution can provide sufficient food for sustenance during the next three decades. But the frightening power of human reproduction must also be curbed; otherwise the success of the green revolution will be ephemeral only."[21]

We have seen catastrophic food shortages occur many times over history in national and subnational settings: The Great Irish Famine; Cambodia under the successive traumas of the Vietnam War and genocidal rule; and the Horn of Africa affected by prolonged drought and conflict. On a global scale, Borlaug's prediction in 1970 of three decades of "breathing space" was prophetic. With agricultural productivity improvements, the real price of food steadily declined until the turn of the century, apart from the spike of 1972–1975.

Over the past fifty years, we have seen two major global food crises, one in 1972–1975 and another often referred to as the 2007–2008 global food crisis. Observing real food price trends based on the FAO Food Price Index (FFPI), however, it is fair to say that the 2007–2008 crisis continued at least until 2012, except for a sharp but short-lived dip in food prices in 2009 (figure 1.1). If not a *crisis*, it was no less than a sustained period of food price stress. Real food prices averaged across 2017–2019—prior to the effects of COVID-19—remained 28 percent higher than they were in the five years prior to 2007, and they have shown little sign of reverting to pre-2007 levels.[22] Nevertheless, I will refer to this period of elevated prices as the *2007–2008 crisis* while recognizing that it occurred within a longer 2007–2012 global food crisis that itself was just the initial phase of a protracted period of high food prices, most recently exacerbated by supply chain disruptions caused by COVID-19 and the Ukraine–Russia conflict.[23]

Dozens of research papers have been written about each of these periods of high food prices. Those who have analyzed the 2007–2008 crisis had the opportunity to compare and contrast findings with 1972–1975. In 2010 Derek Headey and Shenggen Fan of the International Food Policy Research Institute (IFPRI) published an insightful analysis of the 2007–2008 crisis, *Reflections on the Global Food Crisis: How Did It Happen? How Has It Hurt? And How Can We Prevent the Next One?*[24]

In seeking to understand the causes of the 2007–2008 food crisis, Headey and Fan astutely acknowledge that it was complicated: "The more one assesses this crisis, the more one concludes that it is the result of a complex set of interacting factors rather than any single factor."[25] They do an admirable job, however, of segregating the more compelling evidence-based causal factors from the selective soundbites put forward by politicians and the press. They identify rising energy prices, the depreciation of the U.S. dollar, low interest rates, and investment portfolio adjustments as the most direct culprits of sharp food price rises. These factors are, of course, related to a number of broader macroeconomic trends that were not directly associated with agriculture and food. Higher energy prices affected the cost of food production and transport. Higher energy prices also created a surge in demand for biofuel alternatives, thus diverting potential food and land from the global supply pool.

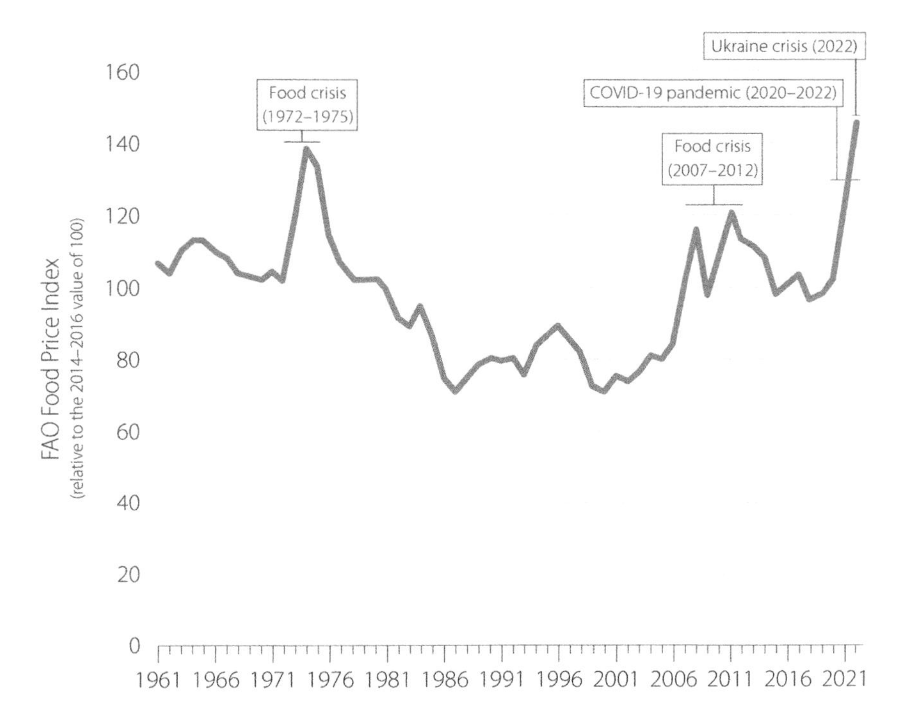

FIGURE 1.1 The FAO Food Price Index using real prices, 1961 to 2022.

Source: Food and Agriculture Organization, "World Food Situation," FAO, accessed May 6, 2022, http://www.fao.org/worldfoodsituation/foodpricesindex/en/. Data for 2022 includes the first four months only. The FAO Food Price Index (FFPI) is used to track the monthly change in international prices of a basket of key food commodities: cereals, vegetable oils, sugar, meat, and dairy products. These commodities represent about 40 percent of gross agricultural food commodity trade and were chosen for their high and strategic importance in global food security and trade. The FFPI is calculated as the average of five commodity group price indices weighted by the average export shares of each group using 2014–2016 as 100 percent. See FAO, *Food Outlook—Biannual Report on Global Food Markets: June 2020*, Food Outlook, 1 (Rome: FAO, 2020), doi.org/10.4060/ca9509en.

Other causal factors were important for specific commodities. Rice prices were particularly volatile, causing some large exporting countries to ban exports and some sensitive importers to make large precautionary purchases. Drought in Australia reduced wheat supply, pointedly bringing climate change into the crisis-focused dialogue. Headey and Fan found little evidence to support popular reports that commodity speculation in futures markets was important. While investor behavior may have exacerbated price increases, Headey and Fan concluded that it was unlikely to have been a causal factor.

There were other long-term influences that are likely to have created the setting for a crisis rather than being a direct causal factor. Overall neglect of agricultural investment, as a result of three decades of broadly declining real food prices, was often mentioned as a factor, not least by me. Like many in the field of agriculture, I believed we should not let a "good crisis" go to waste. It was quite a novelty to be asked to speak with journalists from the *New York Times* and the *Economist* and accompany them on fact-finding trips to help them understand the nature of the global food crisis and the options for response and recovery. During that time, especially during 2007–2012, stories of the food crisis, especially in Africa, were in high demand.

Another often-mentioned factor shaping food demand and prices was the diversion of food crops to animal feed, especially in countries undergoing a rapid dietary transition. There is no doubt these long-term trends in supply and demand of agricultural commodities provided fertile ground for the higher energy prices, the depreciated U.S. dollar, low interest rates, and investment portfolio adjustments—the real triggers of the crisis—to take effect, and thereafter for secondary factors like protective national policies and perhaps commodity market speculation to exacerbate the effects.

Headey and Fan draw comparisons with the 1972–1975 crisis. The profiles of the two crises were similar in scale and scope. Price changes of the main food commodities and fertilizers were comparable. As with the 2007–2008 crisis, the primary causes of the food price spike in 1972–1975 were rising oil prices, the devaluation of the U.S. dollar, and market shocks on both the supply and demand sides; the United States was a major player in terms of export markets and food aid; and the food-importing countries and low-income consumers, most notably in Africa, were hardest hit by the price spikes. Professor C. Peter Timmer, an authority on global food security, argues that the 1972–1975 crisis was rooted in a severe weather shock that reduced global grain production, though exacerbated by global financial turmoil and policy actions by the United States and the Soviet Union.[26] Timmer notes that the OPEC decision to embargo oil exports to the United States and Europe came after the initial food price spike. Higher oil prices did raise fertilizer prices, however, and likely helped extend the high food prices through 1975 and into 1976. Timmer concluded from both crises that government policy interventions are needed. These include consistent long-term investment to improve agricultural productivity and managing strategic reserves of food grains.

RESPONDING TO THE CRISIS

I am personally more familiar with the 2007–2008 crisis than the events of 1972–1975 and will share some observations from that experience. Secretary-General

Ban Ki-moon of the United Nations established the High-Level Task Force on Global Food and Nutrition Security (mercifully assigned the shortened acronym HLTF) in 2008. After moving from Nairobi to New York in mid-2009, I was asked to serve on the HLTF Senior Steering Committee, a technical group that supported an assembly of more than twenty heads of UN agencies, development banks, and other international organizations working on food and nutritional security. The HLTF was masterfully guided by David Nabarro, a British physician and seasoned UN official, who would be awarded the World Food Prize in 2018 for his contributions to food security and improved nutrition.

Working as an adviser to Jeffrey Sachs, then the secretary-general's special adviser on the Millennium Development Goals (MDGs), I had the good fortune to participate in a number of HLTF meetings at UN Headquarters and observe an unprecedented effort to improve coordination of agencies as the food crisis progressed and generated food insecurity in various parts of the world. I admired the commitment, patience, and political acumen of the secretary-general and Nabarro as they herded, rather than directed, the assembled agency leaders toward better outcomes for hundreds of millions of people suffering from the high food prices. That said, three things bothered me.

First, the HLTF appeared to have little financial leverage to deploy centrally and collectively; ultimately, it was up to the individual HLTF members to allocate resources as they saw fit. Each agency had its own budget and plan and, to a degree, competed with the other agencies for the attention and financial support of donors. Flexibility to respond creatively and boldly to the unfolding crisis was limited.

Second, there were no representatives of the private sector at the UN. That seemed unfortunate. Could our mission be better served by having leaders from international firms in logistics, supply chains, telecommunications, marketing, and others joining us at the secretary-general's "high table"? Yes, there were various committees in place to promote dialogue with the private sector. But I observed little in the way of direct collaboration and joint problem-solving with business.

Third, the focus of almost all the discussions was emergency relief and not development. Little, if any, attention was directed to the means for avoiding a future food crisis or a transition to a more stable and sustainable food-secure world. To that end, Sachs and I vigorously advocated for a new global fund on food security that would enable low-income countries to finance improvements in agriculture. Several years earlier, Sachs had successfully championed the establishment of a multilateral fund for global health—the Global Fund to Fight AIDS, Tuberculosis, and Malaria. By 2020 the Global Fund was mobilizing and investing more than US$4 billion a year to support disease-eradication programs run by local experts in more than a hundred countries.[27] Surely, it was time to do the same to achieve global food security?

While it would be unfair to refer to Ban Ki-moon, David Nabarro, and Jeffrey Sachs as "prophets of doom," they and their colleagues in the HLTF were effective in creating a much-needed sense of urgency to invest in agriculture and food, drawing on the increased awareness and anxiety that emerged from the 2007–2008 crisis. Their efforts contributed in large measure to the Group of Eight's (G8's) L'Aquila (Italy) Joint Statement on Global Food Security (July 2009), the opening paragraph of which drew attention to the causes and scale of global food insecurity:

> We, Heads of State, Government and International and Regional Organizations convened in L'Aquila, remain deeply concerned about global food security, the impact of the global financial and economic crisis and last year's spike in food prices on the countries least able to respond to increased hunger and poverty. While the prices of food commodities have decreased since their peak of 2008, they remain high in historical terms and volatile. The combined effect of long-standing underinvestment in agriculture and food security, price trends and the economic crisis have led to increased hunger and poverty in developing countries, plunging more than a further 100 million people into extreme poverty and jeopardising the progress achieved so far in meeting the Millennium Development Goals. The number of people suffering from hunger and poverty now exceeds 1 billion.[28]

Countries represented at L'Aquila committed to mobilizing US$20 billion over three years "through this coordinated, comprehensive strategy focused on sustainable agriculture development." In fact, over US$22 billion was pledged, one-third of which was additional to existing commitments. Further progress was made at the subsequent G20 meeting in Pittsburgh, where leaders called on the World Bank to establish a multilateral trust fund to scale up agricultural support to low-income countries.[29]

From these efforts emerged the Global Agriculture and Food Security Program (GAFSP). Since 2010 GAFSP has financed projects worth US$1.7 billion in forty-five low-income countries, US$1.3 billion of which has been grant financing.[30] By late 2021 GAFSP had directly reached more than 16 million rural people, including 6.3 million women. While serving on the GAFSP Technical Advisory Committee, responsible for technical evaluation of national proposals, I observed firsthand the enormous potential for this financing vehicle to improve agriculture and food security, especially in low-income countries. The problem then and now, however, is that the scale is way too small to have any meaningful impact on food availability and access in individual countries, let alone on our goal of universal food security.

The food crisis of 2007–2008 was a wake-up call—a reminder of the "inequality" of demand and supply that Malthus had trumpeted more than two hundred years earlier. But despite some promising early actions, such as the L'Aquila Joint Statement and GAFSP, and increased attention of the international finance institutions on agriculture and food, I do wonder whether we have turned off the alarm and gone back to sleep.[31] The HLTF was effectively disbanded in 2015. Donor contributions to GAFSP have plateaued over the past few years; in any case, US$1.7 billion over ten years is a drop in the bucket.

Implementation of the much-heralded L'Aquila commitment of US$22 billion over three years was slow at first; less than half of the committed funds were disbursed within four years of the announcement.[32] The 2019 Biarritz Progress Report, however, released by the French G7 presidency to track donor commitments and disbursements, gave a more positive assessment, reporting that the G7 had in fact disbursed a total of US$24.4 billion by January 2018.[33] Some creative accounting and the absence of a serious disbursement deadline can make any pledge achievable in the long term.

With memories of the high prices, food riots, and L'Aquila fading, the attention of the international community shifted from food security to a broader set of concerns, including climate and the environment, culminating in the 2030 Agenda for Sustainable Development (hereafter the 2030 Agenda) and the Sustainable Development Goals (SDGs) that were approved by 193 nations in September 2015.[34]

THE 2030 AGENDA

On January 1, 2016, the SDGs formally replaced the Millennium Development Goals as the global framework for action on development. SDG 2, "End hunger, achieve food security and improved nutrition and promote sustainable agriculture," replaced MDG 1, "Eradicate extreme poverty and hunger"—more specifically MDG Target 1.C, "Halve, between 1990 and 2015, the proportion of people who suffer from hunger."

The MDGs had served as the global development framework from 2000 to 2015. MDG Target 1.C focused on hunger and made no reference to food security, malnutrition, or agriculture. By 2015 Target 1.C had been more or less achieved. The proportion of undernourished people—a measure of dietary energy deprivation I elaborate on in chapter 7—in the less-developed regions had fallen by almost half: from 23.3 percent in 1990–1992 to 12.9 percent in 2014–2016.[35] By the end of 2015, however, FAO reported that almost 800 million people in the developing world remained undernourished.[36] At best, the larger job of *ending* hunger was only half completed.

With the articulation of SDG 2 in 2015, the global narrative shifted from *halving* hunger to *ending* hunger, the time frame now extended to 2030 with additional, equally ambitious calls to achieve food security, end malnutrition in all its forms, and promote sustainable agriculture. Elsewhere in the SDGs were calls to cut food losses and waste (SDG 12) and reduce the environmental damage caused by food systems in oceans (SDG 14) and on land (SDG 15).

The origins of the 2030 Agenda and the SDGs can be traced to the early contributions of Carson, Meadows, and others in the 1960s and 1970s. International concerns about environmental sustainability, in relation to agriculture, food, and other sectors, gathered pace with the publication of the report by the World Commission on Environment and Development in 1987. *Our Common Future* (also known as the Brundtland Report) introduced a vision of intergenerational equity that was adopted by the Rio Earth Summit in 1992 and provided an important conceptual anchor for the SDGs.[37]

At the UN World Summit on Sustainable Development in Johannesburg in 2002, the international community redefined the term *sustainable development* to encompass three interdependent and mutually reinforcing pillars: economic growth, social inclusion, and environmental protection. In 2009 UK chief scientist Sir John Beddington prophetically warned of a "perfect storm" of food shortages, water scarcity, and insufficient energy resources, leading to public unrest, international conflict, and mass migration.[38] The importance of the food–environment–health nexus was further reinforced through the Rio+20 Summit in 2012, which provided the foundation of the 2030 Agenda and the SDGs.

While generally a positive document focusing on a better world and how to get there, the 2030 Agenda expresses a state of the world that would not have surprised the likes of Malthus and Crookes. It is salient, however, to note the absence of strong language around hunger and food security. The narrative of the 2030 Agenda is focused on the social and environmental ills facing our planet, illustrated in the following passage:

> We are meeting at a time of immense challenges to sustainable development. Billions of our citizens continue to live in poverty and are denied a life of dignity. There are rising inequalities within and among countries. There are enormous

> disparities of opportunity, wealth and power. Gender inequality remains a key challenge. Unemployment, particularly youth unemployment, is a major concern. Global health threats, more frequent and intense natural disasters, spiralling conflict, violent extremism, terrorism and related humanitarian crises and forced displacement of people threaten to reverse much of the development progress made in recent decades. Natural resource depletion and adverse impacts of environmental degradation, including desertification, drought, land degradation, freshwater scarcity and loss of biodiversity, add to and exacerbate the list of challenges which humanity faces. Climate change is one of the greatest challenges of our time and its adverse impacts undermine the ability of all countries to achieve sustainable development. Increases in global temperature, sea level rise, ocean acidification and other climate change impacts are seriously affecting coastal areas and low-lying coastal countries, including many least developed countries and small island developing States. The survival of many societies, and of the biological support systems of the planet, is at risk.[39]

Hunger is a product of all the maladies described in paragraph 14 of the 2030 Agenda. Yet hunger seems to have drifted from the center stage. Had the 2030 Agenda been crafted in 1960 or 2010 or 2022, I do wonder whether food security would have been more prominent. Fortunately, since 2015 there was a surge of interest in food and nutrition, particularly in the context of environmental sustainability. The COVID-19 crisis and the Ukraine–Russia conflict have also focused our attention on the vulnerability of food supply chains and the differential impacts of crises on different social and economic groups (see chapter 14). In September 2021 the United Nations convened the Food Systems Summit "to launch bold new actions to transform the way the world produces and consumes food, delivering progress on all 17 Sustainable Development Goals."[40] The timing of the summit provided impetus to acknowledge the shortcomings of the global food system and reignite the quest for universal food security.

TODAY'S PROPHETS AND PRIORITIES

During 2019 three important reports were published on the parlous state of the global food system, the vulnerabilities and risks therein, and the urgency to act: the EAT–*Lancet* Commission report "Food in the Anthropocene" (January 2019), the World Resources Report *Creating a Sustainable Food Future* (July 2019), and the Intergovernmental Panel on Climate Change (IPCC) *Special*

Report on Climate Change and Land (August 2019).[41] The arrival of COVID-19 in 2020 brought a flurry of thoughtful analyses on the effects of the pandemic on food systems and food security, most notably reports of the High Level Panel of Experts on Food Security and Nutrition (HLPE).[42]

The EAT–*Lancet* Commission, cochaired by Professors Johan Rockström and Walter Willett, comprised more than thirty leading scientists across relevant disciplines and sectors. Their goal was to reach a consensus on what defines a healthy and sustainable diet. They called for a "Great Food Transformation" around five strategies: shifting to healthier diets; reorienting agriculture toward producing healthier food; sustainably intensifying agriculture; improving governance of land and oceans; and halving food losses and waste. The authors argued persuasively that the "data are both sufficient and strong enough to warrant immediate action." In thinly veiled neo-Malthusian tones, the commission expressed a sense of urgency: "Without action, the world risks failing to meet the UN Sustainable Development Goals (SDGs) and the Paris Agreement, and today's children will inherit a planet that has been severely degraded and where much of the population will increasingly suffer from malnutrition and preventable disease."[43]

Among its recommendations, some stakeholders and observers harshly criticized the EAT–*Lancet* Commission's position on what constitutes a healthy diet. The commission's call for reduced consumption of animal-based products, albeit accepting a *flexitarian* diet that allowed for some animal protein, was not received well by livestock industry advocates. But the science-based synthesis of knowledge on food and diets was rigorous and constitutes a strong supportive case for healthy diets, an essential component of an integrated framework for investment toward a food-secure world (see chapter 12).

The World Resources Report (WWR) was the outcome of a multiyear partnership of international development agencies led by the World Resources Institute. The report noted that rising global population and increased incomes across the developing world will increase overall food demand by more than 50 percent by 2050, while demand for animal-based foods will likely increase by nearly 70 percent. Thus, the WRR argued, "the world urgently needs to change the way it produces and consumes food."[44] While agriculture needs to meet the world's food demands, at the same time serving as an engine of economic growth, it must do so with a lighter environmental footprint. To that end, the WRR proposed a twenty-two-item "menu for a sustainable food future," divided into five "courses":

- Reduce growth in demand for food and agricultural products.
- Increase food production without expanding agricultural land.
- Protect and restore natural ecosystems and limit agricultural land-shifting.

- Increase fish supply (from both wild fisheries management and aquaculture).
- Reduce greenhouse gas (GHG) emissions from agriculture.

The IPCC *Special Report on Climate Change and Land* focused more sharply on land use and GHG emissions. The report recognized the crucial role of productive land in achieving food security and the role of agriculture in reducing GHG emissions. It noted that climate change exacerbates land degradation. And when land is degraded through soil erosion or desertification, it becomes less productive, thus reducing the soil's ability to absorb carbon. We therefore see a downward spiral contributing to poverty, hunger, migration, and conflict. In response, the authors offer a range of options in support of mitigation, adaptation, combating desertification and land degradation, and enhancing food security. As with the EAT–*Lancet* Report and the WRR, the IPCC recommends policy action across the global food system, including reduction in food loss and waste and changes in dietary choices, which enable more sustainable land-use management, lower GHG emissions, and enhanced food security.

The COVID-19 pandemic and the Ukraine–Russia conflict have exposed and amplified existing vulnerabilities of food systems around the world. The HLPE report on the impacts of COVID-19 on food and nutrition security and sustainability identified implications for six dimensions of food security: availability, access, utilization, stability, agency, and sustainability.[45] In response to COVID-19 impacts, the HLPE called for transformative policies on better social protection systems for the most vulnerable consumers, workers, and producers, and more resilient systems of production and distribution.

The analyses and recommendations of the EAT–*Lancet* Commission Report, the WRR, the IPCC *Special Report on Climate Change and Land*, and the many fine reports of the HLPE are well researched and effectively communicated. They are timely, evidence-based resources for agencies and individuals dedicated to improving food security. From Malthus to Borlaug to the IPCC, one message is common: business as usual is not an option if we are to achieve a food-secure world. What began as simple mathematics pointing to a widening gap between food supply and demand is more nuanced today in terms of understanding the causes and consequences of food insecurity, the complex and contextual array of interacting and complementary solutions, and questions around resilience and sustainability that are being revealed through research and practice.

Malthus and his fellow prophets of doom had important points to make. Had Malthus lived to observe the Great Irish Famine, the mid-twentieth-century food shortages in Asia, or the global food crisis of the early twenty-first century, he may well have remarked, "I told you so." Over more than two centuries, however, science and practice have taught us that a sharp focus on the extent and health of our land—a supply approach—is necessary, but not sufficient, if we are to achieve food security for all people, at all times. It is more complicated than that, and the solutions must reflect and address that complexity. As I will show in this book, there remains much scope to improve the productivity of land and make more effective use of its products. And we are managing to flatten the population curve, at least on a global scale. We can reasonably expect a population of around ten billion by 2050 and eleven billion by the end of the century. I hope to show that Malthus's "gigantic inevitable famine" is avoidable and a reversal of environmental degradation is feasible. Through science and technology, effective public policies, strong institutions, education reforms, and transformational leadership, a food-secure world is within our reach.

2

GREEN REVOLUTIONS

William S. Gaud, then United States Agency for International Development (USAID) administrator, is widely credited for having first used the term *Green Revolution* in a speech delivered to the Society for International Development at the Shoreham Hotel in Washington, D.C., on March 8, 1968: "It is not a violent Red Revolution like that of the Soviets, nor is it a White Revolution like that of the Shah of Iran. I call it the Green Revolution."[1]

Reporting unprecedented harvests from Asia, Gaud drew contrasts with alternative color-coded depictions of revolutionary change. Referring to the record yields of rice and wheat across Asia, Gaud declared, "We are on the verge of an agricultural revolution," one that is "as significant and as beneficial to mankind as the industrial revolution of a century and a half ago."

Gaud served under Lyndon Baines Johnson, thirty-sixth president of the United States. On a sunny, late October day in 1966, President Johnson visited the International Rice Research Institute as part of a three-day trip to the Philippines.[2] Accompanied by President Ferdinand Marcos and First Lady Imelda Marcos, Johnson ventured into IRRI's research fields to see for himself the progress being made in developing new, high-yielding varieties of rice. One of the genetic lines Johnson was shown had the designation IR8-288-3: a sturdy, short-statured variety that IRRI would approve and release a few weeks later as IR8, a major milestone in Asia's Green Revolution.

Johnson went on to give a rousing speech to the assembled IRRI staff and dignitaries:

> I am glad to know the Institute is prepared to make these seeds available to all nations, to *all* nations, whatever their politics and ideologies. Because, if we are

to win our war, and the only important war that really counts, if we are to win our war against poverty, and against disease, and against ignorance, and against illiteracy, and against hungry stomachs, then we have got to succeed in projects like this, and you are pointing the way for all of Asia to follow. And I hope they're looking. I hope they're listening. And I hope they're following.[3]

Later that day Johnson flew to Cam Ranh Bay, an American military base in Vietnam. By 1966 he was escalating the war effort and had decided that agricultural assistance, especially though rice production—the main staple and livelihood of the Vietnamese people—would be an important, if not decisive, complement to direct military action. Johnson was convinced that "pacification" through an abundance of rice would ultimately win over the hearts and minds of a rural population caught in an ideological battle about a pathway to a better life. He pressured USAID to promote IR8 throughout the Mekong Delta.[4]

FIGURE 2.1 A visit to the International Rice Research Institute on October 26, 1966. *From left*, Peter Jennings (IRRI rice breeder, *standing*), Henry ("Hank") Beachell (IRRI rice breeder), Robert Chandler (IRRI director general), President Ferdinand Marcos, and President Lyndon B. Johnson.

Credit: Urbito T. Ongleo—Image collection of the International Rice Research Institute (https://www.irri.org).

My old friend and former IRRI colleague Tom Hargrove shared his experience as a U.S. Army adviser in the Mekong Delta in 1969–1970 and his return visit twenty years later in a book: *A Dragon Lives Forever: War and Rice in Vietnam's Mekong Delta 1969–1991, and Beyond*.[5] Hargrove joined IRRI in 1973 as a communications expert, seven years before me. I recall many a long coffee break in his office, listening to him reminisce about various experiences in Vietnam, the glories of his alma mater Texas A&M University, and the vital importance of having steak very well done.

Hargrove worked as an agricultural adviser in Chương Thiện, a Viet Cong (VC) stronghold and classified by the U.S. military as one of the two least secure South Vietnamese provinces.[6] On the front line of pacification efforts of President Richard Nixon, by then Johnson's successor, Hargrove described how IR8 was officially named Lúa Thần Nông (Rice of the Farming God) but was better known as Honda Rice, implying a good crop of IR8 would produce enough rice to buy a motorcycle.[7]

Rice would play a major role in American efforts to win local support. Airplanes dropped leaflets on North Vietnam pronouncing, "IR-8, South Vietnam's Miracle Rice. All Vietnamese can enjoy this rice when peace comes."[8] The distinctive new rice variety provided a symbolic yet visible dividing line been the past and the present, between poverty and prosperity, and between hunger and plenty.

The influence of *rice diplomacy* was not lost on the VC. While they initially resisted IR8 as an imperialist plot of the Americans, they ultimately secured IR8 seeds, proclaimed it their own, and distributed the variety to areas under their control.[9] IR8 had emerged as a hot item in the Cold War. When a food commodity like rice is so central to life, it becomes the focus of political discourse and sometimes a weapon of war.

Earlier in 1966, less than six months into his presidency, Ferdinand Marcos had visited IRRI and been so impressed with the possibilities of the new high-yielding rice varieties that he called on the institute leadership to come to Malacañang Palace on June 11 and present their results to the Philippines cabinet and heads of agencies, with the press in attendance.[10] IRRI director general Robert Chandler presented Marcos with a 2 kilogram (kg) bag of IR8-288-3 and committed another 50 metric tons (t) for immediate distribution in the Philippines. On June 12, 1966, front-page headlines in the *Manila Bulletin* announced: "Marcos Gets Miracle Rice," a term that would stick with IR8 and the generations of high-yielding varieties that emerged from IRRI's research in the following decades.

By 1968, as preemptively announced in William S. Gaud's March 8 speech, the Philippines was self-sufficient in rice, a proud achievement for both Marcos and

IRRI. Marcos had campaigned the previous year on the three Rs—rice, roads, and 'rithmetic—the third "r" related to construction of school buildings.[11] However, with its high population growth rate, coupled with typhoons and disease outbreaks, the Philippines struggled and generally failed to sustain full self-sufficiency in the ensuing years and to the present day.

Many development policy experts argued and continue to argue that 100 percent self-sufficiency is not an economically rational policy—grow what you grow well, and import the rest. However, because of the dominance of rice in Asian diets, the domestic politics of rice in the Philippines and in most other Asian countries, and a thin international market, where only about 10 percent of global production is internationally traded by a small number of exporters, self-sufficiency would remain a national policy objective for decades to come. To that end, governments in Asia prioritized policies and programs that would raise the productivity of staples like rice and wheat.[12]

NORMAN BORLAUG

It may come as a surprise to learn that Asia's Green Revolution had its roots in Mexico or, as some might argue, in Iowa and will forever be associated with the work of Norman Borlaug.[13] Born in 1914 and raised on a farm in Cresco, a small rural community in northeastern Iowa, Borlaug would become known as the father of the Green Revolution. Through the adoption of his new high-yielding, disease-resistant wheat varieties and similar efforts that this work inspired, Borlaug would be credited for saving as many as one billion lives from starvation. In 1970 he was awarded the Nobel Peace Prize for his role in increasing global food supply. Indeed, the advances in rice production described earlier owe much to Borlaug's path-breaking research to improve wheat yields in Mexico.[14]

In 1944 Borlaug joined the Rockefeller Foundation (RF) in its Mexican Agricultural Program, before taking over the program's wheat research in 1945. This research aimed to breed new wheat varieties with high yield and resistance to stem rust, a fungal disease that had devastated Mexico's wheat crops a few years earlier. Borlaug later crossbred these varieties with short (known in cereal breeding as dwarf) varieties from Japan, leading to the development of relatively short-statured (semidwarf), disease-resistant, high-yielding wheat varieties.[15] He deployed a novel approach called *shuttle breeding*, whereby two successive generations of wheat were grown in different contrasting locations within the same year: a summer crop was grown in the cooler highlands near Mexico City, followed by a winter crop in the warmer conditions of Sonora at sea level in northwestern Mexico.

Through shuttle breeding, Borlaug would cut by half the time needed to develop a new, genetically stable variety, a radical approach that had never been deployed with wheat. These new wheat varieties had wide ecological adaptation and broad resistance to stem rust and were responsive to irrigation and fertilizer application. Borlaug's new varieties, capable of producing 5 to 6 metric tons per hectare (t/ha), transformed wheat productivity in Mexico, enabling self-sufficiency in 1956.[16] By 1963, 95 percent of Mexico's wheat lands grew Borlaug's wheat varieties, and production had increased sixfold since his arrival.[17]

Borlaug's program also placed high emphasis on training local researchers and technicians. He insisted that hands-on fieldwork and a strong connection to farmers and extension were required for success. Borlaug believed that the value of the researchers' work should ultimately be measured by impacts in farmers' fields. During seventeen years of the RF's assistance, hundreds of Mexican researchers were trained, including two hundred at master's and PhD levels, mostly in the United States. In 1961 the National Institute for Agricultural Sciences (known by its Spanish acronym, INIA) was established to continue the work that Borlaug, the Rockefeller Foundation, and their national collaborators had started. This two-pronged approach of introducing new technologies and national capacity building would become a feature of international agricultural development efforts in the following decades. The RF's program in Mexico would ultimately lead to the establishment in 1966 of the International Maize and Wheat Improvement Center (known by its Spanish acronym, CIMMYT).[18]

As the fruits of Borlaug's work were unfolding in Mexico, India and Pakistan were struggling to meet their food requirements. In both South Asian countries, researchers had been evaluating Borlaug's wheat varieties since the early 1960s. But by 1966, the year President Johnson visited IRRI, India's cereal production, consisting primarily of wheat and rice, had fallen for the second successive year and the country needed to import 11 million t of grain. Food aid was mobilized, primarily from the United States, but complemented with imports of the high-yielding wheat seed from Mexico: 200 t in 1966 and 18,000 t in 1967.[19] The spread of these seeds in both India and Pakistan, coupled with better weather conditions in 1967, led to a remarkable turnaround in cereal production in both countries, and thereafter across much of Asia.

In addition to direct dissemination of the imported Mexican varieties, Indian and Pakistani researchers crossed these with locally adapted strains to improve local adaptation and acceptability. The specter of a Malthusian apocalypse had been removed, at least for now. Contrary to the fears of the Paddock brothers,

India *could* be saved! Not by food aid alone—though this bought time and saved lives—but through the science, policies, and institutions associated with the Green Revolution. While Borlaug's wheat was the flagship of this rebuttal to the Malthus hypothesis, Asia's main food staple—rice—was also undergoing a makeover through the work of IRRI.

RICE IS LIFE

I joined IRRI on Monday July 7, 1980, roughly fourteen years after the release of IR8. I had spent the previous three years as an agronomist on an Australian-funded rural development project on the island of Mindanao in the southern Philippines. In those predigital days in remote Mindanao, it was a challenge to access new ideas and technologies. Every few months I would visit IRRI and the neighboring University of the Philippines Los Baños (UPLB) to collect seeds, pick up or photocopy useful publications, and meet with fellow agricultural scientists.

IRRI's reputation for research excellence and its role in the Green Revolution were firmly established as I visited the institute in May 1978. Then, and for many years thereafter, the institute had a practice of announcing daily visitors on a notice board outside its visitors center. On that day, I was listed below the king and queen of Nepal and the chief scientist for the United Kingdom's Ministry of Agriculture, Fisheries, and Food, Sir Charles Pereira, as visitors for the day—an everyday reflection of IRRI's extraordinary standing.

Rice was a key staple in Nepal, and, as in most Asian countries, the heads of state and political leaders of the region saw IRRI as a source of hope and opportunity for improving national food security. Like Presidents Johnson and Marcos and many others, Nepal's rulers wanted to see for themselves. King Birendra was one of many heads of state and senior government officials who were, like me, attracted to what was then the engine room of Asia's Green Revolution.

Pereira was no stranger to IRRI. He was a distinguished British hydrologist who pioneered the scientific study of the relationship between water and crops and had reviewed IRRI's work during the 1970s. According to his obituary in the *Times* of London, Pereira's "ambition was to improve food production at a time in the mid-20th century when starvation and under-nourishment were commonplace among rapidly growing populations."[20] Aside from, but possibly because of, his knowledge of water resources and food production, Pereira was known for his strong views on population, famously and controversially advocating the widespread promotion of contraception at the Nehru Centenary Conference of World Scientific Academies in New Delhi in 1989.[21] Malthus would surely have nodded in approval.

As with Borlaug's work in Mexico, the RF was central to the establishment of IRRI as a focus for rice research in Asia. Warren Weaver and J. George Harrar, senior RF officials, visited Asia in the early 1950s to explore possibilities for the foundation's assistance in rice research and education. In 1954 they prepared a paper for the RF Board of Trustees on the case for establishing an "International Rice Research Institute in Asia."[22] Weaver and Harrar outlined the benefits of multicountry cooperation that would address fundamental problems that were independent of political boundaries. They argued efficiencies of scale could be gained by focusing resources and avoiding duplication of efforts. The Ford Foundation joined this effort, and in 1960 IRRI was established in the Philippines, initially as an office in Manila and later relocating to a site at Los Baños, 72 km to the south, on the campus of UPLB. IRRI's research fields, currently 252 ha, are formally operated under a lease from UPLB.

As outlined earlier in this chapter, IRRI's achievements during its first decade emerged from the development and spread of IR8 throughout Asia. The focus of IRRI's research was clear: develop high-yielding varieties that were short, stiff-strawed (meaning firm-stemmed varieties that would not bend or break with heavy grains), fertilizer responsive, and early maturing.[23] The IRRI rice breeders sought a fundamental change in the architecture of tropical rice varieties. In the tropics, rice varieties were tall (usually taller than a meter at harvest) and weak-strawed and tended to fall over with improved soil fertility. However, by crossing these tall varieties with short, stiff-strawed, temperate-zone varieties, IRRI's rice breeders developed a new plant type that was adapted to the tropics and responsive to fertilizer and irrigation. These new varieties had a higher *harvest index*, meaning that more energy and nutrients ended up as grain and less as stem and leaves. In addition, the Green Revolution varieties were earlier maturing than the traditional rice and enabled farmers to grow two crops in succession, provided they had access to irrigation.

During the dry season of 1966, Surajit K. (S. K.) De Datta, a young Indian agronomist who joined IRRI in early 1964, was the first to demonstrate IR8's extraordinary yield potential in response to fertilizer applications, good water management, and effective weed control.[24] At a time when the national rice yield in the Philippines was 1.3 t/ha, De Datta obtained experimental yields of around 10 t/ha. He would go on to serve twenty-eight years at IRRI. His textbook *Principles and Practices of Rice Production* (1981) remains the premier text on the world's most widely consumed food crop.[25]

IR8, and a succession of high-yielding varieties that followed, transformed rice farming and food security in Asia with impacts extending to Latin America. Along with Borlaug's wheat varieties, IR8 had averted the Malthusian apocalypse, for now, at least. In the most widely cited assessment of the impact of these Green Revolution varieties, Robert Evenson and Douglas Gollin distinguished between an early Green Revolution period (1961–1980) and a late one (1981–2000).[26] Contrary to popular perceptions, they found that these new varieties of rice and wheat contributed more to productivity improvements in the late Green Revolution than in the earlier two decades. They argued that the Green Revolution should be viewed as a *long-term* increase in cereal productivity rather than a one-time jump in production in the late 1960s. As new varieties were developed and released by IRRI, CIMMYT, and their national research partners, supported by national policies encouraging food security, productivity gains continued, ensuring food security for most of Asia until the early years of the twenty-first century.

New varieties of rice and wheat were critical to improvements in productivity. By 1980 about 40 percent of the total cereal area in Asia was planted with Green Revolution varieties; by 2000 the coverage had increased to 80 percent. Adoption of these new varieties was accompanied, however, by increased fertilizer use and modest expansion of irrigation. Between 1967 and 1982 fertilizer use grew by more than 10 percent per year, while the area of irrigated land grew by 2 percent a year over the same period. Government support accelerated the availability of these Green Revolution technologies through aggressive extension campaigns, input subsidies (fertilizer, power, and water), cheap and accessible credit, and market price stabilization. Longer-term investments in research and infrastructure complemented these efforts. Public investment was an essential feature of the Green Revolution. By the early 1970s Asian countries were spending on average about 15 percent of their total government budget on agriculture.[27] While the private sector played an important role in producing and distributing inputs (mainly fertilizer and pesticide) and in processing and marketing surpluses, government investments and supportive policies were critical to the Green Revolution's success.

In production terms, the impact of the Green Revolution in Asia was astonishing. Between 1970 and 1995 cereal production more than doubled, from 313 to 650 million t per year, while population increased by 60 percent over the same period.[28] As a result, cereal availability per person increased by nearly 30 percent, and wheat and rice became cheaper in real terms. It is often overlooked that the main beneficiaries of the Green Revolution were consumers, both rural and urban. Farmers also benefited, but those benefits were dampened by lower grain prices and therefore were ultimately determined by the crop productivity and access to input and output markets.

LIMITS AND PIVOTS

In a comprehensive and insightful retrospective of the Green Revolution, economist Prabhu Pingali examined both its achievements and its limitations.[29] It is perhaps not surprising to see that, having worked at both IRRI and CIMMYT, as well as the World Bank, the UN Food and Agriculture Organization, and the Bill and Melinda Gates Foundation, Pingali recognized the positive impacts of the Green Revolution on productivity improvement and increased food supply from the world's major staple crops. Consistent with the findings of many others, Pingali concluded that the Green Revolution sharply reduced poverty and hunger, while slowing land conversion that would otherwise be required to meet growing demands. One study, based on simulations using a global economic model, found that 18 to 27 million hectares of pastures and forests—mostly in low-income countries—were saved from conversion to agricultural land as a result of the Green Revolution.[30]

The numbers on cereal crop yields and productivity, hunger reduction, poverty reduction, and land-sparing speak for themselves. Even the most enthusiastic supporters of the Green Revolution, however, accept that this productivity-driven, cereal-led approach did not solve all problems for all people. As Pingali explains, "Some areas were left behind, and even where it successfully increased agricultural productivity, the [Green Revolution] was not always the panacea for solving the myriad of poverty, food security, and nutrition problems facing poor societies."[31] He describes several limitations of the Green Revolution, which I list here. Some critics have used these limitations to connote an overall failure of the Green Revolution approach. Instead, we should see these unintended shortcomings as forming a new agenda for action and a pivoting of investment priorities:

- **Persistence of hunger.** Even with a limited calorie-based definition of hunger, at least one in ten people on Earth go to bed hungry each night. Notwithstanding irrefutable progress, the Green Revolution has not *solved* the global hunger problem. As Borlaug observed, it has bought us time. This issue is further explored in chapters 7 and 12.
- **Proportionally less benefit to women.** A significant gender gap remains in agriculture. Women have less access than men to agricultural assets,

inputs, and services, such as credit and extension. Closing the gender gap would generate significant gains for the agriculture sector and for society.[32]

- **Slow progress toward improved nutrition.** More than one in five children under five years of age are stunted: a condition that is associated with poorer and largely irreversible cognitive and educational outcomes in later childhood and adolescence, often with dire economic consequences later in life (see chapters 7 and 12). Even in some nations that have achieved cereal self-sufficiency and managed to export surpluses (e.g., India and Cambodia), there is persistent chronic undernutrition.
- **Negative environmental impacts.** Excess use of fertilizers and pesticides in specific settings have caused negative environmental impacts. Land degradation, water resource depletion, biodiversity loss, and greenhouse gas emissions have also been linked to Green Revolution policies. These negative environmental consequences and the scope for practical solutions are addressed in chapters 3 to 6.
- **Limited impact in marginal and riskier production environments.** Productivity gains were generally greatest and earliest in more favorable growing environments, especially those with irrigation (see chapter 4). Productivity improvements, however, are possible in rainfed environments. In the following section of this chapter, I will highlight the case of Cambodia, where remarkable progress was made in a less favorable production environment.
- **Slow progress in Africa.** At least until the turn of the twenty-first century, there was a dominant narrative that Africa had been bypassed by the Green Revolution. Evidence is emerging, however, of significant progress in overall production and pockets of productivity improvement, suggesting a partial or delayed Green Revolution may be a more appropriate assessment. The facts and the myths of this narrative are presented and discussed later in this chapter.

By the time I joined IRRI in 1980, the focus of attention had begun to shift to what were called *less favorable environments*. Five years earlier, the institute's annual report asked the question: "What about the other three?," referring to the then three-quarters of rice farmers who had not adopted so-called modern varieties. These farmers toiled in harsh environments where IRRI's new varieties had not shown any advantage over what the institute described as "hardy but low-yielding local varieties and ancient farming methods."[33]

Irrigation clearly played an important role in improving productivity. In essence, what this did was enable the new varieties to express their yield potential by eliminating, or at least reducing, the damaging effects of water deficits on plant growth. In many parts of Asia, however, as well as in Africa, rice was grown under rainfed conditions and depended on rainfall to meet the crop's substantial water requirements. Rainfed crops were less consistent in their response to fertilizer, which was a core ingredient of the Green Revolution package.

Beyond the biophysical challenges of rainfed environments, where yields were lower and poverty levels were higher, IRRI placed increasing emphasis on the development of national agricultural research systems. The rationale for this approach was that many of the unmet challenges had to do with the local adaptation of technology. For the same reason that IRRI was established in the Philippines, rather than, say, California or Louisiana, it was necessary to adapt and develop new technologies for the less favorable environments in different locations. This challenge was made worse when agriculture—and the national institutions that support it—had been severely disrupted by conflict. Cambodia was one such example.

CAMBODIA CALLS

In the twentieth century few countries experienced the levels of social disruption, human carnage, and physical destruction that befell Cambodia.[34] As the Vietnam War spilled across the border, the country was bombed relentlessly between 1969 and 1973.[35] During that period, the U.S. military flew more than 200,000 sorties and dropped 2.75 million tons of munitions on Cambodia.[36] Despite this extreme use of force, on April 17, 1975, the capital, Phnom Penh, fell to forces of the Communist Party of Kampuchea, known widely as the Khmer Rouge. In the preceding five years, the population of Phnom Penh had doubled to two million as the rural population fled the bombing and the advance of the rebel forces.[37]

Within weeks of overthrowing the Lon Nol government, the Khmer Rouge emptied the capital and other cities, relocating the population to all parts of the country, separating families, and creating organized labor and production brigades. Recognizing the symbolism of empowering rural people, the Khmer Rouge leader, Pol Pot, immediately placed a high priority on increasing rice production. He established an ambitious national target of 3 t/ha that was likely modeled on a similar campaign instituted in China by Hua Guofeng. As early as May 1975 Radio Phnom Penh pronounced: "If we have rice, we have everything."[38] Although Cambodia was a significant rice exporter during the 1960s, average national yields were mainly in the range of 1.0 to 1.4 t/ha, the lowest in Asia.[39] To more than double rice production without external inputs like fertilizer and well-adapted seeds would be a formidable and ultimately an unrealistic task.

Between 1975 and 1978 the Khmer Rouge instituted some extraordinarily draconian and ill-conceived agricultural policies. Within the context of a four-year plan to build socialism in agriculture and other sectors and to achieve the 3 t/ha target, the government moved to replace Cambodia's traditional rainfed rice with high-yielding irrigated rice systems that would, in principle, rapidly increase production. Irrigation canals were constructed on 1 km^2 grids by deploying forced labor, many of whom had been relocated from Phnom Penh and other cities. Within these irrigation grids, one-hectare plots were designed from existing parcels and farmed by communal production groups. The infrastructure, however, was poorly designed and hastily constructed without consideration of water requirements and stream capacities and flows. This caused flooding and required frequent major repairs. Throughout the Khmer Rouge period, production stagnated at around 1 t/ha, and the country plunged into deep food deficit and widespread starvation. The Khmer Rouge rejected what they considered to be modern technology, as illustrated by this quote attributed to a Khmer Rouge leader when emphasizing the overriding importance of their unique brand of Cambodia's socialist revolutionary movement: "Technology is not the decisive factor; the determining factors of a revolution are politics, revolutionary people, and revolutionary methods."[40]

Under the Khmer Rouge leadership, the country, then called Democratic Kampuchea, experienced almost four years of genocide and crimes against humanity, while launching periodic attacks against Thailand and Vietnam. While estimates vary, noted Yale University expert on the Cambodia genocide, Ben Kiernan, assesses that the Khmer Rouge was responsible for the deaths of about 1.7 million people, an extraordinary 21 percent of the population.[41]

In December 1978 Vietnam invaded Cambodia, and by January 1979 it had removed the Khmer Rouge leadership from Phnom Penh. Several years of slow recovery followed, supported by the governments of Vietnam, the Soviet Union, Eastern European Soviet-bloc nations, Cuba, and India, as well as several nongovernmental organizations (NGOs). Because of Vietnam's role in the overthrow of the government, the United Nations—in particular, China, the United States, and the United Kingdom—refused to recognize the new Vietnam-supported regime, known as the People's Republic of Kampuchea. For more than a decade, civil war continued, and Cambodia remained internationally isolated by an official UN doctrine that rejected the nondemocratic regime change, despite the clear, albeit delayed, acknowledgment of the genocidal actions of the Khmer Rouge.

It was within in this complex, bloody, and tragic historical setting that, in 1985, IRRI director general Dr. M. S. Swaminathan initiated a program of international cooperation with Cambodia that would serve as the country's primary source of scientific and technical support to the agriculture sector for more than fifteen years. With his deep knowledge of Indian agricultural research as a scientist and administrator, Swaminathan appreciated the necessity of a strong national research infrastructure. He argued that "only a strong national research system could take advantage of advances in international research" and pointed to the ease with which India embraced and deployed new wheat and rice varieties that launched his country's Green Revolution.[42]

After several months of exchanging letters and telexes, and following a visit to IRRI by two Cambodian agricultural officials, the then agriculture minister and deputy prime minister, Kong Sam Ol, formally approved a mission to Cambodia by a team of three IRRI scientists.[43] With arrangements made by the UN Children's Fund (UNICEF), one of the few international organizations allowed to operate in Phnom Penh, IRRI's first postwar mission to Cambodia took place from January 23 to 30, 1986. The group comprised Dr. Gurdev Khush (renowned IRRI rice breeder and World Food Prize laureate in 1996), Dr. Donald Puckridge (IRRI agronomist and representative in Thailand), and me.

FIGURE 2.2 War-damaged agricultural research buildings in Cambodia (1986).

Credit: Glenn Denning.

Mobilizing funds to support the IRRI–Cambodia program presented unusual challenges. The lack of diplomatic recognition of Cambodia, outside the Soviet-bloc countries and India, made IRRI's traditional donors reluctant to contribute. However, I encouraged Swaminathan to approach the government of Australia, recognizing its geopolitical interests in the region. Through a good friend and colleague—Fabian Sweeney, a former political activist in the Australian Labor Party—I was aware that the then Australian foreign minister, Bill Hayden, had a strong personal interest in supporting the people of Cambodia and reengaging the government in Cambodia, a position that the United States strongly opposed.[44] Through a series of discreet meetings and correspondence with Australian officials in Canberra and Manila, including a ten-minute discussion between Swaminathan and Prime Minister Bob Hawke during the latter's visit to Manila in May 1986,[45] agreement was reached to establish a somewhat-clandestine IRRI–Indochina Program that would include Cambodia, Vietnam, and the Lao People's Democratic Republic (Lao PDR, also known as Laos). Later this program would become more transparently the Cambodia–IRRI–Australia Project (CIAP), while separate country-focused projects continued in Vietnam and Lao PDR. Between 1987 and 2001 Australia contributed almost US$25 million to support Cambodia through CIAP.[46] I coordinated IRRI's work in Cambodia, Vietnam, and Lao PDR from 1986 until I departed the institute in 1998.

The first decade of IRRI's program of research cooperation in Cambodia was comprehensively documented in CIAP project leader Harry Nesbitt's *Rice Production in Cambodia*.[47] This volume highlighted some of the unique challenges encountered in improving production in a country dominated by diverse, rainfed rice ecosystems. Most of Cambodia's farmers were clearly part of IRRI's "other three": those three-quarters of rice farmers who subsisted in the harshest production environments. Major areas of emphasis included rice varietal improvement, soil fertility management, farming systems research, integrated pest management, mechanization, and postharvest improvement.[48] Donald Puckridge published a richly detailed, anecdotal history of IRRI's Cambodia program, titled *The Burning of the Rice*, in 2004.

Under Swaminathan's guidance, IRRI's collaborative program in Cambodia embraced the importance of local ownership and leadership as the foundation of sustainable capacity development. To that end, two critical programmatic priorities were established at a very early stage: a massive training of Cambodian agricultural scientists and practitioners, and the establishment of a national research institute to be known as the Cambodian Agricultural Research and Development Institute (CARDI). On reflection, the development model was not unlike the Rockefeller Foundation's program in Mexico forty years earlier.

The report of our first mission in January 1986 noted that the biggest constraint facing the country was the shortage of trained human resources and that local capacity development should be given the highest priority.[49] The team noted that only twenty of the three hundred graduates of the prewar University of Agriculture remained in the country; the rest had died or had fled. Later, Nesbitt estimated that only forty of the four hundred qualified agriculturalists remained in Cambodia immediately after the war. Through a variety of capacity-building opportunities, which included in-country training, international courses at IRRI and elsewhere in the region, and master's and PhD programs in several countries, Cambodia was able to rapidly build its capacity in rice research and extension. By 2001 a total of 1,700 Cambodians had been trained with IRRI's support.[50]

Our team also observed in January 1986 that "there are no facilities for development and evaluation of technology" in Cambodia.[51] Most of the infrastructure had been destroyed. The IRRI mission received a request from the Ministry of Agriculture to assist in establishing a rice research and training center. Following an evaluation of different sites, complex and difficult negotiations with landowners, the government, and donors, and many years of functioning as the de facto national rice research institute, CARDI was formally established in 1999 and inaugurated by Prime Minister Hun Sen in 2000.[52] Fully consistent with Swaminathan's vision of the vital role of a national research institution, Hun Sen stated in his inauguration speech: "CARDI should also have an important role to play in networking with regional and international research institutes and centers to strengthen the cooperation in research and exchange of experiences, outcomes and technical information of scientific value."[53]

Dr. Swaminathan had the unique perspective of having served as leader in both a national agricultural research institution—the Indian Council of Agricultural Research—and an international agricultural research center—IRRI. He appreciated that a strong national research institution is a prerequisite for international cooperation. In his own words: "If we wish developing countries to progress in agriculture, we must help them to build strong National Agricultural Research Systems. The stronger the NARS, the greater is the benefit of the International Agricultural Research Centres." Moreover, Swaminathan ventured: "This will confer long-term benefits and also help strengthen the morale and capability of national scientists, working on national salaries."[54]

Perhaps the most striking example of international cooperation under this program was in conserving and repatriating Cambodia's unique diversity of rice varieties. The massive dislocation of people, disruption of farming, and distortion of policies left Cambodia's farmers with only remnants of their diverse, uniquely adapted, traditional rice varieties after the Khmer Rouge period. The Ministry of Agriculture and NGOs approached IRRI for assistance in locating the "lost" traditional varieties. In response, IRRI repatriated 766 traditional varieties that had been collected in December 1972 and January 1973 and safely stored in the International Rice Genebank at IRRI headquarters in the Philippines.[55] IRRI and Cambodian scientists purified, evaluated, and reintroduced many of these varieties to meet the needs of Cambodia's diverse rainfed conditions, where modern varieties developed elsewhere for more favorable environments were found poorly adapted.

One Cambodian variety, Phka Rumduol, was chosen four times as the World's Best Rice at the Rice Trader World Rice Conferences (2012, 2013, 2014, and 2018).[56] CARDI developed and released Phka Rumduol in 1999 with the support of CIAP. The important task of conserving Cambodia's rice varieties continues today under CARDI's leadership, in cooperation with IRRI. The larger strategic role of genetic conservation and gene banks in advancing food security is detailed in chapter 5.

The long-term impact of IRRI's cooperation with Cambodia is difficult to quantify. Nevertheless, it is appropriate, in the light of the preceding historical sketch, to at least reflect on the pattern of rice production in Cambodia from 1961 to the near present (figure 2.3). From a prewar production level of almost 4 million t, annual rice production fell to around 1 million t immediately before and during the Khmer Rouge period because of the flawed policies and technologies described earlier. National production in 1979 was just 540,000 t. By 1985, six years after the Khmer Rouge was removed from Phnom Penh, the country's rice production had recovered to about 2 million metric tons per year (t/yr), primarily through a threefold expansion of planted area. From just 2 million t in 1985, production has subsequently increased to 11 million t/yr in 2020 (a 5 percent annual growth rate), through a combination of further doubling of the harvested area and a more than doubling of yield per hectare to 3.8 t/ha, exceeding the elusive target set by the Khmer Rouge.[57]

Even as Cambodia's population approached seventeen million in 2020, the nation was generating enough surplus rice to emerge as a significant exporter. In 2020 the country officially exported 700,000 t of milled rice, mainly to China and the European Union. Fragrant (jasmine) rice—including CARDI's Phka Rumduol—made up 80 percent of those exports. A further 3 million t of unmilled (paddy) rice was exported informally to neighboring countries. The total value of these formal and informal exports exceeded US$1.26 billion.[58]

The experience of Cambodia—replicated on a more limited scale in neighboring Lao PDR—marked a very different phase of Asia's Green Revolution. Much of the surplus production emerged in rainfed rather than irrigated land. And it

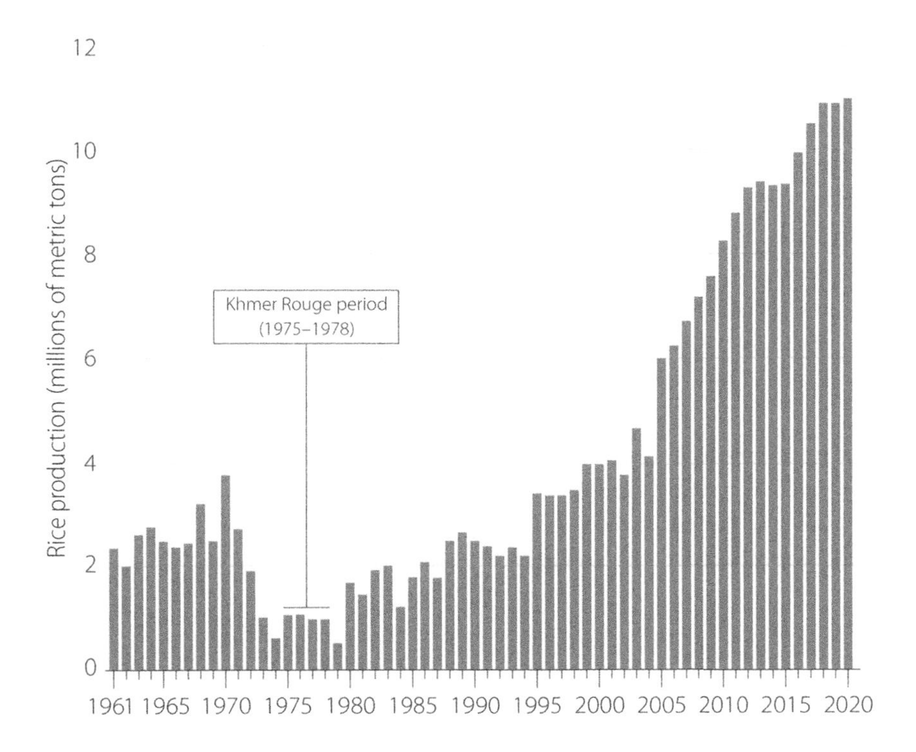

FIGURE 2.3 Rice production in Cambodia, 1961–2020.

Source: FAOSTAT, http://www.fao.org/faostat/en/#data/QCL.

was not IRRI-bred varieties that were making the difference. Instead, most of Cambodia's additional rice came through the reintroduction and further improvement of its traditional rice varieties, coupled with improvements in crop management. While the yield improvements were modest compared with those in Vietnam and Indonesia, where most of the rice is grown under irrigated conditions, adopting improved rice-production practices helped boost rural incomes and reduce poverty in the country.

During my eighteen years with IRRI, I had the good fortune to travel and work in dozens of rice-growing countries, mostly in Asia. In August 1983 I visited Bhutan—the first of more than twenty missions to the country—to support the government in establishing the country's first research program on rice.[59] IRRI colleagues and I also designed and implemented a program in Lao PDR of similar scale to the Cambodian project; it was led by John Schiller and Lao colleagues and had financial support from the Swiss government.[60] As part of the

IRRI–Indochina Program, I partnered with Professor Vo-Tong Xuan, then the vice rector of Can Tho University, to reenergize a long-standing program of cooperation with Vietnam.[61] While the experience gained and lessons learned from Cambodia have stood out among the others, some important success factors emerge as relevant today as we ponder the recovery of nations and regions devastated by war and neglect:

- the transformational potential of science and technology
- the contributions of national institutional and human resource capacity
- the power of international collaboration
- the ingenuity and resilience of local farmers

In June 1998, as I headed to Kenya for the next phase of my journey in understanding food systems and food security, these factors would guide my vision and actions. I was reminded, however, that Africa is different from Asia, and that I should not expect an African Green Revolution.

A UNIQUELY AFRICAN GREEN REVOLUTION

"We are here together to discuss one of the most serious problems on earth: the plague of hunger that has blighted hundreds of millions of African lives—and will continue to do so unless we act with greater purpose and urgency."

So began an address by UN secretary-general Kofi Annan before an audience of several heads of state, dozens of ministers and other senior officials, and assorted development experts and diplomats at the UN Conference Centre in Addis Ababa, Ethiopia, on July 5, 2004.[62] I was charged with organizing the event on behalf of the UN Millennium Project in partnership with the government of Ethiopia. Held on the fringes of the annual African Union Summit to maximize high-level participation, the event's goal was to highlight, as its title event stated, *Innovative Approaches to Meet the Hunger Millennium Development Goal in Africa*. The outcome was a clarion call for action that effectively launched a new effort to tackle hunger in Africa. In Annan's words, "Let us generate a uniquely African green revolution—a revolution that is long overdue, a revolution that will help the continent in its quest for dignity and peace."[63]

The secretary-general's speech starkly acknowledged that Africa faced an existential crisis: it was the only continent where food availability and child malnutrition were worsening. While agricultural productivity had advanced in Asia, Latin America, and the Middle East, there had been no green revolution in Africa.

Drawing on the findings of the UN Millennium Project Hunger Task Force and other recent reports, Annan acknowledged that the continent had lagged Asia and other regions.[64] Farmers were not adopting the high-yielding Green Revolution varieties that had spread throughout much of Asia. And there were very good reasons why they did not.

For a start, most of Africa's agriculture depends on rainfall and is therefore particularly vulnerable to climatic shocks. In his address, Annan described African soils as being depleted of nutrients. At that time, farmers in sub-Saharan Africa applied on average about 10 kg of fertilizer per hectare, just 6 percent of the rate applied by Asian farmers.[65] Annan also highlighted the important role of women in African farming and the failure of institutions to recognize their needs for finance, training, and land rights.

One of the most common questions I heard at that time and still hear today is: Why did Africa not have its *own* green revolution? It is an important question to ask, because if we know the answer, we can begin to invest in correcting the problems that had caused African agricultural productivity to stagnate during the second half of the twentieth century, in sharp contrast to Asia.[66] Unfortunately, there is no simple and elegant answer. In my review of the literature, my discussions with experts, and more than a decade of living and working in Africa, I can offer several influential factors. But before I do, we should address a more fundamental question: *Should* there be a green revolution in Africa?

On the latter question, my answer is an unreserved *yes*. Even in the lead-up to Kofi Annan's Addis Ababa speech, there was considerable back-and-forth with the secretary-general's office on whether to even use the term *green revolution*. Some pundits argued that this term had "baggage" and would not fly with the NGO and academic communities who were influential stakeholders in African development. The InterAcademy Council, also appointed by the secretary-general's office, called for numerous "rainbow evolutions" that acknowledged diverse farming systems and institutions in Africa, "rather than a single Green Revolution."[67] The latter's critics derisively described it as *industrial agriculture* and inappropriate for the continent. It was a narrative that many critics repeat today as a dichotomy—to me, a false dichotomy—between two believed-to-be opposing models of agricultural development: industrial agriculture and agroecology, the latter sometimes referred to as regenerative agriculture.

As we dig deeper into the criticisms, it becomes clear that such negative perceptions of the Green Revolution are locked in a rhetorical 1960s time warp with little recognition of (1) its extraordinary impact on food supply, food security, and poverty reduction, especially in Asia, and (2) the advances made to improve environmental sustainability and local adaptation of technologies associated with a second incarnation of the Green Revolution, referred to by some as a "doubly green revolution" or an "evergreen revolution."[68] It was therefore a victory of sorts that the term *green revolution* survived the editorial red pen of the secretary-general's office and, even more satisfying, that Kofi Annan characterized it as "a uniquely African green revolution," one that would tackle challenges that are unique to the continent.[69] With low agricultural productivity and high levels of hunger and malnutrition, nothing short of a revolutionary approach would be needed.

To return to the question of why the (Asian) Green Revolution had not reached Africa, here are some of the key reasons why Africa (specifically sub-Saharan Africa) did not experience crop productivity improvements at the same time and to the same extent as Asia, Latin America, the Middle East, and North Africa.

Agro-ecological and demographic factors

- **Arable farming land in sub-Saharan Africa is 96 percent rainfed, increasing vulnerability to drought.**[70] In contrast, Asia is predominantly irrigated. An expanded discussion of these contrasting conditions is found in chapter 4.
- **African soils are depleted of nutrients.** Successive crops grown on often inherently poor soils without replenishment led to a decline in soil fertility. World-renowned soil scientist Pedro Sanchez described soil fertility depletion as the fundamental biophysical cause of stagnant per capita food production in Africa.[71]
- **African agriculture is diverse.** There are several different staple foods, including maize, sorghum, millets, rice, cassava, yams, plantains, highland cooking bananas, and teff. Livestock production is also important in single or mixed farming systems across Africa. In contrast, rice and wheat dominate the food systems and food security priorities of Asia. Therefore the quest for an IR8-type solution with widespread impact through a single crop or variety would be futile.
- **Public health challenges remain widespread.** While both malaria and HIV/AIDS were present in Asia, their spread within the agricultural population and the effects on institutional capacity were less damaging.

Economic and political factors

- **Structural adjustment and the dismantling of government support mechanisms to agriculture, including credit, input subsidies, and price stabilization.** In contrast, Asian governments stepped up their support to agriculture to aid the Green Revolution.

- **Poor physical infrastructure, especially the roads and rail network.** Typically, transport networks served resource industries, such as mining, and were not designed to support major agricultural areas.
- **Urban bias.** Maintaining low food prices entrenched urban support for governments.
- **Poor governance and conflict.** These issues are both causes and consequences of poor agricultural productivity.

Of course, these constraints should not be viewed in isolation. Poor transport systems and the absence of credit or subsidies for small-scale farmers (typically farming 2 hectares or less and known as smallholders) mean that fertilizer is expensive, and farmers are less likely to replenish nutrients in their soil.[72] Rain-fed farming is risky and unlikely to attract the interest of banks. The diversity of farming systems makes research on African staples more challenging. It is best to see these factors as part of a complex of interacting circumstances that were suppressing African agriculture. To succeed, a uniquely African Green Revolution would have to address this complexity.

During the year prior to the Addis Ababa event, the UN Millennium Project Hunger Task Force had developed a practical framework for action on hunger with the tagline, "It can be done."[73] The Hunger Task Force identified the need to assist Africa's impoverished smallholders in gaining access to agriculture inputs—primarily fertilizers, high-yielding seeds, and small-scale water management equipment—all within a comprehensive strategy for rural development. In his Addis Ababa speech, Annan captured the essence of a uniquely African Green Revolution that would reflect an understanding of the complex of causal factors outlined above. He pointed out that the knowledge for positive change exists, but "what is lacking, as ever, is the will to turn this knowledge into practice."[74] Annan proposed several interventions, derived from rigorous analysis and experience of the UN Millennium Project and others:

- investments in improving soil health
- water and irrigation investments, coupled with improved access to safe drinking water
- improved seed varieties, including the potential for biotechnology
- roads connecting rural communities to markets
- microfinance to provide access to credit
- small-scale rural electrification and improved cooking fuels
- community-based agricultural extension workers to improve farm management and spread innovations

- community-based health workers to deliver basic health services
- effective antimalarial control

While Annan's speech in July 2004 provided much-needed visibility and impetus, the genesis of the African Green Revolution can be found much earlier. African governments promoted fertilizer use during the 1970s and early 1980s through several interventions, including direct fertilizer subsidies, government credit programs, centralized fertilizer procurement and distribution, and deployment of marketing boards.[75] Improvements in maize productivity were demonstrated in Kenya, Zimbabwe, and Zambia during the 1980s.[76]

Japanese industrialist Ryoichi Sasakawa enlisted Norman Borlaug to bring Asia's Green Revolution to Africa. From 1986 to 2003 Sasakawa Global 2000 and the Sasakawa Africa Association established test plots and demonstrations across the continent to promote new crop varieties, increased fertilizer use, and better crop-management methods.[77] Consistent with experience in Asia, results showed that local production could be easily doubled or tripled. But these surpluses often resulted in localized supply gluts, postharvest losses, and depressed prices. It became clear from this experience that a narrow focus on production was not suited to African conditions, where infrastructure, markets, and financial systems were poorly developed.

Despite these efforts by Sasakawa and others, agricultural productivity in Africa remained stagnant as we entered the twenty-first century. Cereal yields were a little over 1 t/ha.[78] While Sanchez had identified poor soil fertility as the major culprit, the ability to replenish soil fertility was constrained by the risky nature of rainfed agriculture, poorly adapted crop varieties, costly fertilizer due to high transport costs, and the absence of policy incentives and strong institutions to promote technology adoption. These constraints had been overcome in much of Asia, but they remained widespread in Africa. A different, more comprehensive and integrated approach to agricultural productivity improvement would be needed.

During the decade leading to 2015, marking the end of the Millennium Development Goals era, there was a clear stepping up of effort toward the realization of Kofi Annan's uniquely African Green Revolution. While by no means a complete list, the following events were significant:

- **2006.** The Bill and Melinda Gates Foundation and the Rockefeller Foundation founded the Alliance for a Green Revolution in Africa (AGRA) "on the belief that investing in agriculture is the surest path to reducing poverty and hunger in Africa."[79] Upon completing his second term as UN secretary-general, Kofi Annan became the founding chair of AGRA.

- **2007.** The World Bank released its annual *World Development Report 2008* with the focus on agriculture, recognizing that a productivity revolution in smallholder farming would provide the basis for economic growth in agriculture-based countries.[80] The World Bank acknowledged sub-Saharan Africa's unique agriculture and institutions, and that it would require a different approach from Asia's Green Revolution.
- **2009.** The G8 met in L'Aquila, Italy, and the G20 met in Pittsburgh, United States, reinforcing high-level global commitment to achieving food security while highlighting the special needs of Africa, making specific reference to the African Union's Comprehensive Africa Agriculture Development Programme (CAADP) as a vehicle for supporting national food security strategies.[81]
- **2010.** The Global Agriculture and Food Security Program was launched in response to the G8 and G20 commitments. After a decade of operations, GAFSP had channeled US$1.7 billion to sixteen million recipients in forty-five countries, with more than 60 percent of funds going to Africa.[82]
- **2014.** The African Union's Malabo Declaration on Accelerated Agricultural Growth and Transformation for Shared Prosperity and Improved Livelihoods included a commitment to end hunger by 2025, through doubling agricultural productivity, reducing postharvest losses by at least half, and reducing underweight and stunting to 5 percent and 10 percent, respectively.[83] Through the Malabo Declaration, the African Union recommitted to increasing investment in agriculture to reduce poverty.
- **2015.** The African Development Bank prioritized agriculture as one of its High 5 priorities: Feed Africa. The Feed Africa strategy aims "to transform African agriculture into a globally competitive, sustainable, inclusive and business-oriented sector, creating wealth, generating employment, and improving quality of life."[84]

AFRICA DELIVERS

Contrary to perceptions in the international development community, the green revolutions in Africa (2004–2019) and Asia (1966–1981) showed strikingly similar results during their early phases of implementation.[85] In terms of growth in cereal production, sub-Saharan Africa modestly outperformed Asia during the fifteen years following their Green Revolution launch dates (2004 and 1966, respectively) (figure 2.4).

With a baseline of 2004—the launch year of the uniquely African Green Revolution in Addis Ababa—cereal production across Africa had increased by 60 percent after fifteen years, as a result of an increase of 33 percent in crop area harvested and a yield increase of 20 percent (comparing averages for 2002–2004 and 2017–2019).[86] Limiting the focus to sub-Saharan Africa, the production increase over the same period is 76 percent, resulting from a 37 percent increase in crop area and a 28 percent increase in yield. Taking 1966—the year IR8 was released—as the starting point for the Asian Green Revolution, the comparable production increase over the first fifteen years was 62 percent, with an 8 percent increase in crop area harvested and a yield increase of 50 percent (comparing averages for 1964–1966 and 1979–1981). While Asia's Green Revolution came almost entirely from yield increases, Africa's Green Revolution required significant contributions from both land area expansion and yield per hectare.

Africa's most widely grown staple crop is maize. Between 2004 and 2019, across Africa, maize production increased by 83 percent, combining a 51 percent increase in crop area harvested with a 21 percent increase in yield (comparing averages for 2002–2004 and 2017–2019). Narrowing the focus to sub-Saharan Africa, maize production increased by 95 percent through increases in both area (53 percent) and yield (27 percent).

Maize production in Malawi increased by a similar proportion (79 percent) over the same fifteen-year period, coinciding with the operations of the nation's farm input subsidy program (see chapter 9). In contrast to the rest of Africa, however, the production increase was mainly due to a 62 percent increase in national average yield, with just a 10 percent increase in harvested area. In Malawi, a country with limited land resources and a high population density, the increase in maize production mirrored Asia's experience with cereals from 1966 to 1981. Malawi provides a more classical example of Green Revolution intensification and demonstrates the potential for increasing production on existing cultivated land under rainfed conditions, without the need to remove existing forests or extend into more marginal areas.[87]

Despite the 62 percent improvement in yield over the past fifteen years, Malawi's average maize yield remains around 2.0 t/ha, far below the global average of 5.7 t/ha (2017–2019). Maize yields in the United States, also grown mainly under rainfed conditions, averaged 1.6 t/ha between 1866 and 1936, prior to the introduction of hybrid varieties.[88] Since then, yields have steadily risen to an average

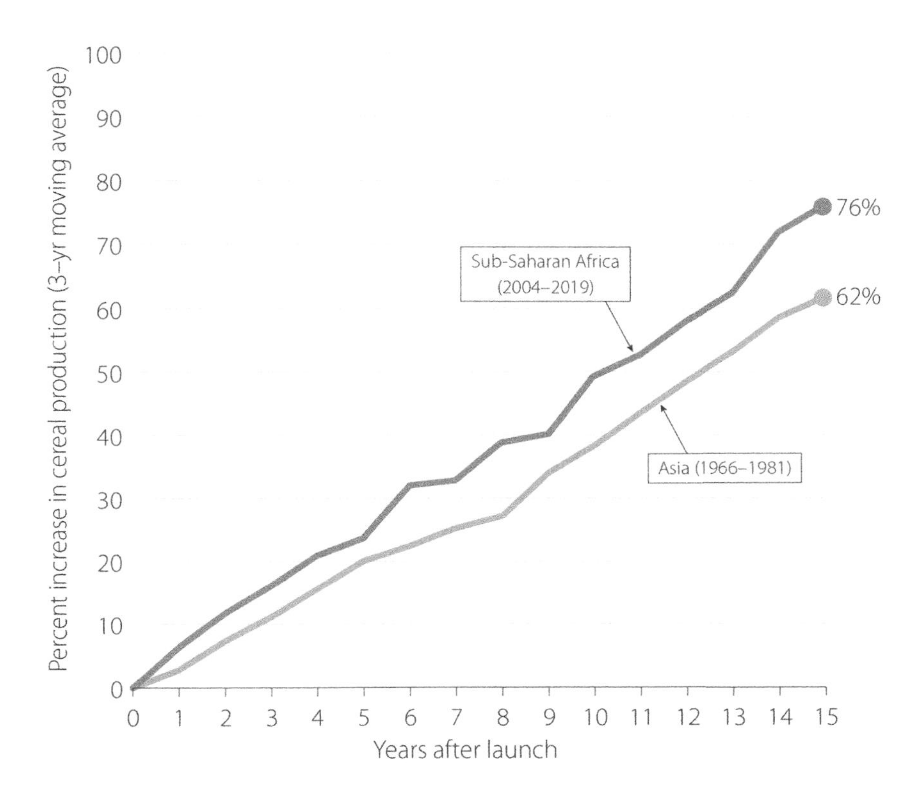

FIGURE 2.4 Increase in cereal production in sub-Saharan Africa and Asia during the first fifteen years of the green revolutions in each region (2004–2019 and 1966–1981, respectively).

Source: FAOSTAT, http://www.fao.org/faostat/en/#data/QCL.

of 11.7 t/ha in recent years (2017–2021).[89] With access to improved varieties and fertilizer, smallholder farmers in the Earth Institute's Millennium Villages Project—across several African countries—obtained maize yields in the range of 5 to 7 t/ha.[90]

There clearly remains scope for at least a doubling of production in Malawi and elsewhere in Africa on existing land.[91] Thomas Jayne and Pedro Sanchez have called for greater intensification of existing land in view of mounting pressure from a rapidly growing population and the environmental costs of further land area expansion. They call for productivity improvements through increased fertilizer use efficiency with greater precision in recommendations, promotion of fertilizer blending facilities, and increased use of organic nutrient sources, coupled with the development of higher-yielding, locally adapted seeds.[92] Just as

Evenson and Gollin observed in Asia, the second phase of the African Green Revolution may prove even more productive than the first.[93]

The early successes and lessons learned from Africa's Green Revolution provide us with a clear and compelling conclusion that a successful green revolution is feasible in Africa. Despite reliance on rainfed farming systems, recurrent droughts, poor infrastructure, and underfunded public institutions and services, Africa has responded impressively to Kofi Annan's call in 2004 for a uniquely African Green Revolution. Malawi, Ghana, and Ethiopia, among others, have shown the way with the contribution of a number of key factors:[94]

- strong and committed national leadership
- applied research to develop and adapt improved technologies
- improved infrastructure to connect farmers to input and output markets
- modern extension approaches to promote technologies
- finance to ensure technologies are widely adopted by farmers
- market support and price stability to encourage investment
- an enabling environment for the private sector

The uniquely African Green Revolution is now underway. Pessimism is giving way to production.[95] But Africa continues to have the highest prevalence of hunger: 21 percent of the population were undernourished, and almost 60 percent faced moderate to severe food insecurity in 2020, while population is still increasing by 2.6 percent per year.[96] Hunger has been on the rise in most African regions since 2015.[97] Drought and conflict are key factors and cannot be addressed by production alone. The disruption of supply chains resulting from the COVID-19 pandemic and the Ukraine-Russia conflict present further challenges to food security in Africa, where most nations remain net food importers.

As is evident from other regions, a successfully executed green revolution approach is but one part of a comprehensive and context-specific food security strategy, as I will elaborate in part III of this book. In addition to sustainable productivity improvement, that strategy should include enhanced market infrastructure for smallholder farmers, improved postharvest stewardship, convergence toward heathy and sustainable diets, and social protection for the most vulnerable. Increased public and private investments throughout the food system will be needed in line with the bold ambitions of the UN's Sustainable Development Goals and the African Union's Agenda 2063, and the growing challenge of climate change.[98] No doubt, Kofi Annan would urge us all to act with even greater purpose and urgency in the quest for our ultimate objective: the end of hunger and the achievement of universal food security in Africa.

The Asian Green Revolution represents one of the greatest achievements in human history. Faced with a looming Malthusian apocalypse in the 1950s and 1960s, scientists, policy makers, and farmers worked together to double cereal production between 1970 and 1995, while population increased by 60 percent over the same period. As a result, cereal availability per person increased by nearly 30 percent, and wheat and rice became cheaper for consumers. The successes of the Asian Green Revolution provided a forceful rebuttal to the Malthus hypothesis. Poverty, hunger, and land conversion to agriculture were all reduced. Africa's recent progress in improving agricultural productivity is giving rise to optimism that the continent may not only meet its own demands but contribute to food security in other regions of the world. But green revolutions are not the panacea for achieving a food-secure world. As Norman Borlaug remarked in his 1970 Nobel lecture, we have been given "breathing space." The challenge we face now is to use that breathing space to invest in a more holistic approach to food systems transformation as a pathway to a food-secure world.

PART II

Knowledge

3

SOIL AND LAND

During the 1930s farmlands across the Great Plains of the United States were subjected to severe and unrelenting drought. Over the previous two decades, encouraged by high demand for wheat from Europe, farmers plowed up vast areas of native grassland. Without soil moisture and the anchoring roots of native prairie grasses, millions of tons of topsoil were swept away by the winds, darkening skies as far away as New York City. The farmlands most directly affected were referred to as the Dust Bowl, a term that would later be synonymous with a decade-long period of land degradation, economic ruin, and social upheaval.[1] Millions of Great Plains farmers and settlers abandoned their lands and migrated west to California and other destinations in search of work. The Dust Bowl is widely regarded as one of America's—and the world's—worst human-created environmental disasters.

When President Franklin D. Roosevelt came to office in 1933, he immediately began programs to provide relief to those people affected by the disaster and to promote soil conservation techniques. In 1935 the U.S. federal government created the Soil Conservation Service (now known as the Natural Resources Conservation Service) to combat soil erosion. In a letter to governors dated February 26, 1937, Roosevelt recognized that "isolated demonstration projects cannot control erosion adequately." He urged states to create soil conservation districts and to recognize that caring for the soil was a matter of national security: "I need not emphasize to you the seriousness of the problem and the desirability of our taking effective action, as a Nation and in the several States, to conserve the soil as our basic asset. The Nation that destroys its soil destroys itself."[2]

In December 1985, on a flight from Port Louis (Mauritius) to Lusaka (Zambia), I peered out my window at another degraded landscape dissected by a winding, rusty-red river. Unlike America's Great Plains, this was a rugged hilly terrain, green but not forested, reminiscent of Mindanao (the Philippines), where I spent three years in the late 1970s. I was looking down at the island nation of Madagascar, a land famous for its exceptional biodiversity and its high concentration of endemic species.[3] Madagascar in the 1980s was one of the poorest countries in the world, mired in poverty and food insecurity.

Several years later, preparing for the first of several missions to Madagascar with the International Rice Research Institute, I read Philemon Randrianarijaona's illuminating paper "The Erosion of Madagascar." Opening with the words, "The Malagasy Republic is being washed away," Randrianarijaona vividly described the results of erosion, with large, deep gullies, known in the Malagasy language as *lavakas,* scouring the hillsides: "These *lavakas* . . . appear as huge cirque-type ravines several hundred meters in diameter whose walls extend in all directions to form characteristic finger-shaped gullies. They are the point of departure for huge mud-flows carrying away soil and sediments—estimated at several thousand tons from each ravine—contributing in spectacular fashion to the silting of riverbeds and their red color."[4]

Widespread soil erosion was the consequence of heavy rains falling on exposed soil. Recent research suggests that *lavakas* may be a feature of the natural landscape and therefore not entirely a product of increased human activity.[5] There is little doubt, however, that Madagascar's red, iron-rich lateritic soils were laid bare through several decades of "slash and burn" (*tavi*) methods, whereby farmers cut and burned forests to grow rice, maize, cassava, and other food crops. After two or three seasons at most, as soils decline in fertility, *tavi* farmers move on to other areas, cutting and burning more forests, inevitably on steeper, less accessible slopes. Other households raised cattle and burned their lands to encourage growth upon the arrival of the rainy season. The result is some of the highest soil erosion rates in the world. As much as 250 metric tons of topsoil per hectare are lost from these fields each year.[6] These sediments are carried downstream, causing siltation and flooding.

The Betsiboka River is the country's largest river system, draining about 50,000 km^2, entering the sea near Mahajanga in northwestern Madagascar. Erosion of lands within the Betsiboka River basin results in sediments in the estuary in the order of 15 to 50 million metric tons per year, producing spectacular plumes extending into the Mozambique Channel that have been observed from the International Space Station.[7] Studies of the seafloor off the coast of northwestern Madagascar found increased deposition of organic-rich terrestrial sediments and a decline in biodiversity since the 1950s.[8] Researchers concluded that intensified tropical cyclones on deforested farmlands may have contributed to the abrupt increase in soil movement from land to sea.

Madagascar remains one of the poorest countries in the world: 75 percent of the population live below the international poverty line, according to the World Bank;[9] deforestation is occurring at alarming rates;[10] agricultural productivity is low;[11] and chronic undernutrition is high.[12] While many factors contribute to Madagascar's perfect storm, including a history of political instability, poor governance, and frequent catastrophic weather events, improvement in food security and quality of life cannot be sustained until the process of soil degradation is reversed. Understanding the nature and role of soils and land is literally the foundation for achieving a food-secure world.

AN INTRODUCTION TO SOIL SCIENCE

We should begin with a working definition of soil. The Soil Science Society of America (SSSA) defines soil as "the layer(s) of generally loose mineral and/or organic material that are affected by physical, chemical, and/or biological processes at or near the planetary surface and usually hold liquids, gases, and biota and support plants."[13] Soil begins at Earth's surface and extends down to the parent rock. It can range from a few centimeters to several meters in depth. Soil is the product of the interaction of that rock with climate, water, and biological activity, conditioned by its location in the landscape, and altered over time by a range of other natural or human activities, just as we saw from those examples from the American Great Plains and Madagascar. Water and wind can erode soil that has been formed on underlying rock, as well as deposit soil formed elsewhere. Volcanic eruptions can deposit nutrient-rich ash on soils, explaining the high productivity of agriculture in areas with a history of volcanic activity, such as the island of Bali and the slopes of Mount Kilimanjaro. And farmers can modify soil with their actions such as irrigation, terracing, plowing, or fertilization.

Soils exist within a landscape. *Land* is a broader concept that includes vegetation, above-ground animals, and even buildings. Soils affect the landscape through the type of vegetation they support. And the landscape affects the soils lying therein. For example, deeper, more fertile soils are normally found in lower and less steep positions along a slope. Later in this chapter, I examine farming practices, soil conservation, and ecosystem services within the context of land and landscapes. But first, it is important to deepen our understanding of soil.

Although I majored in soil science in my undergraduate degree in agricultural science at the University of Queensland, I am not what one would call a card-carrying soil scientist. I have been most fortunate during my career, however, to have worked with three of the world's greatest soil scientists: Nyle Brady, Dennis

Greenland, and Pedro Sanchez. Brady was IRRI director general when I joined the institute in 1980. Prior to IRRI, he was professor of soil science at Cornell University and coauthored the classic textbook *The Nature and Properties of Soils*, now in its fifteenth edition, with Professor Ray Weil, another giant in the field.

Dennis Greenland was deputy director general of IRRI and served as a mentor for much of my early to midcareer. A former professor of soil science at the University of Reading (UK), he had a lasting impact on soil science through his seminal work in Africa on soils under shifting cultivation.[14] While at IRRI, in addition to being patron of the IRRI cricket team, which I captained, Greenland advanced understanding of the sustainability of rice-farming systems.[15] He was always quick to explain that the "best farmers" were invariably on the "best soils."

Notwithstanding the lessons from my IRRI mentors and colleagues, I owe most of my practical education in soils to Pedro Sanchez, whose work in soil fertility in Africa I cited in chapter 2. With a career spanning more than five decades, Sanchez is a leading authority on tropical soils. He worked extensively in Latin America, Africa, and Asia, and I have traveled with him in the field in all three continents. In 2002 he was awarded the World Food Prize, the foremost international award honoring contributions in agriculture and food.

As director general of the World Agroforestry Centre and as director of tropical agriculture at Columbia University's Earth Institute, Sanchez was my direct boss for more than a decade. Throughout those years and since, he has been a friend, colleague, and teacher. In 2019 Sanchez published the second edition of *Properties and Management of Soils in the Tropics*, a detailed yet readable 666-page tome that will be a valued reference for academics, researchers, and practitioners for decades to come.[16] It is one of those rare books that you can set by your bed, open randomly at any page, and find something interesting and useful to read. In this chapter, I draw extensively on the knowledge and experience of Pedro Sanchez, from his book and countless personal conversations over the past twenty-five years.

TOWARD A COMMON LANGUAGE

We can begin with the notion that an improved understanding of soils will enable better decisions by farmers, their advisers, and policy makers on how to manage soils for sustainable food production. The study of soils requires a common language, starting with systems of classification based on observable and measurable characteristics. There are two widely used soil classification systems: the United

States Department of Agriculture (USDA)—National Cooperative Soil Survey (Natural Resources Conservation Service [NRCS]) Soil Taxonomy and the United Nations Food and Agriculture Organization (FAO) World Reference Base (WRB) for Soil Resources.[17]

Soil Taxonomy was initially developed to serve U.S. farmers but is now used more widely. The system enables differentiation of soils based on easily observable quantitative characteristics, along with those measured by standard, reproducible laboratory methods. According to Sanchez, Soil Taxonomy has made a major contribution to the understanding of soils in the tropics. It divides the soils of the world into twelve soil orders and is a hierarchical system of six levels, from soil order down to soil family, the latter being the finest level of differentiation.

Sanchez illustrated the system of classification with a soil family within the Oxisol order, which is the most extensive of the twelve soil orders and covers 25 percent of all land in the tropics.[18] Under the Soil Taxonomy system, Sanchez described this particular soil family as a "fine, kaolinitic, isohyperthermic Rhodic Acrudox." The final syllable *ox* told us that this soil was in the Oxisol order. All that precedes *ox* explains exactly what type of Oxisol we are observing. In this case, the soil has 35 to 60 percent clay (*fine*), with kaolin being the dominant clay mineral. The soil is found in the hot tropics (*isohyperthermic*). It has a deep red color (*Rhodic*), indicating high iron oxide content. The soil has a *udic* soil moisture regime (the *ud* syllable in *Acrudox*), indicating the subsoil is not dry for more than ninety cumulative days in a year. And finally, the soil has a very low effective cation exchange capacity and a pH (acidity) level of less than 5 (the *Ac* syllable in *Acrudox*), both of which are important in determining the ability of soils to retain and release nutrients to plants. It may come as no surprise that the inventors of Soil Taxonomy hired a linguist to help them design the system.[19]

The other main classification system, the WRB, began as an FAO initiative in 1962 to develop a soil map of the world. The objective was to develop an inventory of the world's soil resources with a common legend. Many national soil maps existed at that time, and while there was some common terminology, there would be obvious benefits from "speaking" a common language and sharing information and experience. The WRB has two categories: reference groups and subgroups. There is a degree of correspondence with the USDA–NRCS Soil Taxonomy. While not nearly as differentiated as the Soil Taxonomy taxa, the WRB groups are more widely used in international development circles. There are about thirty reference soil groups under the WRB classification system. Even at the suborder level, WRB remains a blunter instrument than Soil Taxonomy in describing and differentiating soils.

Development practitioners working on agriculture and food security may well ask: How can *we* make use of these systems? Unless you have a doctorate in soil science and/or years of experience as a soil taxonomist, you are unlikely to be able to independently identify and classify soils to even the Soil Taxonomy soil order or the WRB soil reference group level. It can be valuable, however, to find out what, if any, soil maps exist at whatever geographical level you are working at. Knowing the dominant soil orders or soil reference groups gives you a first approximation of what the major soil-related challenges to productivity might be, and where potential for soil improvement exists.

For example, Oxisols are usually red or yellow in color, deep, well-drained, acid, and low in fertility. Long neglected by agricultural scientists, these soils have emerged as highly productive when their fertility is enhanced with the addition of lime (to correct the acidic pH) and phosphorus (the major limiting nutrient).[20] Over the past thirty years, large areas of Oxisols in the *cerrado*, a large savanna ecosystem in Brazil, have been deforested to produce soybeans, corn, and cattle for export. As a result, the *cerrado* is also one of the most threatened natural ecosystems in South America.[21] Drawing on the Brazilian experience, we can begin to understand the constraints and opportunities of the Oxisols that cover about 14 percent of Africa's land area, mostly in the Congo Basin, yet remain undeveloped.[22]

A third system, the Functional Capability Classification (FCC) system (originally known as the Fertility Capability Soil Classification System [FCC] but later renamed with the same acronym), was developed in the 1970s and 1980s to focus on soil attributes that are most important to plant growth.[23] The classification system describes the physical, chemical, and biological soil attributes that are known to be important in determining plant growth and therefore agricultural productivity. Drawing on the FCC system and other soil attributes highlighted by Sanchez and others, I will now elaborate on several key factors that shape the productivity and potential of agricultural soils.

SOIL FERTILITY

A key concept in soil science is soil fertility. Sanchez defines soil fertility as "the capacity of soils to supply essential nutrients to plants."[24] The SSSA similarly focuses on nutrient supply, defining soil fertility as the "quality of a soil that enables it to provide nutrients in adequate amounts and in proper balance for the growth of specified plants or crops."[25] Soil fertility, defined in this way, can be viewed as part of a broader and increasingly used concept known as *soil health*, described by John Doran and Michael Zeiss as "the capacity of soil to function as a vital living system, within ecosystem and land-use boundaries, to sustain plant and animal productivity, maintain or enhance water and air quality, and promote plant and animal health."[26]

It may be helpful to think of parallels to human health and well-being. Both humans and plants grow best with good nutrition, the former being the focus of chapter 7. In both cases, good nutrition is necessary for a healthy and productive life. But it is also accepted that a healthy diet and good nutrition must be combined with mental health, physical fitness, and absence of diseases to achieve an overall state of good health and well-being. Except for mental health (as far as we know), that is true for plants too. Many chemical, physical, and biological features of soils affect plant nutrition. In the case of both soil health and human health, there are important, complex interactions between nutrition and overall health. Acknowledging these interactions, we will focus first on soil as a supplier of essential plant nutrients.

Seventeen elements are essential for plant growth. Three of these—carbon, oxygen, and hydrogen—are absorbed mainly as the core ingredients of photosynthesis and respiration: carbon dioxide (CO_2), oxygen (O_2), and water (H_2O). The other fourteen are mineral nutrients taken up from the soil. These mineral nutrients are divided, somewhat arbitrarily, into macronutrients and micronutrients based on the amounts required by plants. Though there are ongoing arguments about exactly where the boundaries lie, and accepting considerable variation in the relative amounts needed, we can divide the two groups of essential mineral nutrients as follows based on the typical content of plant leaves:[27]

- **macronutrients**—nitrogen, potassium, calcium, magnesium, sulfur, and phosphorus
- **micronutrients**—chlorine, iron, manganese, boron, zinc, copper, nickel, and molybdenum

Tens of thousands of scientific papers and reports have been written about the nutritional requirements of food plants. Millions of soil analyses and field trials have been undertaken to better understand soil fertility, generally with a view to identifying which among these essential nutrients is deficient and how best to correct those deficiencies with a view to improving productivity.

Based on four decades of work as an agronomist in Asia and Africa, I have drawn the conclusion that the two most important plant nutrients are *nitrogen*

(N) and *phosphorus* (P). By importance, I mean that the extent of deficiencies and the response to replenishment in these two nutrients is more widespread than any of the others. These two nutrients are also the primary sources of environmental damage that I will outline later. In no way do I suggest that other nutrients are not critical for production of some plants in some situations. In practice, "the most important nutrient" is the one that is deficient. I have seen rice crop yields in the Philippines add several metric tons per hectare from adding just a few kilograms of zinc fertilizer. Sandy coastal soils often respond well to the addition of potassium. And Pedro Sanchez has argued for greater attention to sulfur deficiency. In *Properties and Management of Soils in the Tropics*, he devotes separate chapters (13–15) to three nutrients: nitrogen, phosphorus, and sulfur.

Before elaborating on nitrogen and phosphorus, I will highlight three important principles of soil fertility management: Liebig's law of the minimum, the synchrony principle, and nutrient cycling.[28] Each of these has important implications for the replenishment, maintenance, and improvement of soil fertility.

Liebig's law of the minimum

Liebig's law of the minimum, popularized by Justus Freiherr von Liebig (1803–1873), a German chemist, states that plant growth will be constrained by the availability of the most deficient essential nutrient.[29] Consider the starting point as a rich fertile soil that generates a wheat yield of 3 t/ha. Each year, nutrients are removed from the soil by the plant and converted to grain, which is then removed from the field. Because this removed grain contains many essential soil nutrients, including a large proportion of the N and P taken up from the soil, there is a net loss of nutrients from the soil, and wheat yield will decline. When a nutrient level drops below what is needed to fuel the biological processes that enable a 3 t/ha yield, it is said to be deficient and yield will decline, sometimes drastically. Only by replenishing that deficient nutrient will we see a return to the original yield level. When seeking to increase yields on a degraded soil, the reverse process is illustrated in figure 3.1, inspired by an illustration from the Sanchez book.[30]

In this hypothetical example, a crop responds to applied nitrogen up to the point where phosphorus supply from the soil is inadequate to support additional growth. Once phosphorus is applied, the crop responds further until it is constrained by another nutrient—perhaps potassium (K), and then a micronutrient such as zinc (Zn).

The law of the minimum can be extended beyond nutrients to factors of production, for example, water availability. If water is limiting growth, no amount

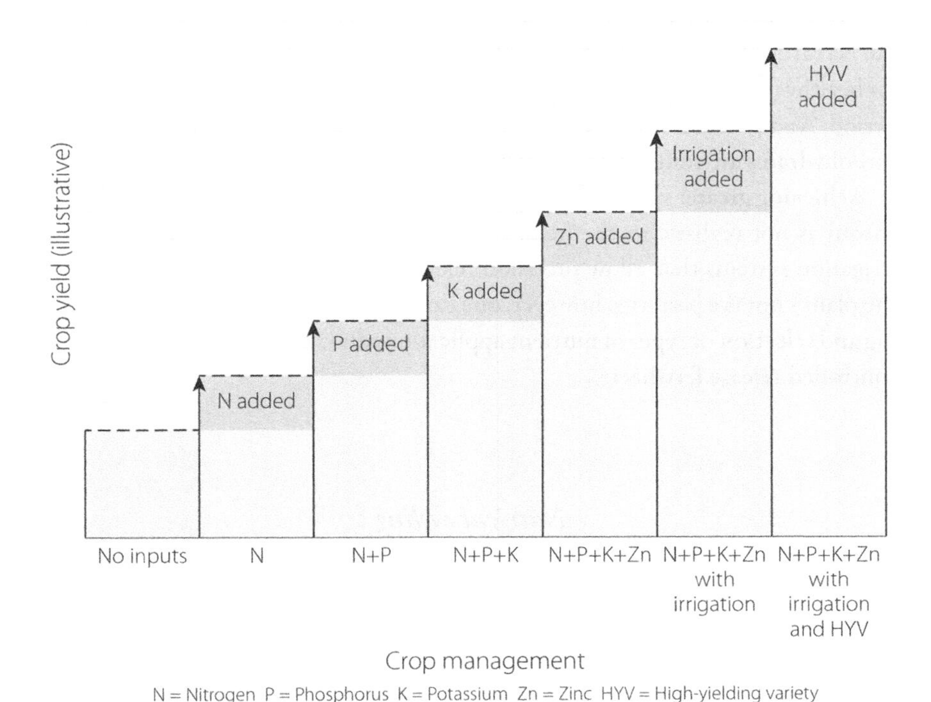

FIGURE 3.1 Liebig's law of the minimum illustrating the crop yield response to progressive additions of individual nutrients, irrigation, and a high-yielding variety.

of fertilizer will compensate for this, and vice versa. Ultimately, yield will be limited by the genetic potential of the crop type or variety. The most important implication of the law of the minimum is that nutrition and other inputs to crop growth need to be in balance.

The synchrony principle

The synchrony principle calls for aligning soil fertility management to the growth pattern and, therefore, the nutrient uptake of plants. In a perfect world, nutrients would be made available to plants on a just-in-time basis. To do otherwise results in periods when nutrients are in surplus or deficit in relation to plant needs. Depending on the soil type, excess nutrients may be lost through leaching, emissions to the atmosphere, and transformation into an unavailable form. As dictated by the law of the minimum, critical nutrient deficits can constrain growth. Different nutrients are needed in different quantities at different stages of growth. For cereal crops like rice and wheat, early application of phosphorus is important for healthy root growth. Nitrogen is needed for photosynthesis in the rapid growth period. And potassium is especially critical during the later stage of growth when carbohydrates are translocated to the grains.

Achieving greater synchrony will improve nutrient use efficiency. Perfect synchrony is not realistic in the field, though it is attempted through some drip irrigation systems that allow precision release of nutrients.[31] By understanding the plant's uptake patterns, however, one can improve synchrony by careful timing and selection of types of nutrient applications and, to a lesser extent, through controlled-release fertilizers.

Nutrient cycling

Nutrient cycling is the third fundamental principle of soil fertility that recognizes the dynamic nature of mineral nutrients in the soil and the atmosphere. Nutrients take on different molecular forms in response to complex chemical, physical, and biological processes. Plant growth leads to the uptake and accumulation of nutrients and their transformation into complex organic forms. Decomposition leads to the breakdown of compounds into forms that can potentially be taken up by plants, lost to the atmosphere, leached from the root zone, or removed through erosion. The molecular form of the nutrient will determine its availability to plants. Each of the fourteen essential mineral nutrients has a unique nutrient cycle.

THE BIG TWO: NITROGEN AND PHOSPHORUS

Nitrogen

Nitrogen is the mineral nutrient that most often constrains food production in both tropical and temperate regions.[32] It is essential for photosynthesis and for the formation of amino acids, proteins, and DNA for all life. Nitrogen gas (N_2) is the most abundant element in Earth's atmosphere, making up 78 percent of the air we breathe. The gas, however, is in a nonreactive, inert form that cannot be absorbed directly by people or plants. Nitrogen fixation occurs when nitrogen gas is converted into a more reactive form such as ammonia (NH_3), which is rapidly converted into ammonium (NH_4+) and nitrate (NO_3-), the two main forms taken up by plants.

Small amounts of atmospheric nitrogen are fixed by lightning. Much more is produced by bacteria, through a process called biological nitrogen fixation, which I will elaborate on later in the chapter.

The importance of nitrogen in plant growth was well established by the mid-nineteenth century. The value of manuring and the role of legumes in rotations were known, but yield declines were causing concern. John Bennet Lawes, owner of the Rothamsted Estate, near Harpenden in England, established Rothamsted Experimental Station in 1843. Working with Joseph Henry Gilbert, a chemist, Lawes conducted long-term field experiments on plant nutrition and agricultural sustainability, some of which continue to this day. Their research showed the need for abundant supplies of nitrogen for all the crops they tested.[33] To counter the unsustainable extraction of nutrients by crops, European and American farmers relied on dwindling and unstable supplies of guano and Chile saltpeter (sodium nitrate) as sources of nitrogen. An alternative source of nitrogen had to be found.

As mentioned in chapter 1, scientists in the late nineteenth and early twentieth centuries saw the potential to transform atmospheric nitrogen into forms that would be available to plants. During the first decade of the twentieth century, German chemist Fritz Haber was able to demonstrate the feasibility of synthesizing ammonia from atmospheric nitrogen and hydrogen gas under high pressure and temperature. Haber's countryman Carl Bosch subsequently developed the means of industrial scale production, which began in Oppau, Germany, in 1913. Haber and Bosch won Nobel Prizes in Chemistry in 1918 and 1931, respectively.[34]

Two dark backstories to the work of Haber and Bosch should be noted for the record. First, the driving force of Germany's interest in the process was the increasing difficulty in sourcing raw materials for explosives. Indeed, the Haber–Bosch process enabled Germany to be self-sufficient in explosives during World War I. Second, following his research on ammonia synthesis, Haber turned his attention to chlorine gas, which he successfully produced for use against the Allied troops in 1915, with high casualties.[35]

The increased availability of synthetic nitrogen fertilizer, through the expanded commercialization of the Haber–Bosch process, led to rapid advances in food production initially in the United States and Europe and ultimately throughout the world. In pursuit of increased food production, through industrial fertilizer production and legume production, we now convert more atmospheric nitrogen into reactive forms than all of Earth's terrestrial processes combined.[36] Urea is the most common nitrogen fertilizer. The distinguished scientist, policy analyst, and author Vaclav Smil attributed the unprecedented population growth during the twentieth century to the industrial synthesis of ammonia. Smil observed: "The world's population now has enough to eat (on average) because of numerous advances in modern agricultural practices. But human society has one key chemical industry to thank for that abundance—the producers of nitrogen fertilizer."[37]

Offsetting these advances in food production (along with increased production of animal feed, textiles, and timber) are a set of now well-documented negative environmental impacts of increased nitrogen use. At a global level, nitrogen (and phosphorus) flows to the biosphere and oceans are among nine *planetary boundaries* (a concept defining environmental limits for safe human existence) that have been threatened by human activity.[38] Increased nitrogen fertilizer use has had many and varied impacts:[39]

- groundwater contamination through leaching of excess nitrate from agricultural land, reducing drinking-water quality and causing adverse human health outcomes—exceeding safe limits in many parts of the world
- eutrophication of freshwater and coastal ecosystems, creating vegetation and algal growth, oxygen depletion, and *dead zones* that cause the loss of livelihoods and threaten food security
- biodiversity and ecosystem-wide changes from deposition of nutrients and acidification on land and water—reducing biodiversity, which may increase vulnerability to shocks and stresses (like drought and pests)
- global warming through emissions of nitrous oxide and methane—two potent greenhouse gases (see chapter 6)

On balance, it is reasonable to conclude that we could not have fed the world population without the industrial and biological fixation of atmospheric nitrogen. But the resulting potent reactive forms of nitrogen have caused damaging environmental impacts. To halt and possibly reverse some of those impacts, it will be necessary to modify nitrogen use practices to reduce losses to the environment.

Nitrogen use efficiency (NUE) measures the ratio of crop yield to applied fertilizer. Pedro Sanchez recommends a more nuanced measure—**agronomic**

efficiency (AE)—that considers the *increase* in yield associated with nitrogen fertilizer application.[40] Our goal must be to increase AE. To do so, we can apply the three principles for improving soil fertility outlined earlier: Liebig's law of the minimum (the need for nutrient balance), the synchrony principle (timing and calibrating applications with plant needs), and nutrient cycling (recycling nutrients through crop rotations, manure use, and composting). Adjusting management practices to reduce nitrogen losses can have the added benefits of lowering application rates and reducing cost to the farmer. Improved efficiencies can also be achieved through the use of slow-release forms of fertilizer such as briquetted urea and polymer-coated urea.[41]

Biological nitrogen fixation is carried out by a specialized group of single-celled organisms that can convert atmospheric nitrogen into ammonia (NH_3), and subsequently to NH_4+ and NO_3–. These microorganisms are said to "fix" nitrogen either as free-living organisms or through a symbiotic relationship with a host plant. In the latter case, the host provides sugars from photosynthesis to the microorganism, which in turn provides usable nitrogen to the host for its growth and development. The most important of these nitrogen-fixing symbiotic relationships in agriculture is the association between legumes (members of the Fabaceae—also known as Leguminosae—family) and bacteria from the *Rhizobium* genus and other related genera (collectively referred to as rhizobia).

Rhizobia live freely in the soil. When they come into contact with the roots of legumes, they cause the formation of nodules that house the bacteria. Under the right conditions—an ample supply of phosphorus and some key micronutrients, particularly molybdenum—these bacteria start their work of converting atmospheric nitrogen into a form usable to the host legume. The nitrogen circulates within the plant and is used to form amino acids and proteins. This explains why food legumes like soybean, peanut, beans, peas, and chickpeas are so highly valued as a source of protein, especially where animal sources are unavailable, expensive, or culturally unacceptable.

In addition to their direct importance in human nutrition, legumes are a key ingredient in animal nutrition, both as a pasture species (e.g., clover) and as a feed grain (e.g., soybean). Also, legumes are deployed to add nitrogen to cropping systems, when grown in rotation or intercropped with cereals, primarily through the decomposition of the legume root nodules. The World Agroforestry Centre, headquartered in Nairobi, Kenya, working with national partners, has undertaken decades of research and development on improving soil fertility through the use of leguminous trees and shrubs in Africa and elsewhere around the world.[42]

I cannot hide my enthusiasm for legumes as a key ingredient of sustainable food systems. From my earliest professional experience in Bali—planting forage legumes between rows of cassava—to introducing mung beans to upland maize systems in Mindanao, to planting soybeans on raised beds in rice paddy fields at IRRI, and most recently in promoting nitrogen-fixing trees and shrubs in Africa at the World Agroforestry Centre, legumes have been part of my toolkit. Legumes, through their unique symbiotic relationship with rhizobia, bring nitrogen into the food system, creating more nutritious grains for human consumption, and more nutritious pastures, fodders, and feed grains for livestock, while improving soil fertility as their stems and roots decompose and serve as a green manure for other crops.

As in almost every aspect of food systems science, policy, and practice, there is always a *however* to add. There are a few caveats to record before we advocate legumes as the solution for all places and all times:[43]

- **The nitrogen from legumes does not come for free.** Yes, the farmer is using legumes to capture nitrogen gas from the atmosphere. However, this requires an investment in land and labor to grow legumes. The lack of rapid, spontaneous adoption of nitrogen-fixing trees and shrubs in Africa was often related to the extra labor required to cut and incorporate legume biomass into the soil. When we plant trees for soil fertility improvement, it is also important to assess any opportunity cost in relation to alternative land uses.
- **Nitrogen fixation by legumes is most productive when the plants are otherwise well nourished.** As noted earlier, nitrogen fixation requires phosphorus and some key micronutrients. When these are deficient, or if there is drought or other stresses, Liebig's law of the minimum kicks in and nitrogen fixation is limited. Therefore we might need to consider adding other nutrients, irrigation, and pest management as additional costs associated with fixing atmospheric nitrogen.
- **Not all legumes fix atmospheric nitrogen.** A case in point comes from Malawi, where a tree legume called *Senna* was promoted widely but did not improve soil fertility. Although some legumes and some other tree and shrub species can improve soil fertility by concentrating nutrients in their leaves and absorbing nutrients from deep in the soil, we cannot assume that all legumes improve soil fertility.[44]

Notwithstanding these three considerations, there is little doubt that legumes play a vital role in improving human and animal nutrition and in sustaining soil fertility, especially nitrogen levels. We shall return later to the role of legumes within the context of sustainable intensification (chapter 9).

Phosphorus

Phosphorus is generally considered the second most important plant nutrient in terms of the extent of deficiencies and the need to replenish supplies. This element is a component of the two main nucleic acids (DNA and RNA) and ATP, the energy-carrying molecule. All plants need phosphorus for photosynthesis, so this nutrient is essential for virtually every process for growth and development. Phosphorus deficiency slows and stunts growth of all plants. As noted, the nitrogen fixation ability of legumes is curtailed when phosphorus is in short supply.

The global phosphorus cycle is very different from that of nitrogen. There is negligible phosphorus in the atmosphere. Phosphorus in soils comes from the weathering of rocks that contain phosphorus, which is released naturally over a geological timescale. With the advent of agriculture, crops relied on the phosphorus in soils to meet their nutritional needs. In traditional farming systems, nutrients including phosphorus were returned to the soil usually in the form of animal manures. Over time, with removal of grain, soils became depleted of phosphorus, as they do with other nutrients when replenishment is inadequate.

Working at his Rothamsted Estate over 150 years ago, John Bennet Lawes was one of the earliest to highlight the importance of phosphorus in agriculture. On a visit there in 2014, I was able to observe a machine that Lawes developed to crush bones, presumably of domestic animals. When treated with sulfuric acid, the phosphorus in the bones became more soluble and available to the turnips grown at Rothamsted. The product—later named superphosphate—was patented in 1842, which enabled Lawes to establish the world's first commercial fertilizer factory.[45] These days, natural deposits of rock phosphate have replaced bones as the source of phosphorus.

Phosphorus deficiencies are common in Oxisols and Ultisols, two soil groups found widely in the tropics. Sanchez notes that, in large areas of Latin America and Africa, phosphorus is the first nutrient deficiency that needs to be corrected, even ahead of nitrogen.[46] Complicating matters, many soils in the tropics have the ability to adsorb phosphorus, greatly reducing its availability to plants, especially under acidic conditions. That is why, in developing these lands, phosphorus applications should be accompanied by the addition of lime (a soil additive made from ground limestone) to raise the soil pH and improve availability of phosphorus, and by the use of species and varieties of crops and pastures that are adapted to this otherwise infertile environment.[47]

Sustainability concerns about widespread phosphorus use in agriculture have been raised for decades. Phosphorus is often referred to as a nonrenewable resource. However, Sanchez has pointed out that tectonic uplift over geological time raises phosphate-containing rocks at a rate similar to that of current phosphorus extraction rates.[48] In practice, our ability to sustain phosphorus fertilizer applications to agricultural land will depend on our ability to locate and extract those resources. For the foreseeable future, it remains prudent to view phosphorus as functionally nonrenewable. These concerns are further exacerbated by the high concentration of phosphorus production in a few countries; China alone produced almost half of the world's phosphorus in 2019. The U.S. Geological Survey estimates that Morocco and the Western Sahara have more than 70 percent of the world's known reserves.[49]

As in the case of nitrogen, we do a poor job of managing this critical resource. Using the planetary boundaries concept, phosphorus use in agriculture has crossed an environmentally dangerous boundary causing massive damage to freshwater and ocean resources. As with our management of nitrogen, we need to improve our efficiency in using phosphorus, drawing on the principles of Liebig's law of the minimum, the synchrony principle, and nutrient cycling.

The question of *peak phosphorus* arises. Estimates of available supplies range from enough for several decades to centuries, depending on the price placed on phosphorus. On the balance of evidence, Sanchez concluded that running out of phosphorus is not a concern for the twenty-first century.[50] In an interview in 2013, he declared: "Once every decade, people say we are going to run out of phosphorus. Each time this is disproven. All the most reliable estimates show that we have enough phosphate rock resources to last between 300 and 400 more years."[51] Nevertheless, it is clear—and Sanchez agrees—that more will need to be done in managing our phosphorus resources more efficiently and increasing efforts to recycle and recover phosphorus from coastal marine sediments.

PHYSICAL CHARACTERISTICS

Beyond the chemistry of soils and the influence of soil fertility on plant growth, it is important to understand the physical characteristics of soils. In this section I focus on two critical and closely related soil parameters: soil texture and soil structure.

Soil texture

Soil texture is an important physical determinant of soil fertility and plant productivity. Soils are made up of particles of different size (diameter): sand (0.05–2 millimeters [mm]), silt (0.002–0.05 mm), and clay (<0.002 mm). Particles larger than 2 mm are not considered in determining texture. Think of a beach sand as one extreme of soil texture. Almost all particles of a beach sand are large and gritty. When dry, the sand slips through your fingers; when wet, a sandy soil is crumbly. For the other extreme, there are clay soils, composed mostly of the smallest soil particles. These soils are slippery and stick to your shoes when wet and rock hard when dry.

Most soils are a combination of sand, silt, and clay. The percentage of each determines the soil texture and is represented by a textural triangle.[52] Various combinations give rise to descriptions of soils such as loam, sandy loam, silty clay, and sandy clay loam. While these textural classifications can be determined easily in the laboratory, I learned to determine texture in the field through what is known as the ribboning method.[53] This involves forming a ball of soil in the palm of your hand, adding moisture, then using your thumb and forefinger to make a ribbon. The longer the ribbon, the higher the clay content. Although it takes practice to get a consistent result, the soil ribboning method is a useful first approximation of soil texture.

Soil texture matters because it determines the ability of a soil to hold water. Soils that are relatively high in sand tend to drain easily. Pour water onto a beach sand and it quickly disappears, and on a sunny day, the surface quickly dries. Do the same on a clay soil and the water may run off the surface or very slowly seep into the soil. Clay soils retain moisture better than sandy soils. That can be an advantage or a disadvantage. Too much water leads to a condition known as waterlogging, which can adversely affect plant growth. Sandy soils, on the other hand, are more susceptible to drought.

Texture also affects the ability of a soil to retain nutrients. Clay particles in the soil serve to retain nutrients, while a soil with high sand content has poor capacity to hold nutrients. Soil leaching describes the process whereby water passing through a soil removes nutrients from the topsoil. Fertilizer applications to sandy soils can lead to low agronomic efficiency and contamination of ground water.

Soils can be more complicated than they look. Some clays hold on to nutrients so tightly that they are largely unavailable to plants. This is the case with the Oxisols that were described earlier. Those soils are high in a clay (kaolin) that binds (fixes) phosphorus. Only when the acidity is reduced by applying lime, coupled with applying inorganic phosphate fertilizer, does the phosphorus become more accessible to plant roots.

Soil structure

Soil structure describes the arrangement of constituent textural particles, organic matter (living and decomposing tissues of plants, animals, and microorganisms), and moisture into clumps or aggregates. Those aggregates are formed through wetting, drying, and biological activities and are influenced by farming operations such as plowing, irrigation, and mulching. While there is no single measure of soil structure, the ideal structure for most production systems is one that permits soil aeration, moisture retention, drainage of excess water, and ease of plant root penetration. Stability of aggregates is an important soil attribute. Organic matter is necessary for binding aggregates.

Tillage is the physical manipulation of the soil to support the establishment of a crop or pasture. Often referred to as plowing or land preparation, tillage can also be undertaken by hand using a hoe, as is commonly observed in traditional subsistence farming systems. Repeated tillage can destroy soil aggregates, expose organic matter to decomposition, and degrade the structure of soils, resulting in increased susceptibility to erosion. This process, in large part, describes the genesis of the Dust Bowl. Later in this chapter I will introduce the practice of conservation agriculture in reducing this form of land degradation.

On the very same lands that created the Dust Bowl, the use of a new generation of farm machines, vividly captured by Charles Mann in his essay on "Our Good Earth," is again threatening the productivity of America's fertile but fragile prairie soils: "Big, heavy machines like harvesters mash wet soil into an undifferentiated, nigh impenetrable slab—a process called compaction."[54]

Compaction is a process that crushes the soil aggregates, decreasing pore spaces and thereby reducing aeration, water infiltration, and root penetration. It can be caused by the passing of heavy machinery or trampling by animals. Some soil types can become compacted through tillage even without the pressure from machinery, a process called hard-setting. Either way, compacting seals off the soil surface, which reduces the emergence of seedlings, reduces the infiltration of water, and results in increased runoff and soil erosion. Adjusting tillage practices to reduce the destructive pressure of heavy machinery and grazing animals, increasing the use of mulch, and fallowing to allow regeneration of grasses and trees have all been shown to be useful in rehabilitating compacted soils.[55]

ORGANIC MATTERS

An immense number and diversity of organisms affect the overall productivity, sustainability, and resilience of soils. A healthy soil will be rich in microorganisms like bacteria and fungi. You will also find larger species like earthworms, ants, and millipedes. Healthy soils harbor thousands of species, making soil the most complex ecosystem component in the natural world and home to a large proportion of the world's genetic diversity.[56]

The living component of the soil plays a major role in decomposing organic matter, enabling nutrients to be released and recycled, and therefore affecting soil health and plant growth. Sanchez distinguishes between the broader term *organic matter* and the more specific *soil organic matter* (SOM). The former refers to above- and below-ground litter, crop residues, manure, dead animals and the like, while the latter—SOM, also known as humus—more specifically refers to organic material below the soil surface that has been partially or fully transformed.[57] Most organic matter and therefore most of the soil's carbon is found within 30 to 50 centimeters (cm) of the soil surface. SOM, more broadly, contains organic forms of essential nutrients like nitrogen and phosphorus that are released and taken up by plants. A component of SOM is soil organic carbon (SOC), which is one of the most important soil properties:[58]

- SOC provides soil-binding properties that help aggregate soil and improve water holding capacity. Aggregate formation helps reduce soil erosion.
- SOC provides cation exchange capacity (CEC), a property that allows soils to retain certain nutrients like potassium and calcium. With low CEC, these nutrients can be quickly lost through leaching.
- SOC provides an energy source for microorganisms. The balance of carbon and nitrogen is important in determining nutrient availability. In the case of rice and other cereals, incorporating too much straw relative to soil nitrogen content leads to nitrogen deficiency and poor yield.
- SOC serves as a carbon reservoir or *sink*, and its management has emerged as a critical component of global warming mitigation.

The soil carbon story is complex. It turns out there are different pools of SOC in soils.[59] The more active fraction is responsible for nutrient release to plants. A different, passive fraction is responsible for sequestering carbon, binding aggregates, and retaining nutrients. Across these various properties, there is consensus that soils with high SOM and SOC are generally desirable.

This topic leads us inevitably to the contested issue of "organic versus inorganic," sometimes communicated as "organic versus industrial" and then, taking a further step, "sustainable versus industrial." What exactly are we talking about? Of course, that depends on what we define as organic. At one level, all crops, all livestock, and all food derived from them are organic. The real issue driving the debate, however, is how closely the production systems are aligned with natural ecosystem functions as opposed to systems that farmers manipulate. Of particular concern to many protagonists of organic agriculture is the use of artificial inputs like inorganic fertilizer and manufactured pesticides. Some argue that organic systems are more sustainable and therefore preferred to production systems that deploy artificial, manufactured inputs. We will return to this in chapter 9 when we discuss the strategy of sustainable intensification. But here, I shall focus on what this issue means in the context of soil fertility management.

Let me start with three inconvenient truths:

- All agriculture involves the human manipulation of genes and the environment. Once we as a species began the move from hunting and gathering to farming around twelve thousand years ago, we no longer depended solely on natural ecosystems. All forms of agriculture are therefore *artificial.*
- Agricultural systems are not closed. Except in the most traditional subsistence systems, nutrients are removed from farms and taken to the points of consumption, increasingly cities, where more than half of the world's population lives. Therefore, over time, there is a net removal of nutrients from food production systems.
- To sustain food production and productivity of the land, those nutrients must be replenished in the proper amounts to ensure healthy growth and development of plants and animals that constitute our diets.

As we saw in chapter 1, Malthus, Crookes, Ehrlich, and others were concerned that agricultural production could not meet the demands of a growing population in the nineteenth and twentieth centuries. The work of Lawes and Gilbert at Rothamsted was prompted by declining productivity of lands that today might be described as organic systems, reliant on animal manures. The depletion of guano and saltpeter deposits in the second half of the nineteenth century led to development and expansion of industrial fertilizer production, especially for nitrogen and phosphorus.

The popularity of inorganic fertilizers, then and now, has a lot to do with packaging and convenience. Inorganic fertilizers deliver nutrients in a concentrated form. It would require 1 metric ton of high-quality dried cattle manure to supply the nitrogen contained in one 50 kg bag of urea (46 percent N)—the most widely used nitrogen fertilizer. The same amount of manure would be required to deliver the phosphorus contained in a 50 kg bag of triple superphosphate (20 percent P).[60] When it comes to convenience, few farmers would opt for the transport and application costs associated with manure use. As described earlier, an alternative organic approach like nitrogen-fixing trees or green manure crops that are grown primarily for their nutrient contributions provides other challenges in terms of the opportunity costs of land and labor. Drawing on an excellent in-depth synthesis of research and practice comparing organic and inorganic approaches, Sanchez draws several important conclusions:[61]

- Plants do not distinguish between the different sources of a specific nutrient like nitrogen or phosphorus. Sanchez has often told me: "Plants don't care."
- Soils do "care" about the source, however. When nutrients are applied in organic forms, like animal manure or decomposing plant matter, they bring carbon and other nutrients that can improve chemical, physical, and biological properties described earlier.
- Based on these two perspectives, the most effective approach in settings where both organic resources and fertilizers are available is to combine the two. This approach, coupled with the use of locally adapted crop varieties, is referred to as *integrated soil fertility management*.[62]
- Neither organic nor inorganic applications cause environmental damage when applied and managed properly, but both approaches (individually or combined) can cause harm when applied incorrectly and in excessive amounts.
- Inorganic fertilizers do not "poison soils" or "make crops hungry" when applied correctly. These are myths with no scientific basis. In fact, the opposite is true. By growing crops *without* adequate nutrient replenishment, lands become degraded to the point of being unsuitable for cropping.
- Organic farming alone is unlikely to meet the world's food requirements. A comprehensive meta-analysis compared the yield performance of organic and conventional farming systems globally. Overall, organic yields are typically lower than conventional yields. The differences are highly contextual, however, depending on the system and site characteristics, and

ranged from 5 percent reduction to 34 percent reduction when the conventional and organic systems were most comparable.[63] Yield reductions of any level would likely require additional land clearing with negative impacts on biodiversity and ecosystem services.

- Organic farming has a role in meeting the preferences of high-income consumers who are willing to pay a premium for an assurance that no inorganic fertilizers or other chemicals were used. There are high levels of uncertainty and disagreement on the health benefits of organically produced food. Based on a meta-review of research by Verena Seufert and Navin Ramankutty, the only clear health benefit of organic foods is the reduced contamination from pesticide residues.[64]

Some of these statements will come as an affront to supporters of organic and other alternative and regenerative approaches that are increasingly attracting the attention of urban elites, hobby farmers, and miscellaneous armchair activists. But like it or not, the conclusions I have drawn reflect the conclusions of peer-reviewed science. That said, drawing experience from other contentious areas like climate change, agricultural subsidies, and genetically modified organisms (GMOs), all of which I address in later chapters, we can conclude that ideological motivations cannot be ignored and often carry the day with policy makers. The line is often drawn uncompromisingly as follows: organic agriculture is good; industrial agriculture is bad.

The mere use of the term *industrial agriculture* is often intended to invite a negative view from audiences. For example, a report published in 2020 by the Biovision Foundation for Ecological Development and IPES-Food derided the Bill and Melinda Gates Foundation (BMGF) for allocating 88 percent of its agricultural research and development resources to "supporting industrial agriculture and/or increasing its efficiency." This report noted that BMGF funding went mainly to "increasing the efficiency of or developing new vaccines" for livestock systems—central to smallholder livelihoods in Africa. The report acknowledged that BMGF was making significant allocations to research leading to "reductions in pesticide use, improved plant varieties, reduced water use and reduced synthetic fertilizer." In a baffling stream of logic, the authors argued for a shift in funding priorities toward an agro-ecological approach based on ecological concepts and knowledge, with less reliance on external inputs.[65]

The irony is that sub-Saharan Africa, the primary focus of the report, is by and large organic or, at best, low-input by default. Throughout the continent, low crop yields are associated with low fertilizer use.[66] Average fertilizer use in 2018 of around 20 kg/ha—compared to 170 kg/ha in South Asia and 294 kg/ha in East Asia and the Pacific[67]—is a major reason why sub-Saharan Africa is food insecure.[68] Sanchez concludes: "Statements by policymakers promoting organic farming in smallholder farms (most of which are nutrient-depleted) as the solution for Africa are not only wrong but irresponsible."[69]

I recall Dr. M. S. Swaminathan speaking about his home country: whatever the experts say about India, the opposite is also true. Adapting this wisdom to the debates contrasting organic agriculture with the use of fertilizers and other manufactured inputs, we may well say with confidence: whatever the experts say about organic agriculture (and for that matter, industrial agriculture), the opposite is also true. Context matters. Science matters. People matter. And it is high time to find space for blended approaches that draw on the scientific and pragmatic truths of organic and agro-ecological approaches, and on the more conventional intensification systems that have fed the world for the past century.

CARING FOR THE LAND

"All the richer and softer parts of the soil having fallen away, and the mere skeleton of the land being left."[70] This quote is an apt description of upland farming areas I have seen in western Mindanao, the Côte Sud region of Haiti, the slopes descending to Lake Victoria in western Kenya, the *tavi* fields of eastern Madagascar, most of Timor-Leste, and, I will throw in, modern-day Greece. These were in fact the words of Plato, in 360 BCE. In describing the mythical land of Atlantis, he painted a landscape where "a single night of excessive rain washed away the earth and laid bare the rock."[71] While the impacts of a single night of rainfall seem a tad exaggerated in a Mediterranean climate, the gullies formed and the muddy waters flowing from a single storm event regularly observed in the tropics and subtropics are evidence of the destructive force of soil erosion.

Soil erosion occurs when there is a "breakdown, detachment, transport, and redistribution of soil particles" in the landscape.[72] In effect, soil is moved by water or wind. What Plato was describing was water-driven erosion enabled by the force of gravity. It occurs on sloping land. The steeper the slope, the faster the water flows, and the greater the erosion. The result of the redistribution is that some parts of the landscape lose soil, while elsewhere soil accumulates. Soils are typically shallower and more degraded on upper slopes. On lower, gentler slopes and on floodplains, the soil is thicker and more fertile. That explains why, over centuries, we have seen civilizations emerge based on the agricultural productivity of great floodplains like those of the Ganges, Yellow, and Mekong Rivers in Asia. Egypt, too, benefited from the silt-laden flows of the Nile before the construction of the Aswan Dam.

From a practitioner's standpoint, we should view soil as a nonrenewable resource. It is true that soil is produced through the weathering of rocks. But this occurs at a very slow rate. Professor David Montgomery of the University of Washington estimated a median soil production rate of about 0.02 mm/year across 188 locations.[73] In other words, it will take 50 years to produce a millimeter of soil and 500 years to produce a centimeter of soil.

Fundamental to understanding the consequences of erosion is a simple equation reproduced by Montgomery based on earlier work by M. J. Kirkby:

$$Tc = Si/(E - P)$$

where

Tc is the critical time to erode completely through a soil profile,
Si is the initial depth of the soil,
E is the rate of erosion, and
P is the rate of soil production.

Under native vegetation, Montgomery found that erosion was more or less in balance with the soil production rate. In principle, therefore, *Tc* was infinite. Under conventional agriculture (i.e., plowing of fields), the median rate of erosion across 448 locations was about 1.53 mm/year, which is equivalent to about 18 metric tons/hectare/year (t/ha/y). At this rate of erosion, a 1-meter-deep soil would last more than 650 years; a soil of 10 cm depth would be gone in about 65 years. The rate of erosion varies greatly, however, reaching more than ten times the median rate in the most vulnerable locations.[74] As mentioned earlier, the sloping lands of Madagascar can lose up to 250 t/ha (or about 2 cm depth) of topsoil each year. If we apply Montgomery's equation for a soil of 1 m depth, we could expect that soil to disappear in 50 years.

Montgomery concluded that "if agricultural erosion rates remain far beyond rates of soil production, global society will eventually be compelled to either adopt agricultural methods that sustain the soil or face increasing competition over a shrinking agricultural land base."[75] To design and deploy more sustainable methods of soil and land management, we must first understand the primary determinants of water-driven soil erosion. These are:[76]

- **Rainfall intensity and duration.** The erosive force of rainfall is greater with storms of higher intensity and duration.
- **Soil type.** Heavy clay soils (because they resist detachment) and sandy soils (because water runoff is slow) are the least susceptible. Soils with a high silt content are the most erodible; they are easily detached and tend to crust, resulting in faster runoff of water.
- **Steepness and length of the slope.** Steeper and longer slopes result in faster runoff. Gravity does the work.
- **Vegetation cover.** Plants protect the soil from the impact of raindrops and slow the flow of water down slopes. Roots also help bind soil and allow infiltration of water, thereby reducing runoff.
- **Protective structures.** Terraces and contour banks, often planted with trees and cover crops, reduce the steepness and length of slopes. Check dams, comprising stone and gabions placed in gullies, can slow water flow and trap soil.
- **Tillage practices.** Increased tillage results in greater soil pulverization that increases susceptibility to erosion. Conservation agriculture, including no-till farming, is practiced to reduce erosion.

For thousands of years, mountain communities have constructed terraces as a way of cultivating steep hillsides while conserving the soil. The Ifugao rice terraces in the Cordillera mountains of the Philippines—a World Heritage List cultural site since 1995—have provided food security for these mountain communities for more than two thousand years. Narrow rice terraces carved from hillsides, reinforced with stone and mud, enable the ponding of water and the growing of traditional rice varieties that are adapted to the cooler temperatures of these high altitudes. These elaborate rice farming systems represent a mastery of engineering and can be observed in other parts of Asia, including on the island of Bali, in the Chinese province of Yunnan, and the mountainous foothills of the Himalayas.

Unfortunately, these magnificent landscapes are under threat. Migration to urban areas has resulted in shortages of labor needed to maintain the terraces. In addition, the high productivity of rice in the lowlands—a result of the Green Revolution described in chapter 2—makes farming the high-quality, yet low-yielding local varieties unprofitable in comparison. Moreover, climate change appears to be

FIGURE 3.2 Terracing of sloping land, coupled with the protection of forests, conserves soil, water, and biodiversity in Bhutan (1988).

Credit: Glenn Denning.

causing havoc with the reliability of traditional water sources. On a recent visit to Bali's Jatiluwih Rice Terraces, another World Heritage site, it was apparent that eco-cultural tourism may be the last hope for the survival of these engineered landscapes that have effectively conserved soil, water, and cultural resources for millennia.

In more recent years, on the lower and gentler slopes, there has been growing interest in a set of practices that are collectively known as *conservation agriculture*. A central component of conservation farming is minimum or, more commonly, no-tillage (no-till farming). Instead of plowing the soil before planting, farmers sow seeds directly into the undisturbed ground. To do so, they may use any of a variety of implements—ranging from a multirow mechanical seeder that slices open the ground and delivers and covers the seed in a single operation to a simple pointed planting stick used by subsistence farmers to form a hole for the seed, which is then dropped in and covered by soil.

Aside from including reduced tillage, there is no universally agreed definition of conservation agriculture. A second key component of the conservation agriculture package is mulching, which is the practice of covering the soil with residues (e.g., straw) from a previously harvested crop or growing and incorporating live mulches, preferably nitrogen-fixing legumes. This practice reduces soil exposure to the impact of rain and slows water movement across the soil surface, thereby reducing erosion. Mulching has the advantage of conserving soil moisture, which may be lost with plowing. Leguminous mulches also contribute to improving soil fertility. Mulching can also suppress weeds.

A third practice associated with conservation agriculture is crop rotations to diversify the sequence of species grown on a field. Crop rotations have the advantage of breaking cycles of pests and diseases. Incorporating legumes in the rotation can improve soil fertility. Most agronomists, however, regard this as good practice generally, and not unique to conservation agriculture.

One of the purposes of plowing is to reduce competition from weeds. For large, commercial farmers practicing no-till farming, there has been widespread uptake of genetically engineered, herbicide-resistant crops coupled with herbicide use. Smallholder farmers in low-income settings are more likely to rely on mulches and hand-weeding. Efforts to promote the spread of conservation agriculture in Africa have been constrained by the absence of herbicide-resistant crop varieties, which remain banned in most countries. One promising innovation is the roller-crimper that cuts and incorporates crop residues and mulches. Developed for application to organic agriculture at the famed Rodale Institute in eastern Pennsylvania, the equipment has been modified and adapted for use in Africa by the Howard G. Buffett Foundation.[77]

Opinion is divided over the potential for expanding conservation agriculture in the developing world, where the highest soil erosion rates are found.[78] Bernard Vanlauwe and colleagues have identified fertilizers as necessary for raising the productivity of conservation agriculture systems and thereby increasing the availability of crop residues as mulch.[79] Ken Giller argues that some of the claimed benefits of no-tillage farming (including carbon sequestration) remain unproven, especially in developing country settings. Giller highlights the labor constraints, limited access to herbicides, and absence of organic resources to serve as mulches, concluding that "an across-the-board recommendation of conservation agriculture is misplaced."[80]

Conservation agriculture emerged from a desire to improve our care for the land. In arid and semiarid areas, the main issue is wind erosion. The experience of the Dust Bowl demonstrated the fragility of land when disturbed by plowing and accompanied by dry conditions and winds. Overgrazing and fires also render the land more susceptible to wind erosion. The most immediate effect is that fertile topsoil is removed. In particular, the removal of clay, silt, and organic matter damages soil structure and biological activity, reduces water holding capacity, and lowers overall productivity.[81] A related impact is that dust is formed and transported, resulting in respiratory diseases and economic disruption thousands of kilometers away.

Essentially the same set of factors that influence water erosion affect wind erosion, though instead of rainfall intensity and duration, a key driver is drought intensity and duration. It is likely that higher temperatures and increasing frequency of extreme weather events with climate change will increase the vulnerability of farmlands to wind erosion.[82] In addition, soil type, topography, vegetation, structures, and tillage practices all influence the rate of wind erosion. Michael Duniway and colleagues from the U.S. Geological Survey in Utah and the University of Colorado conducted a comprehensive review of wind erosion in the undeveloped, uncultivated drylands of the southwestern United States. They concluded that wind erosion can be mitigated by three factors: cohesion of soil particles—crusts, aggregates (clods), and moisture; presence of nonerodible objects in the soil, such as rocks and plant parts; and surface roughness to create turbulence and reduce wind velocities, for example, plant canopies. Translating this knowledge into practice for application to farming, we arrive at three broad strategies that are well supported in a range of settings:[83]

- For cropping systems, keep the soil covered with crops, crop residues, or mulches; avoid bare exposed soil; plant in strips; use no-till conservation agriculture.
- For livestock systems, avoid overgrazing by managing stocking rates and feed resources.
- Plant trees and shrubs as windbreaks or shelterbelts to reduce wind speed.

There is no doubt that the visual spectacle of Plato's "skeletal" landscapes induces a desire to act with purpose and without delay. The World Food Prize recipient in 2020, Professor Rattan Lal, has researched land degradation for five decades. In a contribution to *Advances in Soil Science* in 1990, Lal observed: "The literature is full of horror stories. From the level of technical information presented, however, it is often difficult to judge whether an author is 'crying wolf' or the threat to natural resources and the environment is genuine."[84]

Yet, even then, the evidence was sufficiently compelling for Lal to conclude that erosion is a universal phenomenon in large part "driven by human needs, greed, shortsightedness, poor planning, and cutting corners for quick economic returns." And this applies equally well to the unsustainable extraction and depletion of essential soil nutrients, along with other forms of soil and land degradation described in this chapter. The technical know-how exists to protect and replenish our soil and land, but the application of that knowledge requires policies and the leadership to act. Indeed, as President Roosevelt wisely observed: "The Nation that destroys its soil destroys itself."

4

WATER RESOURCES

Understanding the role of water in achieving food security begins with the most fundamental chemical process in the global food system: photosynthesis, the light-driven mechanism used by green plants (and most algae and some bacteria) to convert carbon dioxide and water into carbohydrates and oxygen. These carbohydrates become the energy source for growth and development of food plants, which, along with the oxygen produced, support livestock and aquatic production systems. Water is essential for photosynthesis and other physiological functions for crop growth and development.

Water shortages impede plant growth. Excess water—waterlogging—can also reduce growth by interfering with oxygen supply to plant roots, preventing uptake of nutrients. There is great genetic variation among different species and varieties of food-producing plants in their ability to withstand water deficits or excesses. What is crucial to understand is that water is one of the fundamental building blocks of all food systems. Agriculture accounts for 70 percent of global freshwater withdrawals.[1]

In this chapter I focus primarily on the role of water in food production. At the same time, I recognize that food systems have water footprints that extend well beyond the production stage of food value chains. In addition, while beyond the scope of this book, water for direct consumption and for sanitation and hygiene is a prerequisite for effective food utilization and food security.

In 2019 the World Resources Institute (WRI) released an assessment of water security to help decision makers understand water risk and identify possible solutions amid concerns regarding water availability, water quality, climate change, and growing user demand.[2] One of the water risk indicators was *baseline water stress*, an index that measures total water withdrawals in relation to available renewable surface and groundwater supplies. Water withdrawals include domestic, industrial, and agricultural uses. Using a scale of 1 to 5, higher values indicate more competition among users. As is often the case with such reports, all eyes were on the country rankings.[3]

The WRI report revealed that seventeen countries face "extremely high" levels of baseline water stress, where irrigated agriculture, industries, and municipalities withdraw more than 80 percent of available supply on average every year (resulting in a score of 4 to 5). Twelve of the seventeen most stressed countries were in the Middle East and North Africa (MENA) region. Qatar ranked number 1, with a score of 4.97. In fifth place (rising from sixteenth place in 2010), with 4.56, was Jordan, a country I first visited in 2006 and that would serve as my personal introduction to the special challenges faced by water-stressed nations in achieving food security.[4]

In a world of water crises, Jordan deservedly attracts much attention. An article by Elizabeth Whitman in *Nature* in September 2019 was entitled "A Land Without Water: The Scramble to Stop Jordan from Running Dry."[5] Whitman describes the perfect (dust) storm faced by Jordan, given:

- the country being extremely dry to start with—Whitman described it as "a toasted expanse of desert"
- climate change reducing the already meager rainfall (50 mm annual rainfall in most parts) and slowing the already unpredictable aquifer recharge, as well as increasing evaporation loss with rising temperatures
- rapid growth in the population, especially the conflict-induced arrival of refugees from Iraq and Syria, increasing demand and overexploitation of groundwater and other sources
- Jordan's lack of petroleum resources restricting its options for energy-intensive desalinization—a situation not shared by most of its MENA neighbors

A more recent modeling analysis by Jim Yoon and colleagues published in the *Proceedings of the National Academy of Sciences* noted that aggressive action would be needed to achieve some semblance of water security, requiring "an ambitious portfolio of interventions that span supply- and demand-side measures, including large-scale desalinization and comprehensive water-sector reform."[6] Among the measures under consideration was a transfer of groundwater access from agriculture to the municipal (urban) sector, a radical policy option that would affect people's lives and livelihoods.

Jordan clearly occupies the extreme end of the spectrum of vulnerability to water shortage, which has the potential to further destabilize an already fragile political and social setting. In traveling along the Jordanian side of the Jordan River valley during several visits since 2006, I was astonished to observe relatively high-water-consuming, low-value crops like bananas and tomatoes being grown. Surely that would need to change. Jordan and other water-stressed countries must plan for an uncertain future. The good news is that substantial research and development has been undertaken to improve our understanding of water and its role in food systems. In this chapter I summarize current understanding and best practices for the sustainable management of water for improving food security.

THE COLOR OF WATER

We can broadly divide crop-based agriculture into two systems, based on the source of water used by plants: rainfed and irrigated. As the name suggests, rainfed agriculture refers to systems of food (and nonfood) production where crops—annual or perennial—depend on water coming entirely from rainfall. Rainwater infiltrates the soil, reaches the root zone, and is taken up through root hairs and transported through the plant. Irrigated agriculture requires the manipulation of water resources to augment or, in some cases, entirely replace the need for direct rainfall—a necessity in a country as dry as Jordan. Irrigated agriculture may require storage structures, diversion weirs, distribution canals, and pumps to convey water from surface or underground sources to the fields where crops are grown.

In 2004 Malin Falkenmark and Johan Rockström introduced a new paradigm for water resources planning and management, based on the concepts of *green water* and *blue water*.[7] Rainfall is the original source of both green and blue water. The green water resource is water vapor or moisture held *in the soil*, and it can be taken up by crops—a process necessary for plant growth called transpiration—or lost to evaporation from the plant and soil surface. Blue water is liquid water that percolates beyond the plant roots or flows across the land surface. This blue-water resource reaches rivers, lakes, dams, and aquifers that can be exploited for irrigated agriculture, as well as for human and animal consumption and for industrial use.

Blue water has long been the primary focus of investment to improve agricultural productivity and food security. But during the past two decades, high cost and sustainability concerns associated with irrigation investments have focused attention on improving the utilization of green water for rainfed agriculture.

RAINFED FARMING: RISK AND REWARD

Representing about 80 percent of the world's cropland, rainfed farming produces just 60 percent of the world's food.[8] Rainfed agriculture is especially risky for smallholders who depend on production for their food security. In practical terms, it is impossible for farmers to control when it rains or how much it rains. In rainfed conditions, farmers face difficult decisions on when to plant and whether to invest in inputs like seeds, fertilizer, and labor.

As I have learned many times over the years, if you plant early on the first rains and the follow-up rains do not come, you can lose your investments. Delay planting and you may run into drought at the end of the season or suffer from an outbreak of pests and diseases that build up on earlier planted crops. If there is too much early rain, fields may be flooded and waterlogged. On plowed sloping lands, soil erosion may occur, carrying away the seeds and other inputs and ruining the possibility of a healthy crop stand. Later in the rainy season, crops may also be vulnerable to dry spells or high intensity rainfall, each of which can have devastating effects on crop growth and yields. Being a rainfed farmer can be a stressful life.

About 96 percent of arable agriculture in sub-Saharan Africa is rainfed, leaving the continent vulnerable to rainfall variability.[9] In these circumstances, farmers, especially smallholders, are reluctant to invest in purchased seeds, fertilizers, and other inputs, explaining in part why sub-Saharan Africa has experienced low agricultural productivity relative to Asia. This variability, associated with the practice of rainfed agriculture, is experienced on many levels. At the household and community levels, reduced yields and sometimes complete crop failure create food shortages, hunger, and even starvation. Experienced at scale, this can lead to a catastrophe affecting regions of a country, sometimes extending across several countries, and even having global impacts, as we witnessed during the 2007–2008 global food crisis.

So, what scope do we have to reduce vulnerability and improve productivity of green water in rainfed agriculture? There are options.

In water-constrained environments, we should aim to maximize the utilization of green water by rainfed crops through improving water productivity. This can be achieved through three complementary strategies:

- **harvesting**—"the process of concentrating rainfall as runoff from a larger catchment area to be used in a smaller target area"[10]
- **conservation**—reducing on-site water losses to evaporation, percolation beyond the root zone, and blue-water flow across the soil surface (runoff)[11]
- **utilization**—improving the capacity of crops to utilize the available green water (plant water use efficiency)

Countering what they call a "blue-water bias," Rockström and Falkenmark argued in 2015 in *Nature* for greater use of small-scale water-harvesting methods to conserve soil moisture, as found in Kenya and India, and using on-farm dams and ditches to channel runoff into fields, as used in Eritrea, Ethiopia, and Israel.[12] These interventions have improved the stability and productivity of crop yields. With the momentum of the 2030 Agenda and the Sustainable Development Goals, approved by the United Nations in the same year, they argue persuasively for increased investment in capacity, planning, and infrastructure development for water-harvesting systems to advance the untapped potential of rainfed agriculture.

The same set of conservation practices that reduce soil erosion, as discussed in the previous chapter, are important in reducing on-site water losses. In essence, we want to minimize the exposure of soil to reduce evaporation losses, maximize water infiltration and retention, and minimize water runoff. This can be achieved through minimum or zero tillage, using live or dead mulches, intercropping, contouring fields, and terracing. Building up soil organic matter also improves water retention.

A crucial complementary strategy is to ensure the crop makes best possible use of the accessible green water that we have conserved. That result is achieved by ensuring a well-adapted crop type, an adequate supply of soil nutrients, and control of weeds that may compete for both the water and the nutrients. Recall Liebig's law of the minimum from chapter 3. If we increase water availability to the plants, we want to ensure that genetic potential and nutrients are sufficient to meet the crop's growth demands (see figure 3.1). In other words, by applying an *integrated* system of soil and water management, we improve green-water productivity and achieve "more crop per drop."

But how does this work in practice? Consider first the case of Australia, where rainfed (also known as dryland) agriculture has provided the bulk of the country's food production and exports for two centuries. Australia is the world's driest inhabited continent. Farm sizes are large and growing, often exceeding 2,000 ha. Yields are volatile, with droughts regularly extending for several years. Yet even in these circumstances, Australian rainfed farms have remained viable and competitive without direct government subsidies. The key to success has been improved agricultural productivity through reduced costs and increased yields, mainly by adopting innovations such as minimum or zero tillage, crop stubble retention, improved varieties, weed control, and precision agriculture technologies that better target fertilizer applications. These innovations have resulted in a higher yield per unit of water from rainfall. For example, the water use efficiency of rainfed wheat in Western Australia doubled between 1980 and 2013.[13]

In southern Africa, rainfed farming faces challenges like those experienced in Australia. In Malawi, where farmers mainly grow maize on fields rarely exceeding 2 ha, yields are also volatile. The high risks associated with rainfed farming in low-income rural settings are a disincentive to investment. Farmers typically take a conservative view to avoid catastrophic crop failure that could lead to hunger, debt, and sometimes the loss of assets through forced sales. They invest at a lower level than would be expected based on outcomes averaged over time.[14]

In behaving in that risk-averse way, Malawian farmers forfeit opportunities to take advantage of more favorable seasons. In these circumstances, farmers would be better served by rebalancing hazard management with efforts to capitalize on better rainfall. Institutional support is required. As described in chapter 2 and further elaborated in chapter 9, Malawi's government took a bold step in subsidizing fertilizers to remove some of the risk associated with increased investment. As a result, average yields doubled, meaning that—just like for their Australian counterparts—water use efficiency doubled. Improved climate forecasting, information support, including decision-making tools, and index-based insurance and credit can further remove constraints to increased investment in rainfed agriculture.[15]

These very different examples from Australia and Malawi illustrate the high potential for improving water use efficiency in rainfed environments. Rockström and colleagues provided one of the most powerful analyses I have seen demonstrating the case for investment in rainfed crop production. Drawing on dozens of studies on cereal crops in temperate and tropical environments, they showed that water productivity increased exponentially as cereal crop yields rose from 1 to 3 t/ha. For smallholder farmers, like those in Malawi, a doubling of production from 1 to 2 t/ha reduced the amount of water required per ton of grain by more than 40 percent.[16] Of course, to achieve that doubling of yield, improved varieties, fertilizers, and good crop management are needed. We need to invest in rainfed systems to achieve these major improvements in productivity.

IRRIGATED FARMING: EXPLOITING BLUE WATER

Parākramabāhu I, king of Polonnaruwa (r. 1153–1186), is known for having united the island that we know today as Sri Lanka. He reformed Buddhist practices, encouraged the arts, and sent successful military campaigns to South India and Burma—all impressive achievements. But for an agronomist like me, this Singhalese monarch is best known as the author of the following exhortation to better manage the kingdom's water. He called on his subjects to go beyond rainfed agriculture and tap the blue-water resources—the rivers and dams—of his island nation:

> In the realm that is subject to me there are, apart from many strips of country where the harvest flourishes mainly by rain water, but few fields which are dependent on rivers with permanent flow or on great reservoirs. Also by many mountains, by thick jungle, and by widespread swamps my kingdom is much straitened. Truly in such a country not even a little water that comes from the rain must flow into the ocean without being made useful to man.[17]

Around 20 percent of global cropland is irrigated, but it produces a disproportionate 40 percent of the world's food.[18] Investment in irrigation accelerated in the second half of the twentieth century, resulting in a tripling of the global irrigated area to more than 300 million hectares. The largest increases in irrigation were in East and South Asia and were an important factor contributing to Asia's Green Revolution. Irrigated land, as a proportion of arable land, is highest in Asia at around 48 percent compared with 17 percent in South America and just 4 percent in sub-Saharan Africa.[19]

Irrigation reduces the risks associated with farming. By reducing vulnerability of food plants to rainfall variations, farmers are encouraged to invest more in other inputs like improved seeds and fertilizer. Irrigation also enables an increase in cropping intensity: more crops can be grown sequentially. In the case of rice production in Asia, most rainfed areas allow for a single rice crop to be grown during each rainy season, sometimes followed by a shorter-duration crop that requires less water. With the provision of reliable irrigation, double cropping becomes possible. In the humid tropical regions of Southeast Asia, double cropping of rice is common in irrigated areas. In South Asia, double cropping more often entails a rice crop grown during the monsoon season and a wheat crop grown during the cooler and drier winter months.

FIGURE 4.1 Irrigation allows farmers to invest more in inputs like improved seeds and fertilizer, and to increase the number of crops planted each year.

Credit: Image collection of the International Rice Research Institute (https://www.irri.org).

The rice–wheat rotation is a common feature across the Indo-Gangetic Plain, covering more than 13.5 million hectares.[20] Agricultural land occurs in sediments eroded from the Himalayas that formed a flat, fertile plain that extends across Pakistan, northern India, southern Nepal, southern Bhutan, and Bangladesh. The Indo-Gangetic Plain is the largest surface water irrigation system in the world, redistributing water from the Indus and Ganges Rivers to millions of smallholder farmers through a canal network 100,000 km in length. Groundwater abstraction—a process of removing water from below-ground aquifers through wells or bores—has emerged as a driving force of agricultural intensification in this vital regional food bowl. There are an estimated fifteen to twenty million tube wells, representing a quarter of the world's groundwater withdrawals.[21]

With the Indo-Gangetic aquifer being such an important water resource for food production in the region, there have been vocal concerns regarding the sustainability of these food production systems, especially with the anticipated impacts of climate change.[22] Important new evidence from high-resolution in situ records of groundwater levels, abstraction, and water quality has shown, however, that groundwater supplies are constrained more by contamination than by depletion. Published in *Nature Geoscience*, the study by A. M. MacDonald and collaborators revealed that between 2000 and 2012 the water table fell across 30 percent of the Indo-Gangetic aquifer but was stable or rose in the other 70 percent. Excessive salinity and arsenic levels, caused by groundwater abstraction practices, were shown to constrain potable water use in 60 percent of the aquifer.[23]

The sustainability challenges observed in the Indo-Gangetic Plain are important in their own right because of the effects they have on food supplies and drinking water quality for more than a billion people in the region. Beyond that, we are reminded of the centrality of irrigation in supporting stable food supplies locally, regionally, and globally. The Sustainable Development Goals address this issue in the SDG 2.4.1 indicator, which monitors "the agricultural area under productive and sustainable agriculture," and the SDG 6.4.2 indicator, which tracks "freshwater withdrawal as a proportion of available freshwater resources." We need only look to the Aral Sea disaster to remind us of what goes wrong when we exploit a water resource beyond its capacity.

Located between Uzbekistan and Kazakhstan, the Aral Sea was once the world's fourth largest lake, with a drainage basin of 1.8 million square kilometers. In the 1960s the former Soviet Union diverted two rivers—the Amu Darya and Syr Darya—to provide irrigation, mainly for cotton production. The project, aimed at making the desert bloom, has emerged as one of the world's greatest environmental disasters. By 2006 the (once) freshwater sea's level had dropped 23 meters, area had decreased by 74 percent, volume of water decreased by 90 percent, and salinity of the southern part increased by more than 1,000 percent (figure 4.2). Today, irrigated agriculture in the deltas of the Amu Darya and Syr Darya suffers from upstream irrigation that has caused downstream salinity and, consequently, lower crop yields. Adding to this, strong winds bring salt and contaminated dust from the dried Aral Sea bed, causing degradation of surrounding lands and adverse health outcomes.[24]

The area equipped for irrigation in Africa (across the entire continent, including North Africa) is around 16 million hectares, or just 7 percent of the arable land area. The Malabo Montpellier Panel is a group of leading African and international experts assembled to support evidence-based, high-level dialogue among policy makers to advance agriculture and food security on the continent. In a report entitled *Water-Wise: Smart Irrigation Strategies for Africa* (2018), the panel argued that the potential for expanding irrigation is vast, especially in sub-Saharan Africa.[25] While North Africa has largely reached its irrigation potential, the panel noted that irrigated land in sub-Saharan Africa could grow from about 8 million hectares to 38 million hectares, thereby improving crop

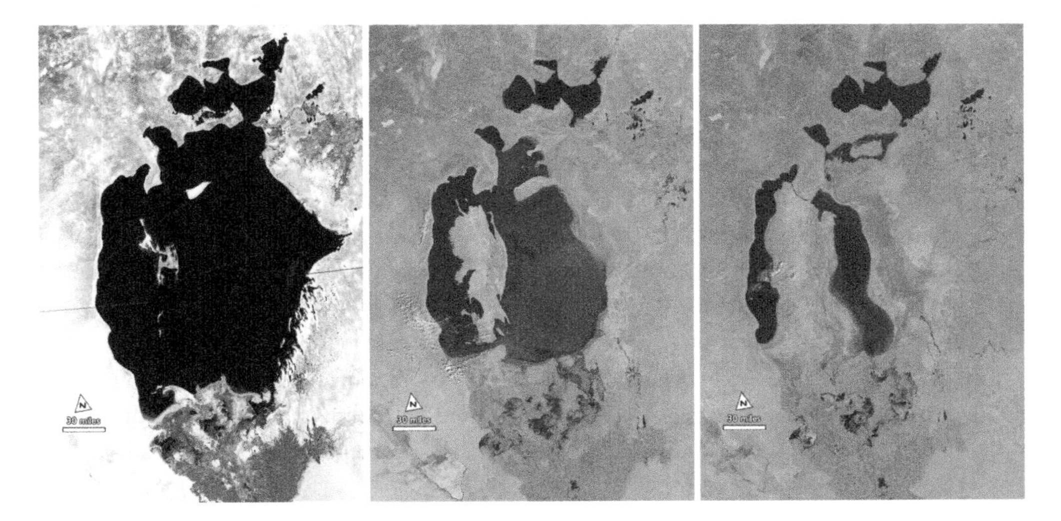

FIGURE 4.2 Landsat 8 images of a shrinking Aral Sea, as shown on NASA's website, https://www.nasa.gov/mission_pages/landsat/news/40th-top10-aralsea.html. Images taken (from left to right) in 1977, 1998, and 2010.

Source: USGS EROS data center, with permission to use in textbooks from NASA, https://www.nasa.gov/multimedia/guidelines/index.html.

productivity (yield per ha), cropping intensity (number of crops per year), and crop diversification opportunities.

Drawing on case studies across the continent, the Malabo Montpellier Panel concluded that governments and the private sector had critical complementary roles to play. Governments needed to prioritize irrigation for long-term investment and provide fiscal incentives and a regulatory environment conducive to investment, while the private sector should lead in the design, development, and dissemination of innovative smart technologies for irrigation. The panel, however, recognized that irrigation expansion in Africa would inevitably produce increased environmental and health risks that plagued other areas—as illustrated in the cases of the Indo-Gangetic Plain and the Aral Sea—and that safeguards were needed to ensure that irrigation investment would lead to sustainable intensification, a topic I return to in chapter 9.

The optimism emerging from the Malabo Montpellier Panel must be tempered by the fact that the irrigated area in sub-Saharan Africa expanded by just 1.3 percent per year (compound annual growth rate) from about 2005 to 2015, primarily through increases in two countries: Ethiopia and Tanzania.[26] In addition, the track record of past irrigation investments in Africa is dismal. A recent study by Thomas Higginbottom and colleagues from the University of Manchester quantified the performance of seventy-nine irrigation schemes, comparing planning documents with estimates of actual achievement based on satellite imagery.[27]

Their findings were sobering. Irrigation coverage was just 16 percent of the proposed area, only a quarter of the schemes delivered 80 percent or more of the planned coverage, and 20 percent were completely inactive. The authors found that, although scheme size and climate are often identified as causes for underachievement, the more likely reasons are (1) emphasis on low-value staple crops, (2) unrealistic expectations driven by optimism bias and political incentives, and (3) lack of technical expertise, local knowledge, and financial resources for long-term maintenance. Yet increased risks associated with climate change continue to be a driving force for investment. At the very least, we must learn from decades of experience.

THE INVISIBLE AQUATIC PRICE TAG

My work as a rice agronomist at the International Rice Research Institute for almost two decades brought me to dozens of rice-growing countries across Asia, Africa, and Latin America. The diversity of conditions under which rice is grown is immense. In Asia, the carefully manicured terraced slopes of Ubud in central Bali were my first exposure to rice. At that time (1975) Bali was in the initial stage of the Green Revolution, where double cropping of early maturing varieties was already taking hold and creating headaches for societies using the traditional water management system known as *subak*. A decade later I visited rice farms in the *dambos* of Zambia. These depressions in the otherwise flat or gently rolling landscape accumulated water each rainy season and allowed a single rice crop to be produced despite the hostile assault of weeds. Around that time I also ventured into the Nile delta, which boasted rice yields of more than 10 t/ha, benefiting from long days of abundant sunshine, cool evening temperatures, high-yielding varieties, and liberal use of subsidized fertilizer.[28]

These diverse production systems embodied water capture, conservation, and utilization. Most rice is produced in *paddies*—terraced fields surrounded by levees or bunds designed to hold water. And the reason is simple: rice consumes a lot of water. Based on field studies with continuously flooded rice, it takes an astonishing 1,000 to 5,000 kg of water to produce *1 kg* of rice.[29] Viewed out of context, however, this estimate exaggerates actual use by the rice plant, since much of the water is lost to evaporation, runoff, and deep percolation. The latter two blue-water "losses" may be reused downstream for agriculture or other purposes, a practice that King Parākramabāhu I would surely have approved. In any case, there is no denying that rice has a high *water footprint* (WF). And over the past two decades, a growing body of research has examined ways to improve the water productivity of rice.

The water footprint measures the amount of water used to produce a good or a service. The concept can be applied to a single primary commodity like rice or to a manufactured good like a pair of jeans or shoes. The WF can also apply to a company, a city, or a country. In essence, it represents the cost of appropriating water to do business. The WF includes the consumption of both green and blue water, as well as the role of *gray water* (wastewater). Professor Arjen Hoekstra developed the WF concept in 2002 while working at the IHE Delft Institute for Water Education in the Netherlands.

Providing the basis for measuring a water footprint is the concept of virtual water: the notion that water used in a commodity's production is *embedded* in that commodity, effectively an invisible aquatic price tag attached to a commodity. Professor Tony Allan, a British geographer, first coined the term, elegantly illustrating the concept as follows: "To produce one tonne of grain, you need 1000 tonnes (cubic meters) of water. Should an individual run out of water, it would be much easier to access one tonne of grain than the 1000 tonnes of water required to produce it."[30]

To be more concrete, Saudi Arabia would rather import a ton of rice than the 1,000 to 5,000 t of water needed to produce it in the country. In his two-page overview of virtual water, published in 1998 and cited more than a thousand times, Allan pointed out that the United States and the European Union exported to the Middle East and North Africa "as much water as flows down the Nile into Egypt for agriculture each year: 40 billion tonnes embedded in 40 million tonnes of grain." He contended that virtual water imports contributed to solving water shortages in the Middle East. Instead of promoting rice or wheat production, governments in water-stressed countries should introduce policies to conserve the limited water resources for household consumption or higher-value products.

Hoekstra and colleagues would go on to publish numerous papers and develop a Water Footprint Assessment tool that quantifies and maps water footprints, assesses the sustainability, efficiency, and equitability of water use, and identifies strategic actions for achieving a more sustainable footprint. In 2017 Hoekstra summarized the development and evolution of WF assessments.[31] He highlighted three reasons why WF assessments are needed:

- Water resource management is increasingly transboundary because of the trade in water-intensive commodities, that is, the movement of virtual water. It is crucial to understand how countries are externalizing their water consumption.
- Water resources renewal is limited, leading to competition for those resources. Competing demands requires accurate accounting for water use.
- Commodities are part of complex value chains wherein water is used at different stages from production to consumption, often extending beyond national boundaries.

Measurements of virtual water and the water footprint provide useful inputs to policy decisions on investments in irrigation and the role of subsidies and taxes in agriculture. Fresh water is often viewed as a free or at least a highly subsidized resource. As such, knowing the water footprint of commodities provides the basis for understanding the extent of the hidden subsidy associated with water use. If all countries had to survive within the envelope of their own water resources, many nations (like Jordan) would inevitably collapse. Trade in virtual water via imported commodities makes it possible for countries with physical water scarcity to be food secure. This phenomenon also opens opportunities for countries with surplus water resources to develop food exports through investments in both rainfed and irrigated agriculture.

As noted earlier in this chapter, sub-Saharan Africa has great potential to exploit both its green- and blue-water resources. As competition for water resources in Asia and the Middle East grows, the priority afforded to agricultural water use in those regions will likely decline in favor of municipal use. By closing the yield gap in rainfed areas and expanding investment in *sustainable* irrigation, many African nations could emerge as significant exporters of food and, hence, profitable exporters of virtual water. As we saw in chapter 3, Pedro Sanchez has argued for increasing nutrient supply to staple crops and makes a compelling case that yields in Africa can triple. But he acknowledges, building on the research of Rockström and colleagues, that those higher yields come from more efficient use of soil moisture: green water.[32]

HOW NOT TO RUN OUT OF WATER

Globally, agriculture (including irrigation, livestock, and aquaculture) is responsible for 70 percent of all freshwater withdrawals from rivers, dams, and groundwater.[33] *Water for Food, Water for Life: A Comprehensive Assessment of Water Management in Agriculture*, published in 2007, provided a critical evaluation of the benefits, costs, and impacts of the previous fifty years of water development, along with the challenges faced and the solutions available.[34] Drawing on no fewer than seven hundred researchers, practitioners, and policy makers, the *Comprehensive Assessment* (CA) was indeed comprehensive, running to 664 pages, including a 37-page summary for decision makers. Asking the most fundamental question—will we run out of water? —the CA concluded: "It is possible to produce the food—but it is probable that today's food production and environmental trends, if continued, will lead to crises in many parts of the world. Only if we act to improve water use in agriculture will we meet the acute freshwater challenges facing humankind over the coming 50 years."[35]

The CA identified eight high-level policy actions to improve water management for achieving global food security. For each action, I have extracted verbatim the most relevant CA text:[36]

1. **Change the way we think about water and agriculture.**

   Instead of a narrow focus on rivers and groundwater, view rain as the ultimate source of water that can be managed. Instead of blueprint designs, craft institutions while recognizing the politically contentious nature of the reform process. And instead of isolating agriculture as a production system, view it as an integrated multiple-use system and an agroecosystem, providing services and interacting with other ecosystems.
2. **Fight poverty by improving access to agricultural water and its use.**

   Target livelihood gains of smallholder farmers by securing water access through water rights and investments in water storage and delivery infrastructure where needed, improving value obtained by water use through pro-poor technologies, and investing in roads and markets.
3. **Manage agriculture to enhance ecosystem services.**

   Good agricultural practice can enhance other ecosystem services. In agroecosystems there is scope to promote services beyond the production of food, fiber, and animal protein. Agricultural production does not have to be at the expense of other services that water provides in rivers and wetlands.
4. **Increase the productivity of water.**

   Gaining more yield and value from less water can reduce future demand for water, limiting environmental degradation and easing competition for water.
5. **Upgrade rainfed systems—a little water can go a long way.**

   Rainfed agriculture is upgraded by improving soil moisture conservation and, where feasible, providing supplemental irrigation. These techniques hold underexploited potential for quickly lifting the greatest number of people out of poverty and for increasing water productivity, especially in sub-Saharan Africa and parts of Asia.

6. **Adapt yesterday's irrigation to tomorrow's needs.**

The era of rapid expansion of irrigated agriculture is over. A major new task is adapting yesterday's irrigation systems to tomorrow's needs. Modernization, a mix of technological and managerial upgrading to improve responsiveness to stakeholder needs, will enable more productive and sustainable irrigation.

7. **Reform the reform process—targeting state institutions.**

Reform cannot follow a blueprint. It takes time. It is specific to the local institutional and political context. And it requires negotiation and coalition building. Civil society and the private sector are important actors. But the state is often the critical driver, though state water institutions are often the most in need of reform.

8. **Deal with trade-offs and make difficult choices.**

Informed multistakeholder negotiations are essential to make decisions about the use and allocation of water. Reconciling competing demands on water requires transparent sharing of information.

Pressures on global water resources have escalated since the CA was published in 2007. The cases of Jordan, the Aral Sea, and the Indo-Gangetic Plain are all sobering reminders of the need to understand the long-term implications of irrigation and to take bold corrective action to reduce negative social, economic, and political consequences. While investments to improve productivity in both rainfed and irrigated systems are required to meet our future food security needs, growing evidence supports a strategic shift in emphasis toward rainfed (green-water) agriculture in some regions, especially sub-Saharan Africa. While acknowledging the need to reduce water withdrawals and decrease the water footprint of some systems and some regions of the world, I see broad agreement among the experts that there is enough water on this planet to feed and nourish a population of ten billion and more. We have the knowledge to greatly improve the management of our water resources. Toward the achievement of universal food security, we must now put that knowledge into practice.

5

SEEDS OF LIFE

On the campus of the International Rice Research Institute in Los Baños, Philippines, lies the largest and most diverse collection of rice varieties in the world. Known formally as the International Rice Genebank, this high-tech cold storage facility stores more than 132,000 different accessions of rice.[1] Most cultivated varieties in the Genebank are of *Oryza sativa*, the most commonly grown and widely consumed species of rice. Another species in the Genebank, *Oryza glaberrima*, originated in West Africa. While not as widely grown, this rice type has been used to breed new varieties suitable for African conditions. Some varieties being stored are no longer grown by farmers, many of whom opted for new, higher-yielding varieties or exited rice farming altogether. The Genebank also houses wild rice species—relatives of rice, within the genus *Oryza*, that were not domesticated but continue to grow in the wild. These, too, are endangered by land-use change and habitat loss.[2]

For six decades, IRRI scientists have been collecting and storing rice seeds that could be used in crossbreeding programs that would ultimately lead to what scientists refer to as *improved varieties*.[3] Peter Jennings, IRRI's first plant breeder, started the institute's rice collection upon his arrival at IRRI in October 1961. Jennings requested seeds from his many contacts around the world, and by the end of 1962 he had amassed almost seven thousand accessions from seventy-three countries and territories.[4] He and Te-Tzu "T. T." Chang, a rice geneticist from Taiwan, studied the genetics of rice in search of characteristics that would increase rice yield in the tropics. In late 1962 Jennings and his team crossed a tall Indonesian variety called Peta and a short-statured variety from Taiwan called Dee-geo-woo-gen. The researchers then examined the second-generation offspring of the crosses (known by breeders as the F2 generation) for patterns of trait inheritance.

Jennings described what happened next in an interview with IRRI historian Gene Hettel:

> We looked down the rows. Something had happened! It was an epiphany! I never had an experience like that in my life—before or since. There were tall plants and there were short plants [among the several thousand plants of the F2 generation], but there were no intermediate plants! The short ones were erect, darker green, and had sturdy stems and a high number of tillers. We counted the tall plants and short plants. Essentially, the ratio of tall to short was 3 to 1—obviously a single gene recessive for shortness![5]

The F2 generation turned out to comprise three-fourths tall plants and one-fourth short plants: the classical Mendelian ratio.[6] This finding demonstrated that shortness was controlled by a single recessive gene and therefore could be relatively easily bred into new varieties. Access to the diverse genetic resources assembled by Jennings and Chang enabled crosses that led to the release of IR8 in 1966 and the start of Asia's Green Revolution, as chronicled in chapter 2. IR8—the first of the Green Revolution rice varieties and a prototype for future varieties in terms of plant architecture—had multiple short and sturdy stems (known as tillers) emerging from a single seed (figure 5.1). This trait of having a short, sturdy stem provided the basis for high yields in response to improved soil fertility and abundant water. Taller, spindly varieties would easily fall over under the weight of high grain production. Genetic diversity provided the foundation for increased rice production in the tropics. And many have described the International Rice Genebank as holding the "crown jewels" of the global food system.

CONSERVING AGRO-BIODIVERSITY

After joining IRRI in 1980, I grasped every opportunity to visit the Genebank, which by then was housed in an impressive new two-story building, the Genetic Resources Laboratory, financed by the Asian Development Bank and the government of Japan. The complex was later renamed the N. C. Brady Laboratory, after IRRI's director general from 1973 to 1981, who pioneered the second wave of rice genetic improvements for more diverse, less favorable growing environments.

Opportunities to visit the Genebank usually came while accompanying visiting VIPs—important donors, high-level officials, and the occasional head of state.

FIGURE 5.1 The first Green Revolution rice variety—IR8—was the product of a cross between a tall Indonesian variety called Peta and a short-statured variety from Taiwan called Dee-geo-woo-gen (DGWG).

Credit: Image collection of the International Rice Research Institute (https://www.irri.org).

T. T. Chang, who directed operations of the Genebank, would greet the most important visitors in the hallway leading to the cold storage units, wearing his signature white lab coat except when entering the coldest vaults of the complex, when he would don his fur-necked parka. We would pass through rooms filled with a dozen or more women meticulously sorting seeds, packing them into envelopes (later vacuum-sealed aluminum), and dispatching these precious packages to the refrigerated interior of the building, which was arranged in the style of a "Russian doll" with progressively colder and less accessible interiors. At the heart of the complex was the base collection, where seeds were stored at −20°C with an expectation of keeping them alive and well for fifty years or more.

The historical, cultural, and economical value of what I always thought of and referred to as *IRRI's gene bank*—though in reality *humanity's gene bank*—is impossible to exaggerate. There were no growling Dobermans or armed guards to keep the collection secure from potential marauders, but each time I stepped into the facility, I was conscious of the importance of the seed collection for past, present, and future generations of rice producers and consumers.

T. T. Chang was a kind, fatherly figure who managed his ever-growing collection with great pride and an always-polite protective zeal. Chang would not have his collection's safety endangered by too many visitors. History may well show that these few hundred square meters of floor space would be the most valuable on the planet in terms of safeguarding food security for billions of people for whom rice is the main staple.

The importance of collecting and conserving seeds was first brought home to me when Dr. M. S. Swaminathan asked me to manage IRRI's support to Cambodia, which I described in chapter 2. Cambodia's traditional rice varieties, selected and improved by farmers over centuries, were uniquely adapted to complex and diverse rainfed environments in that region. IR8 and the Green Revolution had not penetrated Cambodia except in a few small pockets of irrigated rice land. The main production variable was water depth and its influence on growth, described in 2004 by Don Puckridge, a Bangkok-based deep-water-rice agronomist, who then represented IRRI in Thailand and, in January 1986, accompanied me on our first postwar mission to Cambodia:

> In the rainfed lowlands [of Cambodia] the topography is not really uniform and a "high" field may adjoin a "low" field. The difference may be only twenty or thirty centimeters, but it has a major effect on crop growth. In low fields the maximum depth of standing water usually exceeds thirty centimeters and is frequently deeper than fifty centimeters. High fields drain more quickly and plants often suffer from drought. In general, low fields are planted with tall "late maturing" varieties that can tolerate the deeper water. Mid elevation fields have fewer drought or flooding problems than other fields, and are usually planted with "medium duration" varieties, particularly those with good eating quality.[7]

Variations in rainfall and flooding patterns across the country resulted in thousands of locally named varieties being used by farmers in specific circumstances. Surveys undertaken by IRRI anthropologists Sam Fujisaka and Richard Lando, the latter with Cambodian colleague Mak Solieng, revealed an impressive diversity and complexity of Cambodia's rainfed rice environments that were reflected in farmers' decision making on the deployment of varieties.[8] Farmers also made choices based on eating and cooking quality. A single farmer would typically grow several different varieties, each adapted to the specific growing conditions of the several fields that made up a family's farm. Fujisaka went on to pursue this theme in several other countries where IRRI worked in the 1980s and 1990s, most notably Lao PDR, Nepal, and Madagascar. He demonstrated that, in heterogeneous rainfed environments, farmers' decision making, including varietal choice and related crop management, was shaped by adaptation to complex risks, potentials, and problems.[9] These observations contrasted with the experience of high-yielding, modern varieties in the more uniform and more favorable irrigated environments where single IRRI varieties like IR8, IR36, and IR64—sometimes referred to as *mega-varieties*—spread to millions of hectares across Asia.

The disruption of Cambodia's farming population, as a result of conflict and Khmer Rouge policies, resulted in the apparent extinction of these uniquely adapted traditional varieties and, to a large extent, the farmers who knew how to grow them. Compounding the genocide, which included the murder of farmers and professional agriculturalists, the Khmer Rouge's population-relocation policies led to people, often inexperienced women and children, farming in unfamiliar environments and struggling to grow crops. The result was widespread crop failures and the consumption of seeds normally kept for sowing the next crop.

Upon the removal of the Khmer Rouge regime, an urgent search began for the seeds of traditional varieties. The good news was that, prior to the most intense phase of the Indochina war, IRRI had sent a team to Cambodia to collect traditional rice varieties. These seeds were stored in the Genebank for about fifteen years before 766 accessions were selected, repatriated, replanted, evaluated, multiplied, and redistributed to farmers as Cambodians sought to reestablish and develop rice farming.[10] The return of Cambodia's traditional rice varieties is arguably the most successful case of *genetic rescue* in human history. I was director of the IRRI–Cambodia project between 1985 and 1998, working with Harry Nesbitt and his IRRI and Cambodian colleagues.[11] Together, we validated the role of genetic conservation and improvement far beyond theory and ambition. Seeds made all the difference. They were the seeds of life after the death and destruction that had characterized the 1970s in this region.

A more recent example of seed recovery and repatriation comes from Aleppo, Syria. The International Center for Agricultural Research in the Dry Areas (ICARDA) had its own gene bank aimed at conserving genetic diversity from the world's dry regions, mainly North Africa, the Middle East, and West and Central Asia. Established in 1985, ICARDA's gene bank in Aleppo held more than 140,000 accessions of major winter cereals (including wheat and barley), food legumes (such as lentils and faba bean), and various rangeland species.[12] As in Cambodia in the 1970s, Syria's civil war threatened these seed resources, but in this case, these were seeds held in trust for the entire region, not just for Syria.

While the physical facility itself was not damaged by the conflict, most of ICARDA's staff departed Aleppo in 2012, and there was a need to redistribute duplicate seeds to safer locations in the region—Morocco and Lebanon—where they could be stored and redistributed for use by ICARDA's researchers and collaborators. Almost all the seeds had been sent earlier for safekeeping to the Svalbard Global Seed Vault, where about 75 percent of accessions from the ICARDA collection are duplicated. Since 2015 ICARDA has retrieved tens of thousands of duplicated accessions from Svalbard for regeneration and characterization.[13]

ICARDA's collection is an integral part of a strategy to improve food security in dryland areas. The gene banks at IRRI, ICARDA, and Svalbard are indeed "banks" open for deposit, safekeeping, and withdrawal when needed. These genetic resources are held in trust for the benefit of future generations under the auspices of the International Treaty on Plant Genetic Resources for Food and Agriculture, which was adopted in 2001 and entered into force in 2004.[14] The treaty has three main aims:[15]

- to recognize the enormous contribution of farmers to the diversity of crops that feed the world
- to establish a global system that provides farmers, plant breeders, and scientists with access to plant genetic materials
- to ensure that the recipients of those genetic materials share the benefits they derive with the countries of origin

The Svalbard Global Seed Vault is the global centerpiece of crop genetic resource conservation. Formally opened in 2008, the seed vault (also known as the Doomsday Vault) is located inside a mountain on an island in the Svalbard archipelago, about halfway between mainland Norway and the North Pole (figure 5.2). It was established as "a fail-safe seed storage facility, built to stand the test of time—and the challenge of natural or man-made disasters," recognizing the vulnerability and strategic importance of the world's 1,700-plus gene banks.[16] The International Rice Genebank is the largest contributor to the seed vault. By early 2022 more than a million seed varieties were stored at Svalbard, which has the capacity to store the seeds of 4.5 million varieties. The seed vault and seed banks around the world are supported by the Crop Trust, established in 2004, which managed an endowment fund of about US$365 million in 2022. The Crop Trust, the Svalbard Global Seed Vault, the International Rice Genebank, ICARDA's gene bank, and the other participating gene banks and genetic conservation programs around the world make it possible for the world's farmers to draw on the diverse genetic resources that have evolved on Earth over millions of years and have been shaped by agriculture over the past twelve thousand years.

FIGURE 5.2 The Svalbard Global Seed Vault is located halfway between mainland Norway and the North Pole. Seeds are preserved for future use by farmers and scientists.

Credit: Crop Trust / L. M. Salazar.

BEYOND GENE BANKS

Which country holds the greatest diversity of banana varieties? Panama? Uganda? Papua New Guinea? The answer may surprise you.

Not all important agricultural plants can be conserved in the form of seeds. For example, the banana plant is normally multiplied through plant tissue—a process known as vegetative propagation. This requirement creates special challenges for the conservation and improvement of sweet and starchy bananas (*Musa* spp.), which are an important source of food security for millions of households in low- and middle-income countries around the world. To safeguard the genetic diversity of bananas, Bioversity International established the International *Musa* Germplasm Transit Centre (known as the ITC) in 1985. Located at the Katholieke Universiteit Leuven (KU Leuven) in Belgium, the collection, comprising more than 1,500 accessions of edible and wild species of banana, is the world's largest collection of banana *germplasm*. Germplasm is the term used to cover all types of living plant material, including seeds, from which new plants can be grown and multiplied. Germplasm carries the DNA codes that define the genetic makeup of a species, variety, or individual plant.

At the ITC, tissue samples of each accession are kept alive in glass vials (known as in vitro conservation) at 16°C. For backup security, samples are also frozen to −196°C, a process called *cryopreservation*. This means that material can be preserved indefinitely and later revived as banana plants when needed. This frozen collection is duplicated at the Institut de Recherche pour le Développement in Montpellier, France. Just as we saw for the seed banks, multiple backup systems ensure the long-term availability of banana germplasm for farmers and researchers alike and reflect the priceless nature of the banana gene pool. These nonseed resources are also conserved and shared through the International Treaty on Plant Genetic Resources for Food and Agriculture.

The storage of seeds or tissues, known as *ex situ* conservation, provides us with samples of the genetic resources collected from farmers' fields at a specific time and place. The genomes of plants (and animals) have evolved as a result of *human* selection and manipulation (including artificial crossbreeding which I discuss later in this chapter) and the *natural* selection pressures of new and changing environmental conditions. Both of these evolutionary processes must continue for agriculture to remain viable as human needs and desires change, and as Earth's environment changes, most notably as a result of climate change and the continuing evolution of pests and diseases.

The process of evolution and adaptation in cultivated fields and natural environments means that, in order to expand the scope of genetic diversity and the opportunities for utilizing that diversity, we cannot rely solely on "frozen snapshots" of the past. Therefore, to complement ex situ conservation, it is also necessary to conserve these resources in situ, thereby allowing evolutionary processes to continue and for diversity to grow with changing human and natural selection pressures.

Mauricio Bellon and colleagues reviewed the state of in situ conservation and its role in future crop adaptation.[17] Bellon, an anthropologist who has worked at both IRRI and the International Center for Maize and Wheat Research (known by its Spanish acronym CIMMYT), has throughout his career highlighted the existence of "coevolving sociobiological systems," which bring together farmer practices and the use of traditional varieties (also known as landraces).[18] While Fujisaka, Lando, and many others captured snapshots of field biodiversity, I see Bellon's work highlighting these systems as *documentaries* in progress: "Evolution of agricultural biodiversity is determined by complex interactions between crop, environment, and humans at a range of spatial scales. Social factors involving the full gamut of interactions from relationships between adjacent and more far-flung communities, to taste preferences and traditional beliefs, ensure that farmers and landraces constitute a complex coevolving sociobiological system."[19]

Bellon and colleagues conceded that the case for in situ conservation requires further evidence of crop evolution under farmer management. While acknowledging that empirical evidence is limited and fragmented, they conclude that evolution of populations in the field can lead to "favorable gene complexes that adapt plants to changes in their biotic and abiotic environments."[20] There is little question that ex situ and in situ conservation are complementary; neither alone is sufficient to safely conserve existing and evolving biodiversity. Wild relatives and traditional varieties of plant (and animal) species can provide potential sources of genetic diversity that enable adaptation and improvement. But if their habitat becomes fragmented, degraded, or transformed (e.g., through land clearing for agriculture or expanded urbanization), these populations may be lost. Gene banks and other ex situ conservation strategies therefore provide insurance against such risks, even if they are limited to the historical genotypes of crop species.

GENETIC IMPROVEMENT

Earlier in this chapter I introduced the work of Peter Jennings, referring to him as IRRI's first plant breeder. A plant breeder is someone who has the knowledge, skills, patience, and sometimes the good fortune to manipulate the genetic makeup of a plant population to produce desirable traits. Wherever genes are involved and traits can be inherited, plant breeders are looking for opportunities for *improvement*. Some commonly sought-after traits are as follows:

- **high yield (more produce per hectare)**, for example, the transformed plant architecture of the Green Revolution's high-yielding varieties
- **better adaptation to biotic stresses**, for example, resistance to or tolerance of pests and diseases that affect the quantity and quality of produce
- **better adaptation to abiotic stresses**, for example, tolerance of high or low temperatures, drought, flooding, and soil fertility deficiencies or toxicities
- **better quality product**, comprising a diverse set of characteristics including nutritional traits, visual appearance, taste, aroma, and storage factors

The origins of plant breeding can be traced back to the first agriculturalists who selected wild plants for managed cultivation and began the process of domestication. Beginning in the Fertile Crescent of Southwest Asia, wild ancestors of present-day crops and animals—wheat, barley, peas, sheep, goats, cows, and pigs—were first selected, tamed, and managed. Hunters and gatherers became sedentary farmers. Jared Diamond describes plant and animal domestication as "the most important development in the past 13,000 years of human history."[21] This process began independently in many parts of the world at different times.[22]

Asian rice was likely domesticated eight to nine thousand years ago, somewhere in southern China and/or northern India, though competing claims regarding the true origin continue to rage. Duncan Vaughan, a British geneticist who collected traditional varieties and wild rice for the International Rice Genebank through the 1980s and 1990s, concluded there is "sufficient archaeobotanical evidence" to support the argument that domestication emerged from a single domestication event in the Yangtze River valley that eliminated grain shattering and thereafter made rice dependent on humans for its survival and spread. African rice was independently domesticated much later, most likely three thousand to thirty-five hundred years ago.[23]

For millennia, generations of farmers purposefully improved their crops and animals through selection based on observed attributes such as production, appearance, and taste. The underlying genetics and the role of inheritance in shaping those characteristics emerged through the contributions of Charles Darwin and Gregor Mendel, among others. For more than a century breeders have used genetics and heredity to understand and manipulate plant and animal traits to meet the preferences of agricultural producers and consumers. As we saw in chapter 2, the Green Revolution in Asia was based on the development of what agricultural scientists usually refer to as high-yielding varieties (HYVs) of wheat and rice. These HYVs had the genetic traits to respond well to improved soil fertility and irrigation, producing higher yields per hectare than traditionally used varieties in these more favorable growing environments.

In the case of wheat, the most successful HYVs emerged from the breeding skills of Norman Borlaug and the shuttle breeding approach he deployed in Mexico. As elaborated in chapter 2, Borlaug's early work focused on genetic resistance to stem rust. For rice, the story began with IR8, which resulted from a cross of two very different varieties from Indonesia and Taiwan. IRRI's breeding efforts began with the work of T. T. Chang, Peter Jennings, and Henry "Hank" Beachell and continued through the 1970s, 1980s, and 1990s under the leadership of Gurdev Khush. In 1996 Beachell and Khush shared the World Food Prize for their contributions to rice breeding that led to improved food security in Asia and around the world.[24]

While I overlapped with Beachell for a couple of years and met with him occasionally, I was most fortunate to work with Khush for the full eighteen years of my tenure with IRRI. He became one of my mentors, helping me understand the importance of varietal improvement and its role in raising yields and achieving food security. Khush's knowledge of the rice plant was unrivaled, yet he was both patient and generous with his students and young professionals like me (during the 1980s) who regularly sought his counsel.

Having grown up on a wheat farm in northern India, Khush took an undergraduate degree at Punjab Agricultural University before obtaining a PhD degree from the University of California, Davis. He joined IRRI in 1967, the year after IR8 was released. While IR8 was very high yielding, it had three serious shortcomings: poor eating quality, poor insect and disease resistance, and relatively long duration to harvest (130 days). Responding to these challenges, Khush and his team developed IR36, which represented a major advance over IR8.

By the early 1980s IR36 had spread to almost 11 million hectares, prompting IRRI to declare this variety "the world's most popular rice."[25] Its resistance to a damaging viral disease, known as grassy stunt virus, came from crossing earlier HYVs with a wild rice species from India, *Oryza nivara*. IR36 was subsequently replaced in most parts of South and Southeast Asia by IR64, a variety that combined excellent eating quality with high yield and robust resistance to most pests and diseases. IR64 was bred from nineteen traditional rice varieties, once again demonstrating the crucial importance of genetic conservation as a cornerstone of varietal improvement.[26]

Through a collaborative program I initiated in 1983, Bhutan's Ministry of Agriculture evaluated hundreds of improved rice varieties during the 1980s as they sought to reduce their growing dependency on rice imports. IR64 proved to be well adapted to the mid-altitude valleys of Punakha and Wangdue Phodrang (at an elevation of 600 to 1,600 meters above sea level) and became the nation's first released rice variety in 1988. An assessment published in 2012 found that IR64 remained popular among farmers in the mid-altitude zone, where it covered an estimated 50 percent of the rice-growing area, because of its high yield and good eating quality.[27] Researchers were concerned, however, about the breakdown of resistance to a damaging fungal disease known as blast. National rice scientists, all trained at IRRI, then crossed IR64 with local varieties, and the first crosses were released to Bhutan's farmers in 1999.

Since the establishment of Bhutan's national research system in the early 1980s, the country has strengthened its capacity to test and evaluate varieties developed in other countries. Of the twenty-three rice varieties officially released in Bhutan between 1988 and 2010, fifteen were introduced from other countries (Japan, Nepal, China, South Korea, Philippines, and India) or from IRRI, while eight varieties were developed through crossbreeding programs undertaken in Bhutan using a combination of exotic and local varieties.[28] Bhutan was also a generous contributor to the International Rice Genebank, contributing hundreds of traditional varieties during the 1980s and 1990s. Many of these varieties had excellent cold-tolerance traits derived from their high-altitude origins. Bhutan was both a beneficiary of and a contributor to the genetic improvement of rice.

These few examples of crop improvement illustrate the continuing efforts by plant breeders to modify the genetic code of food crops to improve adaptation, production, and ultimately, acceptance by consumers. We have seen that the most basic means of improvement is simple human selection and multiplication of plants that demonstrate superior traits. Deliberate crossbreeding, as we saw in the case of IRRI's HYVs, brings diverse genetic pools together, allowing for further selection and improvement through field testing and observation. A special category of plant breeding is the development of hybrid varieties.

HYBRID VIGOR

In popular use, the term *hybrid* may refer to "an offspring of two animals or plants of different races" or even more broadly "something heterogeneous in origin or composition."[29] As such, all the hybridization outcomes of deliberate crosses could be reasonably described as hybrids. In agriculture, however, we generally refer to hybrid varieties as the crosses between two distinct *genotypes* that often produce offspring that grow more vigorously than their parents, a phenomenon known as hybrid vigor.[30]

The most prominent example of hybrid vigor in agriculture is hybrid maize (corn), which is produced by crossing two genetically unrelated parents.[31] The first-generation offspring exhibit hybrid vigor, but the seed they produce (the second generation) does not retain the enhanced traits. Growing these second-generation seeds can result in yield losses of 30 percent or more, reducing and perhaps eliminating any yield advantage in subsequent planting. To benefit from hybrid vigor, farmers would need to purchase new seeds every year, rather than replant seeds collected from the previous harvest.

The science and practice of hybrid maize was developed and commercialized in the United States during the first three decades of the twentieth century.

Enthusiasm of farmers, breeders, and businesses alike led to the release of commercial varieties that quickly replaced traditional *open-pollinated* nonhybrid varieties. By 1960 almost all the maize planted in the United States was hybrid. In the early 1930s U.S. maize farmers were harvesting just under 2 t/ha. By the early 1960s the national yield had risen to around 4 t/ha.[32] The 2021 crop averaged almost 12 t/ha.[33] The development and introduction of new hybrid varieties made a major contribution to increasing maize production in the United States. In a paper published in 2001 chronicling the rise of hybrid maize in the United States, the legendary maize breeder Donald Duvick reported that improvements in hybrids since their introduction were responsible for 50 to 70 percent of the on-farm yield gains, the balance coming from more fertilizer and better weed control.[34]

The government of Malawi launched a hybrid maize breeding program in 1950, following a severe drought in 1949.[35] The first maize hybrids were released in 1958, and by the 1990s they were adopted by about a quarter of Malawi's smallholders. Research indicated a consistent yield advantage of hybrids over local maize varieties at all levels of fertilizer use, including in a drought year.[36] However, while advocating for their use by smallholders in Malawi in the early 2000s, I would receive persistent pushback from those opposed to private-sector involvement in smallholder agriculture and the need to purchase seeds each year.

During the second year of the Malawi subsidy program, voucher recipients were given a choice of hybrids or open-pollinated varieties (OPVs): 2 kg of hybrid seeds or 3 kg of improved OPV. Based on coupon redemption, 76 percent of farmers chose hybrids over OPVs.[37] Like their counterparts in the U.S. Corn Belt, it was clear that Malawian smallholders, farming less than a hectare of land, could see the benefit of hybrids.

During one of my regular visits to Malawi between 2004 and 2010, at the height of excitement and controversy over Malawi's Farm Input Subsidy Program, I was taken aside by an otherwise supportive member of the international development community in Lilongwe. With a tone of disappointment, this senior official remarked, "Glenn, you never told me that you were promoting GMOs in Malawi." My immediate thought was to respond with, "Why would it matter if I was?" But diplomacy got the better of me, and I was able to comfort the concerned bureaucrat with a brief explanation of the difference between hybrids and GMOs.

I explained that hybrids have been with us since the time of Charles Darwin, widely grown in the United States since the 1930s, and commercially available in Malawi since the late 1950s.[38] GMOs, I pointed out, are a more recent innovation. And in any case, none of the maize varieties being promoted or even researched in Malawi were GMOs. My bureaucrat friend was off the hook. He would now be ready if attacks were to be launched in the press or by HQ officials or even the government of Malawi.

Hybrid rice is another notable example of deploying the phenomenon of hybrid vigor at scale. Yuan Longping (1930–2021) is known as the "father of hybrid rice" and was a laureate of the World Food Prize in 2004 for his "pioneering research [that] helped transform China from food deficiency to food security within three decades."[39] After demonstrating yield increases of 20 percent, Yuan and colleagues went on to develop seed-production methods leading to widespread adoption of hybrid rice in China. By the early 2000s about half of China's rice area was planted to hybrid rice, and the resulting production increases fed an additional 100 million Chinese people each year.[40] Yuan's work prompted IRRI in 1979 to embark on developing hybrids for tropical conditions and has inspired its adaptation and adoption in India, Vietnam, and the Philippines. IRRI's research on hybrid rice, led for almost thirty years by my longtime colleague Sant Singh Virmani, continues with governments and private-sector partners.[41]

Yuan Longping's determination and courage to pursue an unconventional approach to rice varietal improvement was driven by his understanding of genetics and heredity. One of Yuan's early influences was Guan Xianghuan, a professor who vigorously opposed the Soviet pseudoscience dogma then being advanced by Trofim Lysenko and followers. Guan and Yuan were instead influenced by the work of Gregor Mendel, the nineteenth-century botanist who established the science of genetics. While Guan sadly took his own life in 1966, Yuan would relentlessly pursue his lifelong mission to apply genetics to advance rice production in China. In 2019 he was awarded the Medal of the Republic, China's highest official honor, by President Xi Jinping.[42] Upon Yuan's passing in May 2021, Xi declared that Yuan "made major contributions to our national food security, agricultural scientific innovation and global food development."[43] Fortunately for China and for the world, science prevailed over doctrine.

ENDLESS FORMS

Through nature's "endless forms most beautiful and most wonderful," Charles Darwin marveled at the diversity of life that had evolved on planet Earth.[44] Through *natural* selection, humans were presented with an extraordinary menu of edible opportunities from the plant, animal, and other biotic kingdoms. The hunters and gatherers reaped those rewards from the wild for the best part of a million years before deciding to settle down and begin farming around twelve thousand years ago.

From that point, purposeful selection—season by season, using the inventive hands and sharp eyes of the farmers themselves, and no doubt influenced by the tastes of household consumers—further shaped a diverse agricultural gene pool that had evolved in farmers' fields. Enter Gregor Mendel, whose experiments with peas began to unravel the underlying science of genetics and heredity.[45] Mendel's principles of inheritance, proposed in 1865, provided the foundation of what we now call *plant* or *crop improvement*, which underpinned the work of Yuan Longping and other notable plant breeders mentioned in this chapter.

Mendel's work was largely ignored until 1900, when Hugo de Vries, Carl Correns, and Erich von Tschermak-Seysenegg were credited with rediscovering Mendel's research on inheritance. Though the precise chronology and correct attribution of the rediscovery are a matter of debate,[46] this improved understanding of heredity and genetics laid the foundation for researchers to develop crop varieties that were no longer based on simple selection and incremental improvements of desired traits. The genetic makeup of plants was now being manipulated, albeit with a kind of random experimentation—a throwing together of genetic diversity—in the hope that new combinations would emerge with desirable traits.

In the 1930s hybrid maize offered two (then) radical propositions. First, new varieties could be created through scientific manipulation of the crossbreeding process. Second, farmers would give up the traditional practice of saving seeds from one crop to plant the following season. As described earlier, the practice of seed saving was of no value when using hybrids. Instead, farmers would now need to purchase seeds each year from seed companies. In the early days of hybrids, farmers did have the option of purchasing the parent seeds from agricultural colleges and creating the hybrids themselves, but, over time, their preference for ready-to-plant seeds was apparent, and the commercial seed industry flourished. Farmers readily accepted hybrid seeds. In the words of Donald Duvick: "Sturdy hybrids convinced skeptical farmers that 'the professors' and their arcane science could do them some good."[47]

As we have seen in this chapter and chapter 2, scientific manipulation of genes through what became known as *conventional* or *traditional* breeding methods had a major impact on food production in almost all parts of the world. Crossbreeding of genetic variants, and the subsequent selection and multiplication of stable new lines, ultimately resulted in new named varieties that could be distributed to farmers. These conventionally bred varieties—principally of wheat and rice—became the stock of the Green Revolution. Other, less common methods were used to change genotypes of organisms. For example, starting in the 1920s, radiation and chemicals were deployed to create genetic mutations, a process known as *mutagenesis*.

After discovery of the DNA structure in 1953, followed by thirty years of improving our understanding of the molecular basis of heredity, it became possible to bypass sexual reproduction and instead transfer specific genes into crop plants. Recombinant DNA (rDNA) technology provided the means to combine the DNA of two different species, creating a new genetic entity in the laboratory that would not normally be found in nature or be possible through conventional breeding.[48] Genomes were in effect *engineered*, hence the term *genetic engineering* (GE). The products of this unconventional gene transfer technology became popularized as genetically modified organisms (GMOs), although it should be obvious to readers by now that *all* products of plant breeding have had their genetic makeup modified.

The most commonly known, most discussed, and most controversial process of GE is where the DNA of one species is transferred to another, a process known as *transgenesis*, resulting in the term *transgenic crop*. An example of the commercial application of this approach is the resistance in maize to insect pests, specifically the European corn borer, through expression of the insecticidal protein Cry1Ab derived from the soil bacterium *Bacillus thuringiensis* (Bt).[49] By using Bt corn with this inbuilt resistance to insect pests, farmers were able to save on pesticide use and improve their bottom line. Similarly, and more recently, a transgenic eggplant (known as Bt eggplant or Bt brinjal) has reduced pesticide use and increased profits for smallholder farmers in Bangladesh.[50]

A related form of GE, known as *cisgenesis*, describes gene transfers between more closely related plants that could, at least in principle, otherwise be achieved through conventional breeding methods. So why use cisgenesis when conventional breeding can be used? First, cisgenesis can reduce the problem of *linkage drag* whereby undesirable genes get lumped in with desirable genes when using conventional methods. Cisgenesis facilitates greater precision in engineering the traits we are seeking in a new variety. Second, cisgenesis more rapidly achieves the final product. Third, cisgenesis also enables the transfer of multiple genes (gene pyramiding or stacking), which can be useful when seeking more durable resistance to pests and diseases or for improving adaptation to complex abiotic stresses like heat, drought, and floods.[51] While cisgenesis is clearly a form of GE, as it transfers a gene to a recipient species, the resulting GMO differs from transgenic plants in that there are no foreign genes from unrelated species.

Henk Schouten, Frans Krens, and Evert Jacobsen from Wageningen University in the Netherlands argue that cisgenesis "respects species barriers, and in this sense differs fundamentally from transgenesis."[52] They conclude that cisgenic plants are as safe as those derived from traditional breeding approaches. We should not conclude from this argument, however, that transgenic plants are inherently unsafe or less safe than cisgenic (or even non-GE) crops. "Respecting species barriers" has not been credibly established as an indicator of product safety.

Much of the confusion, disagreement, and controversy derives from the complexity of the science underpinning GMOs and the rather imprecise and interchangeable use of related terms like biotechnology, GE, and GMOs.[53] To move the discussion forward, I would like to create some working definitions. Let us agree that *biotechnology* is a very broad, process-defining term that includes all types of human-directed, biologically driven transformations that are intended for generating useful products. Bread, cheese, and wine are all millennia-old examples of biotechnology. We can all accept that biotechnology has numerous useful applications, and we should avoid using biotechnology as a substitutable term for GE, which is really one rather narrow category of biotechnology that can be used to produce GMOs. Hereafter, I shall refer to GMOs as products of both transgenesis and cisgenesis. To reduce confusion, I shall refer to *GM crops*—one category of GMO—as crops propagated by seed or other germplasm created through transgenesis, cisgenesis, and other advanced molecular techniques, including genome editing. Nevertheless, I recognize that there is ongoing debate on whether seeds produced through genome editing and some other genetic engineering methods should be legally defined as GMOs. Some authors prefer to use the less politically charged term *GE crops*.[54]

GM crops were commercially introduced in 1996. By 2018, according to the International Service for the Acquisition of Agri-biotech Applications (ISAAA), GM crops were grown on about 192 million hectares in twenty-six countries—around 13 percent of global crop area.[55] The United States, Brazil, Argentina, and Canada were the four top producers of GM crops, constituting 84 percent of the total GM crop area planted. Soybeans, maize, cotton, and canola were the most widespread applications. In 2020 the most widely adopted GM crop traits were herbicide resistance and pest tolerance. Globally, more than three-quarters of the areas planted with soybean and cotton used GM seeds. For maize, about 30 percent of the planted area used GM seeds. Other important GM crops include alfalfa, papaya, eggplant, potato, apple, safflower, pineapple, and sugarcane. Among low- and middle-income countries, India stood out with more than 11.6 million hectares of Bt cotton in 2018. China and Pakistan were other important producers of Bt cotton, with close to 3 million hectares each.[56]

Farmers today are choosing GM crops for the same reasons that American farmers in the 1930s rapidly accepted hybrid maize: it pays off. Farmers have been willing to spend more on GM seeds because of the expected returns. Seed companies invest in research and development because they know there is a market for GM crops that deliver traits that are valued by farmers. Herbicide-resistant crops enable farmers to adopt conservation tillage, including no-till farming, discussed in chapter 3. With fewer passes of machinery, farmers cut their costs and reduce soil damage, including erosion.

ISAAA likes to argue that developing nations are taking up GM crops even more rapidly than developed countries are. To support that case, they consider Brazil and Argentina to be developing countries. But the reality is that adoption of GM seeds—mainly soybean and maize—in these countries remains largely restricted to modern, large-scale, mechanized farming systems that are similar to those found in the United States. High adoption levels that are having an impact on smallholder farmers in developing countries are, by and large, limited to insect-resistant Bt cotton in India, China, and Pakistan, which represents less than 10 percent of global GM crop adoption.

Why have smallholders in the developing world not taken up GM seeds, aside from the relatively small amount of Bt cotton? This is an important question with economic and ethical dimensions.

As we saw with hybrid maize, the private sector, backed up in part by strong, publicly funded research, played a key role in developing improved crop varieties that met farmers' demands. A fact of economic life is that businesses follow the market opportunities. Large-scale farmers, who are well connected to domestic and international markets, were the obvious target for seed companies that invest large amounts in research and development. The adoption of GM cotton by smallholders could also be explained by the market-driven nature of production. Cotton is a cash crop, and even small-scale farmers are willing to invest their cash or borrow to increase their returns, provided there are established markets. Prior to Bt cotton, farmers were already investing heavily in pesticides. Adoption of Bt cotton drastically reduces those costs and improves profits.[57]

For the world's two most important food crops in low-income countries—rice and wheat—there has been almost no penetration of GM seeds. The limited use of hybrids in these two crops (except for hybrid rice in China, where government plays the lead role in seed supply) leaves reduced scope for private sector profits for these crops, even in high-income countries. For nonhybrid varieties, farmers can reuse their own seed or obtain it directly from other farmers.

Investment in GM rice varieties began in the 1970s through an initiative of the Rockefeller Foundation. The Bill and Melinda Gates Foundation has invested heavily in GM rice and other smallholder staples over the past two decades. Public sector and philanthropic research investments are critical where farmers are not well connected to markets and face high levels of risk in recouping and justifying their seed costs. I like to think of these as *pro-poor* GM seed investments. They are directed to crops and regions where traditional seed firms have seen little scope for high profits and market growth. These applications of GM seed are often overlooked in heated discussions on the use of GMOs.

Current examples of pro-poor GM seed development are as follows:

- **virus-resistant cassava** to build resistance to cassava brown streak disease and cassava mosaic disease, which can decimate smallholders' cassava crops in Africa[58]
- **water-efficient maize** to improve the resilience of maize under drought conditions[59]
- **pod borer–resistant cowpea** to counter *Maruca* pod borer, which can cause 70 to 90 percent yield loss in cowpeas, an important source of food and income for Nigerian farmers[60]
- **Bt eggplant** to confer resistance to the eggplant fruit and shoot borer, which is the main constraint to yields and profits to smallholder eggplant farmers in Bangladesh[61]
- **Golden Rice**, which was engineered to produce beta-carotene, a precursor of vitamin A[62]

THE CASE OF GOLDEN RICE

IRRI developed and supported the introduction of Golden Rice to tackle vitamin A deficiency, especially among low-income consumers who depend on rice as their staple food. Vitamin A deficiency (VAD) is the leading cause of preventable blindness in children; it also compromises immunity from common infections such as diarrheal disease and measles. The World Health Organization estimates that 250,000 to 500,000 children lose their sight each year due to VAD; half of them die within twelve months of loss of sight.[63]

Humans need beta-carotene in their food or as a supplement. The body converts beta-carotene into vitamin A. Rice plants have beta-carotene in their leaves and stem but not in the grains. Through GE (more specifically, transgenesis), scientists have been able to deliver beta-carotene in the seed, giving the grains their signature yellow-golden color. Initially the two genes needed for producing beta-carotene in grain were taken from a bacterium and a daffodil; the daffodil gene was later replaced with a maize gene that enabled greater beta-carotene accumulation.[64] This is the same beta-carotene found in vegetables and fruit, and identical to synthetic beta-carotene used in vitamin A supplements.

Golden Rice is intended to supply up to 30 to 50 percent of the vitamin A needs of preschool-age children and pregnant or lactating mothers. IRRI has worked with plant breeders from the Philippines, Bangladesh, and Indonesia to ensure that Golden Rice varieties are comparable to existing varieties in terms of yield, pest resistance, and grain qualities. The only difference will be their better nutritional value. And unlike hybrids, the seeds of Golden Rice can be harvested and replanted. Yet some militant nongovernmental organizations have expressed fierce opposition, including the destruction of field plots in 2013.[65]

The story of Golden Rice is long and complex. The promise of this GM crop in preventing blindness and death was heralded in a July 2000 *Time Magazine* article entitled "This Rice Could Save a Million Kids a Year."[66] From the outset, environmental NGOs like Greenpeace and Friends of the Earth opposed the project with extraordinary hostility. In March 2001 influential author and food activist Michael Pollan, writing in the *New York Times*, referred to Golden Rice as the "Great Yellow Hype."[67] He lashed out at the "biotechnology industry" for an "audacious new advertising campaign to impale people like me—well-off first worlders dubious about genetically engineered food—on the horns of a moral dilemma." It is hard to fathom where the "moral dilemma" lies, but I do see a moral imperative.

Critics of GMOs, like Greenpeace, Pollan, Vandana Shiva, and others have long feared Golden Rice as a Trojan horse that would undermine their uncompromising objections to GM seeds. Here I outline their most prominent arguments against Golden Rice and counterarguments that reflect my perspective:

1. Golden Rice is unproven and unsafe for consumption.

Counterargument: Golden Rice has been thoroughly tested and evaluated through peer review research over the past two decades. In a letter from the U.S. Food and Drug Administration on May 24, 2018, the agency expressed no concerns about the safety for consumers.[68] Regulatory authorities in Australia, New Zealand, and Canada earlier declared Golden Rice to be safe.

2. There are alternatives to Golden Rice in reducing VAD; there is no need for Golden Rice.

Counterargument: Those alternatives have long existed, yet the extent of VAD remains widespread and unacceptable. Despite the optimism of the *Time Magazine* cover story, Golden Rice has never been seen as the sole path for reducing VAD. It must be viewed as complementary to other approaches like increased vegetable consumption and vitamin A supplements. In situations where low-income people cannot afford high-priced vegetables and where supplement programs have limited effectiveness, Golden Rice has the potential to contribute to increased beta-carotene intake.

3. Golden Rice is a plot of multinational biotechnology companies to win favor with the general public and reduce resistance to development and use of GMOs.

Counterargument: Golden Rice was developed using scientific techniques developed by the public sector in partnership with industry. In 2004 Syngenta gave up its commercial rights to profiting from the distribution and sale of Golden Rice seeds. Since then, research on Golden Rice has been undertaken by the public and nonprofit sectors, led by IRRI and several of its national government partners in Asia. If the private sector's reputation is indeed enhanced by its participation, would that be sufficient reason to block its progress?

On December 10, 2019, the government of the Philippines, where 17 percent of children under the age of five are vitamin A deficient, became the first major rice-growing country to authorize the use of Golden Rice for human consumption. After scrutinizing data submitted by the Philippine Rice Research Institute and IRRI, several regulatory committees representing the agriculture, environment, and health sectors concluded that Golden Rice was as safe as conventionally bred varieties and should proceed for seed multiplication and varietal registration.[69] On July 21, 2021, Golden Rice was approved for commercial production, twenty-one years after *Time* heralded its promise.[70] This was a major step forward, and there are prospects that Bangladesh and Indonesia will follow the Philippines in approving commercialization of Golden Rice. IRRI is currently developing high-iron and high-zinc rice with the end goal of releasing a variety containing levels of beta-carotene, iron, and zinc that can help address these multiple micronutrient deficiencies that affect more than two billion people worldwide (see chapter 7).

The case of Golden Rice has highlighted the politically charged, divisive debate on the perceived benefits and costs of GM seeds. More than two decades have passed since the technology was developed by Ingo Potrykus at the Swiss Federal Institute of Technology in Zurich (ETH Zurich) and Peter Beyer at the University of Freiburg in Germany. The science completely supports the efficacy of Golden Rice in delivering beta-carotene cost-effectively to low-income rice consumers without any health or environmental risks. Yet these consumers have been held to ransom by NGOs and activists whose leaders, members, and followers have little to lose personally by taking a nonscience, nonnegotiable, ideological position on Golden Rice and indeed *any* GM product, irrespective of source or target.

On June 29, 2016, a group of Nobel laureates called on Greenpeace and its supporters "to re-examine the experience of farmers and consumers worldwide with crops and foods improved through biotechnology, recognize the findings of authoritative scientific bodies and regulatory agencies, and abandon their campaign against 'GMOs' in general and Golden Rice in particular." With the signatures of more than 150 Nobel laureates across the fields of medicine (more than one-third of the signatories), chemistry, physics, economics, literature, and peace, the letter called on the United Nations and governments to reject campaigns "based on emotion and dogma contradicted by data," concluding, "How many poor people in the world must die before we consider this a 'crime against humanity?' "[71]

In response to the laureates' letter, Greenpeace effectively abandoned its campaign against Golden Rice. A press release on June 30, 2016, denied any intention to block Golden Rice, saying that "accusations that anyone is blocking genetically engineered 'Golden' rice are false," while expressing concerns that Golden Rice would "pave the way for global approval of other more profitable genetically engineered crops."[72] In fact, its statements on food are—aside from some blustering rhetoric—largely consistent with the strategies outlined in this chapter and this book. Greenpeace calls for "sustainable food for all," demanding "a healthy future that ensures our children live in a world with fresh air, abundant forests and a stable climate" while seeking "a better way of eating and producing food" and "a global food system that is fair and sustainable for farmers and food workers."[73] Greenpeace concludes: "Together, we're determined to change the future of food." I could not agree more. To achieve our common vision of a food-secure world, can we explore ways to work *together*?

GENOME EDITING AND BEYOND

Genome editing (also known as gene editing) represents an important advance over earlier methods used to alter an organism's DNA. Genome editing uses enzymes called nucleases to make cuts in the DNA strands at specific locations, enabling removal of existing DNA and insertion of replacement DNA.[74] The editorial analogy of "cut and paste" has often been used to describe this approach to genome manipulation. In plants, however, the process of inserting replacement DNA remains uncommon. In most genome editing of crops, the mutations produced are indistinguishable from mutations that occur spontaneously in nature or through conventional breeding.[75]

The potential to edit genomes has been known since the 1970s, and the methods have become progressively more precise and usable.[76] A revolutionary new approach to genome editing—CRISPR—has gained media attention and is increasingly used for crop and animal improvement, in addition to groundbreaking and sometimes controversial applications to human health.[77] In 2020 Emmanuelle Charpentier and Jennifer A. Doudna were awarded the Nobel Prize in Chemistry for their discovery of these "genetic scissors" that "have taken the life sciences into a new epoch."[78]

The use of CRISPR, coupled with simultaneously emerging technologies like machine learning, means that genetic engineering can be easier, less expensive, and more efficient than previous methods.[79] This approach could make development of crops with new and useful traits more accessible to start-ups and public agencies that do not have the financial capacity or staying power to compete with large biotech firms.[80]

One potential advantage of gene-edited crops, in terms of public acceptance and regulatory hurdles, is that they are technically not GMOs by some definitions. Although advanced molecular methods are used, the resulting product does not necessarily contain genes from a foreign species. For example, in 2016 a nonbrowning white button mushroom (*Agaricus bisporus*) became the first CRISPR-edited organism to be declared by USDA's Animal and Plant Health Inspection Service as free of GMO-related regulation.[81] In this case, CRISPR was used to knock out an enzyme that caused browning. No foreign DNA was incorporated into the end product. Therefore CRISPR-edited plants are potentially free of the regulatory oversight required for transgenic GMOs. As a result, they are not legally defined as GMOs in some countries, including the United States and Canada, provided they do not contain the gene sequences of foreign species. Other countries consider CRISPR-edited products as GMOs or remain undecided on the need for special regulations.[82]

Earlier in this chapter I emphasized the importance of traditional varieties and wild relatives as sources of useful traits, such as pest and disease resistance, climate stress tolerance, and improved nutritional value. Over thousands of years of farmer selection and more than a century of plant breeding, the genetic makeup of the most widely used crop varieties has been narrowed to meet today's producer and consumer preferences. But as we shall see in chapter 6, Earth's climate is changing and so too the varieties we grow must change. With the use of CRISPR, there has been recent progress to unlock useful traits of wild species through a process called *de novo domestication*.[83] Through CRISPR, researchers are now able to incorporate new traits into undomesticated plants in just a few years, creating the potential to generate new better-adapted varieties that otherwise might have taken decades—or even proven impossible to develop—using traditional breeding methods.[84]

With CRISPR technologies gaining popularity for genome modification in crops, concerns have been raised about the risks of off-targeting—the introduction of unintended mutations in the genome. However, in a review of current knowledge on the precision of CRISPR in plant systems, Rothamsted researchers Florian Hahn and Vladimir Nekrasov concluded that genome editing is a very precise tool when applied to plants and off-targeting can be avoided.[85]

Over more than a decade of teaching at Columbia University, I have observed incoming graduate students struggle with the debate over GMOs: Are they useful? Are they safe? Are they necessary? Interestingly, few asked: What exactly are they? Most of my students have little or no scientific knowledge regarding GMOs or, for that matter, agriculture and the global food system. Yet many arrived, and a few left Columbia, with strongly formed negative views about "industrial agriculture" more broadly, and GMOs more specifically. Their ideas were shaped by the media and the public positions held by organizations that many of them respected and supported, like Greenpeace.

After being guided through the narrative that I have shared in this chapter—beginning with the importance of conserving genetic diversity and then exploring ways to unlock that diversity for human well-being—more than a few found themselves confused by the technical complexity of something as simple as seeds. For the past several years, I have hosted an Oxford-style debate on the proposition: GMOs are necessary for sustainable agriculture. Without exception, the debates have been lively and entertaining. While a handful of students remain entrenched in the positions at each end of the debate scale, a large majority conclude that, well, *it depends*.

> *It depends* on who are the ultimate beneficiaries.
> *It depends* on the risks to human health.
> *It depends* on the effects on the environment.
> *It depends* on who controls the seeds.

So *what if* small-scale farmers and low-income consumers are the ultimate beneficiaries of GMOs?

What if there is scientific consensus that the health risks from GMOs are no greater than most conventionally bred food commodities? *What if* GMOs are *healthier* than existing options?

What if there are net benefits to the environment, including less use of potentially toxic chemicals and improved crop adaptation to climate change?

What if GMOs were regulated or even produced by public-sector or public-interest entities with oversight to advance social inclusion and environmental sustainability?

Each of those provisions *can* be met (and are already being met to some degree). As outlined in this chapter, emerging scientific tools are being applied to meet the needs of smallholder farmers for suitable crops with specific genetic traits. All scientifically credible bodies have found that the health risk of food containing GMOs is no greater than that of non-GMO food, and there are cases like Golden Rice where the transformed foods are healthier because of their higher micronutrient content. Genetically engineered traits like drought tolerance in maize have the potential to improve resilience of smallholders in climate-stressed environments. And, importantly, there is growing support from public-interest agencies like the Consultative Group on International Agricultural Research (CGIAR) research centers, such as IRRI and CIMMYT, and supportive philanthropies like the Bill and Melinda Gates Foundation for investing in new, cutting-edge technologies that can bring desired traits to smallholders and low-income consumers.

The scientific toolbox for the genetic improvement of crops and animals is expanding with extraordinary speed. But this phenomenon creates uncertainty and fear in the general public in the absence of a trusted, science-based, regulatory framework, an honest and effective communication strategy, and capacity-building efforts that address the benefits and costs of action and inaction.[86] New breeding technologies are not—and never were—a panacea for achieving universal food security. In a food-insecure world facing climate change and resource degradation, however, we need all hands on deck and all tools at our disposal.

6

CLIMATE CHANGE

On November 1, 1975, I arrived at Ngurah Rai International Airport on the island of Bali, Indonesia. I would spend the next twelve months conducting research toward a master's degree in agricultural science from the University of Queensland. From day one, the heat was on—in more ways than one. The rainy season was already underway in southern Bali, where my field experiment would be conducted—amid the coconut trees of Kelan, then a quiet village located on an isthmus leading south to the rugged and dry Bukit Peninsula. With high humidity and threatening clouds, there was little time to waste before establishing my field experiment. A delay in planting would risk my maize crop running out of water in this relatively arid region of Bali, a challenge made worse by sandy coastal soils that dry out quickly once the dry season arrives.

On day three I traveled several hours to a government research center in Jember, East Java, to collect seed for my long-planned field trial to assess the potential for interplanting maize with a forage legume known as stylo (*Stylosanthes guianensis*); the idea was to produce a high-quality feed for the Bali cattle, a local breed that survived on crop residues and roadside grazing. The following day my experiments were laid out under a well-manicured field of coconuts. By my calculation, the maize seedlings would emerge within four or five days, in perfectly uniform rows, awaiting my inspection and nurturing hands.

It all went according to plan for the first couple of weeks. Then disaster struck. Torrential rains fell one afternoon. Although this was a relatively flat plot of land, water flowed across the fields, uplifting an inch or two of topsoil from some parts and depositing it elsewhere on the farm. My precious maize plants were uprooted, flattened, or smothered. The experimental fields were an unruly mess and unlikely to recover. Stoically, I declared we would replant. I still had enough maize seed leftover from my trip to Jember. But the bewildered look of my host farmer said it all. It was already too late to plant maize in Kelan. Obviously.

Fortunately, my local academic adviser, Dr. I Made Nitis, a professor at Udayana University and a field researcher with two decades of experience, felt my pain and came to the rescue. Nitis recommended that I change my plans and grow cassava instead of maize. Cassava was then the third most widely grown crop in Bali, behind rice and maize. Unlike maize, cassava is a hardy tuber crop, well suited to sandy soils. But even so, the rainy season was well advanced and we would have to choose a special local variety, known as Kesela Del, that thrived in the harsher environment of the nearby Bukit Peninsula. Nitis helped me find the stem cuttings of cassava I needed.

On January 10, 1976, I planted the cassava cuttings and then the stylo seeds, three days later. The stylo flourished, but even the hardy Kesela Del was no match for an earlier than usual dry season and the drought-prone sandy soils of Kelan. No tubers formed. I dutifully took measurements of the leaves and stems, but there would be no dining on the fruits of my initial foray into agronomic field research.

My research plans were ultimately rescued (once again) through some controlled greenhouse experiments within the safe confines of the Udayana University campus in Denpasar. But simply recalling that experience has caused me to relive the anxiety and uncertainty of trying to grow a crop without any control over the weather. In an *average* year, it would have been possible to grow a handsome crop of maize under the coconuts of Kelan. But as I learned the hard way, the probability of experiencing an average year in any given year is precisely *zero*.

I was fortunate that my food security did not hinge on a successful harvest of maize or cassava in Kelan. Smallholder farmers—indeed all farmers— draw on their experience and their understanding of the climate to make judgment calls on what and when to plant, and how much to invest in their crops. The poorer the farmer, the fewer the options. To reiterate an observation that I shared in chapter 4, being a rainfed farmer can be a stressful life, much more so in a changing and unpredictable climate.

END OF AN ERA

Climate—the long-term pattern of weather—has shaped the distribution of crops, livestock, and fisheries around the world. Hundreds of domesticated species have been selected over millennia to grow best in environments defined in part by temperature, rainfall, and sunshine. First through domestication, and then through experimentation and experience, farmers have selected crop types and livestock breeds to meet their consumption and market needs within the biophysical boundaries set by climate and other factors, such as soil type, landscape position, and access to water resources.

Generations of farmers have taken those seeds and animals with them when migrating to new lands. Some of these genetic resources have survived and thrived, while others have perished. As population pressures and market opportunities have increased, farming has inevitably extended into areas that are marginal or beyond the practical limits of adaptation for particular species or varieties, often degrading natural resources, sometimes with disastrous consequences.

The past twelve thousand years or so, known as the Holocene geological epoch, have witnessed the domestication, spread, and improvement of agricultural genetic resources.[1] This era has been defined by a relatively stable climate within which agriculture flourished in most parts of the world. The Holocene's climatic stability enabled societies to domesticate plants and animals within a specific region and then select for improvements over many generations. In his seminal work *Climate Change and the Health of Nations*, Tony McMichael argues that the relative stability of climate, specifically global average temperature, during the Holocene epoch made it less risky to shift from hunting and gathering to a settled system of farming.[2]

This era of what McMichael calls "climatically congenial times" is now over, and humanity has become "a geological force in its own right." For the past two decades there have been calls to recognize a new geological epoch: the Anthropocene. Atmospheric chemist and Nobel laureate Paul Crutzen popularized this term in 2000 to recognize the human impacts on the environment.[3] Human-induced climate change, along with biodiversity loss, is one of two defining characteristics of the Anthropocene.

RICE AND CLIMATE CHANGE

In the early 1990s I noticed some odd-looking structures appearing on the rice fields of the International Rice Research Institute farm in Los Baños, Philippines. These 1 meter by 1 meter Plexiglas enclosures were, in fact, the earliest manifestation of field studies to better understand the effects of climate change on rice supply and the impacts of rice farming on Earth's climate. The research was led by Dr. Reiner Wassmann, a young German scientist, seconded to IRRI from the Fraunhofer Institute of Atmospheric Research.

Interviewed in 2007 for *Rice Today*, Wassmann outlined IRRI's efforts to unravel the complex relationship between climate change and rice production.[4] IRRI's first experiment on the temperature effects on rice was conducted in 1961, just a year after its opening. Research on the effects of high carbon dioxide (CO_2) concentrations on rice growth began in 1971, long before the establishment of the Intergovernmental Panel on Climate Change in December 1988. And in 1990, with support from the U.S. Environmental Protection Agency (EPA), IRRI began research that explicitly focused on climate change impacts, including the effects of ultraviolet B radiation, increased CO_2 levels, and higher temperatures. Plant breeders, physiologists, agronomists, and modelers joined forces to tackle what was then an exciting new frontier of interdisciplinary research for IRRI. Those transparent boxes strategically located in the rice paddies expanded that effort on climate change impacts by novel research on greenhouse gas (GHG) emissions from rice.

While much of IRRI's early attention focused more on the potential dangers of a changing climate to rice production and global food security, research was also underway on methane (CH_4), a potent GHG that is generated in flooded, oxygen-depleted rice fields. Throughout and beyond the 1990s, IRRI took the research beyond its labs and experimental farm in the Philippines to engage in regional and later worldwide networks to deepen understanding of climate and food security.[5] IRRI's research found that CH_4 emissions from rice fields varied across rice-growing seasons, and with differences in water regime, soil properties, and crop-management practices. These findings paved the way for deliberate and practical modification of crop management to reduce CH_4 emissions.

But not all options were win-win. Even in 1995, the editors of *Climate Change and Rice* recognized that "many options to mitigate methane emissions enhance nitrous oxide emissions [an even more potent GHG] or reduce productivity," further concluding that "any strategy to reduce methane emissions must be judged with respect to costs, especially in light of anticipated increases in food demand."[6] This need to consider trade-offs in our efforts to achieve food security, while also adapting to and mitigating climate change, remains with us today.

THE GLOBAL GREENHOUSE

The facts are indisputable. Increased atmospheric GHG concentrations since around 1750 are unequivocally caused by human activities.[7] Thanks to the work of the IPCC and the thousands of scientists who have contributed through their research, we now know that Earth is getting warmer and that this is, in part, attributable to human activity.

The IPCC uses 1850–1900 as its historical baseline period for measuring changes. This baseline represents the earliest period of complete observations for estimating global surface temperature and serves as an approximation for preindustrial conditions. During 2011–2020 global surface temperature was 1.09°C higher than for 1850–1900, with larger increases recorded over land (1.59°C) than over oceans (0.88°C). Compared with 1850–1900, using different models, the IPCC reported that global surface temperatures could increase by 1.0°C to 1.8°C under a very low GHG emissions scenario, by 2.1°C to 3.5°C in an intermediate scenario, and by 3.3°C to 5.7°C under a very high GHG emissions scenario.[8] The IPCC notes "with high confidence," that such large temperature increases would pose great risks to global food security.[9]

Global warming comes from the *greenhouse effect* whereby infrared radiation is trapped by Earth's atmosphere. Of course, that is what made Earth habitable in the first place. The problem now is that emissions of heat-trapping GHGs have increased dramatically since the Industrial Revolution. Three GHGs are mainly responsible: CO_2, CH_4, and nitrous oxide (N_2O). Other GHGs include ozone (O_3), and synthetic gases, such as chlorofluorocarbons (CFCs) and hydrofluorocarbons (HFCs). Water vapor is also a major GHG; its concentration rises with increased global temperature, creating a feedback loop that amplifies global warming.

These different GHGs have different potencies in terms of their impact on global warming. *Global warming potential* (GWP) is the term used to enable comparisons across different GHGs. GWP is a measure of how much energy the emission of 1 t of an emitted gas will absorb over a given period of time, usually one hundred years, relative to the emission of 1 t of CO_2. Using this measure, 1 t of emitted CH_4 is twenty-eight to thirty-six times more potent than 1 t of CO_2. N_2O has a GWP 265 to 298 times that of CO_2.[10]

Agriculture and other components of the global food system have been and continue to be significant contributors to global warming through their GHG emissions. Using the IPCC terminology, Agriculture, Forestry, and Other Land Use (AFOLU) activities generated 13 percent of CO_2, 44 percent of CH_4, and 81 percent of N_2O emissions from human activities globally in the period from 2007 to 2016.[11]

Land clearing and land degradation are the main food-related causes of CO_2 emissions. CH_4 is generated from anaerobic decomposition of carbon compounds under flooded conditions (principally from rice cultivation, as outlined earlier, as well as from dams and reservoirs) along with enteric fermentation by ruminants, and decomposition of food waste from landfills. N_2O is emitted from soils mainly through excessive fertilizer applications relative to crop needs and uptake, and from manure deposition on managed pastures and rangelands. During the decade 2007–2016, East Asia, Europe, South Asia, and North America were all major contributors of agricultural N_2O emissions through large-scale nitrogen fertilizer applications. Recent research points to growing N_2O emissions from emerging economies, particularly Brazil, China, and India.[12] Overall, AFOLU activities contribute about 23 percent of total net anthropogenic emissions of GHGs.[13]

Beyond primary production emissions, it is also instructive to examine the broader food system. Francesco Tubiello, Cynthia Rosenzweig, and colleagues have estimated that the food system contributed one-third of the global anthropogenic GHG emissions (in terms of CO_2 equivalents) in 2018.[14] Of that amount, 20 percent came from land-use change associated mainly with the conversion to agriculture of forests and other natural ecosystems. On-farm production activities, including on-farm energy use, contributed 44 percent of GHG emissions. The remaining 36 percent came from pre- and postproduction activities, including emissions resulting from food transport, supply chains, and waste disposal. These numbers provide insights to how the food system could potentially contribute to climate change mitigation efforts, which will be discussed later in this chapter.

CLIMATE IMPACTS

Climate has been a major factor in shaping the distribution and productivity of agriculture and food systems around the world. As we move from the relative calm of the Holocene to a more volatile Anthropocene, we can expect increasing disruptions to the food system. The IPCC provides us with balanced, science-based evidence on the impacts of climate change on food systems, while acknowledging there may be some localized benefits from increased CO_2 level, through its positive (fertilization) effect on photosynthesis. The direct physical effects of rising GHG levels on food systems are as follows:

- increased temperatures on land and water
- changes in precipitation (rain and snow) patterns, both distribution and intensity

- melting of glaciers and corresponding long- and short-term impacts on downstream river systems
- increased extreme weather events, such as heatwaves, droughts and associated fire risks, floods, and storms
- sea level rise and corresponding intrusion of saline water as well as enhanced flooding in river deltas
- ocean acidification (more specifically related to CO_2 emissions)

These tangible impacts from increasing GHG emissions are already being felt differentially across sectors, regions, and societies. Food systems are vulnerable to these effects, and there are impacts on food security at all levels: household, community, subnational, national, regional, and global. The timescales at which global warming will affect food systems and food security will vary. Some impacts, for example, rising temperatures and sea levels, will be gradual and cause incremental increases in stress. Others will be immediate, extreme, and potentially catastrophic events, such as hurricanes, floods, and fires. Based on assessments by the IPCC and the UN's Food and Agriculture Organization, as well as other prominent peer-reviewed analyses, the changing nature of climate stresses, climate variability, and extreme climate events will negatively affect all four pillars of food security: availability, access, utilization, and stability.[15]

While the El Niño–Southern Oscillation (ENSO) has long existed, climate change has been found to increase the frequency of extreme El Niño and La Niña conditions, intensifying droughts and floods, and shifting hurricane patterns.[16] Evidence from country food balance sheets during the El Niño phenomenon of 2015–2016 shows reductions in food availability and increased food prices.[17] Modeling studies have consistently shown that the productivity of major crops will be reduced in tropical and subtropical regions with increasing global temperature.[18] Crop productivity of high-latitude regions like North Asia or Northern Europe may benefit under future warming up to 1.5°C, though warming benefits beyond 1.5°C are limited. The effects of CO_2 fertilization are often cited as a potential benefit to crop productivity. However, the field-level impacts of increased CO_2, as distinct from temperature effects, remain poorly understood and uncertain.[19]

Overall, global crop and economic models project a median increase of about 8 percent in cereal prices in 2050 due to climate change, thereby increasing the risk of food insecurity and hunger.[20] Food security of low-income households will be particularly affected by higher food prices as a high proportion of their available income is spent on food, especially cereals (to be further discussed in chapter 8). Food-importing nations are vulnerable to supply reductions and shocks, and associated price spikes. Sub-Saharan Africa imported food valued at US$43 billion in 2019. Nigeria is a net agricultural importer of over US$5 billion per year, while Angola, the Democratic Republic of Congo, and Somalia import another US$5 billion per year combined. Within sub-Saharan Africa, the biggest net food importers are resource exporters or conflict-affected states.[21] In addition, China, South Korea, and Saudi Arabia have emerged as large net importers of diversified food products owing to increased consumption spurred by economic growth and an emergent middle class.[22]

Low-income farming households are additionally vulnerable to direct climate shocks through disruption of livelihoods and loss of livelihood assets.[23] Crops are damaged by storms, temperature extremes, droughts, floods, and salinity. In 2008 Cyclone Nargis struck Myanmar, causing massive disruption to agriculture and fisheries. In addition to the direct destruction of standing crops, milling and transport infrastructure were damaged.

Pastoralists are often forced to sell livestock in drought and flooding events. For example, in pastoral communities, such as parts of Somalia, Ethiopia, and Kenya, livestock herds serve as valuable resources that are exchanged for food and other essentials. Several years of drought in the Horn of Africa, however, have greatly reduced livestock populations. Reductions in goats, camels, sheep, and cattle populations in 2017 alone ranged from 20 to 40 percent—and were as high as 60 percent in the worst-affected areas.[24]

The livelihoods of coastal fishing communities are also disrupted by storm and flooding events.[25] The impact can be felt for many years as a result of lost livelihood assets (such as boats and equipment), habitat loss, and damage to marketing infrastructure. Because of the critical role of livestock and fisheries in food security, the quantity, quality, and dietary diversity of food can be affected by these climatic events.

Climate variability and extreme events can also affect the safe utilization of food, increasing the prevalence of acute malnutrition through contamination of drinking water and poor sanitation and hygiene practices. Climate shocks have also been found to increase the prevalence of infectious diseases, disrupt child care and breastfeeding practices, and reduce access to basic health care.[26] These effects are observed at the immediate sites of climate impact and can be exacerbated when people are forced to leave their homes and relocate to camps and other temporary accommodations. Persistence of these effects can then lead to chronic malnutrition.

In sub-Saharan Africa, women play a critical role as providers of food and cooking fuel for households, making them even more vulnerable when flooding and droughts occur. In Malawi's maize crop failure in 2005, I observed women being forced to leave their homes to seek work on commercial farms and in cities. In these circumstances, women were exploited, even turning to prostitution for survival at a time when HIV/AIDS was prevalent in the region.[27]

Climate-related shocks are not a new phenomenon in agriculture and food systems. Farmers have always struggled with inter- and intraseasonal variations in rainfall and temperature. And, as far as we know, there have always been damaging storms affecting food production and distribution. One common factor I have observed from farmers around the world is that they always complain about the weather—it is never as good as it used to be.

Farmers' selections of crop and animal types and broader farming enterprises reflect trial-and-error experience to optimize their returns within the climatic conditions they experience. With that in mind, we should acknowledge that climate adaptation strategies are not new and that improvements in adaptation would be beneficial even if there was no evidence of climate change. The challenge now, with changing weather patterns, increasing variability, and more extreme events, along with increasing population pressure on an ever-degrading resource base, is to understand and adapt to a new set of more intense and more widespread climate-related threats.

LEARNING TO ADAPT

The most effective food system response to the realities of a changing climate is (1) to adapt to the likely effects *and* (2) to contribute to mitigating the causes.

The IPCC defines adaptation as "the process of adjustment to actual or expected climate and its effects."[28] With increasing climate risks, agriculture and other components of the food system will need to transform and adapt to continue to meet societies' demands. The IPCC acknowledges the close relationship between adaptation and resilience: "Adaptation goals are often expressed in a framework of increasing resilience, which encourages consideration of broad development goals, multiple objectives, and scales of operation, and often better captures the complex interactions between human societies and their environment."[29] John Ingram and colleagues at the Food Systems Group at Oxford University have described three strategies for improving resilience: the three Rs,[30] which I have paraphrased as follows:

- **Resist.** This strategy captures interventions that enable the food system to resist climate-related stresses and shocks. For example, CIMMYT is developing drought-tolerant maize varieties for southern Africa.[31] Adoption of these varieties by farmers will reduce the likelihood of catastrophic yield reductions associated with drought. In South Asia, IRRI researchers and national partners have developed and introduced rice varieties with a gene (called *SUB1*) that enables crops to survive submergence of up to two weeks (figure 6.1).[32] In Cambodia, the Asian Development Bank is providing loans to build roads that can withstand most floods and ensure that the country's rice exports are not adversely affected.[33]

- **Recover**. Even the most robust food systems can collapse in the face of climate shocks. In such cases, farmers and other food system actors need the means to recover as quickly as possible following a disruption. Farmers have long deployed crop diversification to reduce the risk associated with climate variability. In southern Malawi, I observed them interplanting maize, pigeon pea, and cassava, each having different degrees of drought tolerance, thereby planning for a range of potential seasonal outcomes. But when disaster strikes, the UN World Food Programme (WFP) and other humanitarian organizations bring food aid

FIGURE 6.1 Experimental fields at the International Rice Research Institute where new varieties are being developed to adapt to climate change impacts, including flash flooding. Patches of surviving rice in the foreground have the *SUB1* gene that enabled these varieties to recover after two weeks of submergence.

Credit: Image collection of the International Rice Research Institute (https://www.irri.org).

and cash to support recovery, increasingly with emphasis on building more resilient production assets. Crop insurance, including weather index insurance, is another way of enabling farmers to get back to work in their fields.[34] In chapter 9 we will see how the government of Malawi controversially used fertilizer subsidies as a strategy to recover quickly from a severe drought that caused nationwide crop failure in 2005. And in chapter 13 we will explore the role of social protection in recovery from shocks, including those related to climate.

- **Reorient.** This is the most far-reaching and, I believe, the most challenging of the three strategies. We are essentially accepting that our best existing efforts to resist and recover from climate-related stresses and shocks are insufficient to achieve and sustain universal food security. Reorientation supports more fundamental transformation of the structure and desired outcomes of the food system. For example, improving resilience may require farmers to go out of business and migrate to seek more lucrative and stable rural or urban livelihoods. Social protection mechanisms may be required to ensure effective and equitable reorientation to more sustainable and resilient food systems.

One large-scale example of reorientation comes from the Mekong River Delta—long considered Vietnam's most productive agricultural region—where salinity levels have increased as a result of reduced freshwater flows, land subsidence, and rising sea level. Farmers who traditionally relied on rice cropping for their livelihoods have shifted in large numbers to a rotational system of growing rice in the rainy season and salt-tolerant shrimps in the dry season (when salt levels are highest).[35] Some areas have converted lands fully to aquaculture. This reorientation has been financially lucrative because of the strong export market for shrimps. There remain concerns, however, about environmental damage (e.g., freshwater pollution) and social consequences (e.g., land grabs) of the replacement farming systems. Several years ago I asked Jordan's minister of water and irrigation what was the future of farming in his country, one of the most water-constrained nations in the world. He replied somewhat cryptically: "We cannot predict the future, only plan for it." With the Jordan River reduced to a trickle—due to water extraction and diversions throughout the Jordan River basin—and aquifers elsewhere in the country being depleted rapidly, one radical policy option would be to sharply reduce the extent of agriculture in the country. Similar existential questions are currently being asked for Australia's Murray–Darling Basin—a catchment area extending across vast swathes of arid and semiarid lands—where reduced water flows from droughts have resulted in catastrophic fish kills, cutbacks in agricultural production, land degradation, and conflicts over water rights.

These three Rs—resist, recover, and reorient—provide a useful framework for improving resilience of the food system to the shocks and stresses of climate change. The three Rs are not mutually exclusive. Reorientation can incorporate greater resistance to shocks and stresses; recovery mechanisms can include incentives for *building back better*. Building resilience thereby emerges as the overarching goal of adaptation, particularly if we extend the scope of resilience beyond biophysical technical solutions to include interventions across the economic, social, political, and cultural dimensions of food systems.

We have discussed resilience and the three Rs in the context of a changing climate and its unfolding impacts on the food system. It is important to note, however, that these strategies, though with different interventions, are equally relevant when examining nonclimate shocks and stresses to the food system, including food price spikes, conflict-induced crises, and pandemics (see chapter 14 on COVID-19).

Famine early warning systems provide an institutional adaptation mechanism for anticipating shocks and stresses, including but not limited to those that are climate-related, that can affect food insecurity on a large scale. These systems were established in the 1970s following the devastating food crises across the Sahel region of Africa. Their objective is to warn national authorities, donors, and humanitarian agencies about emerging food security crises and to enable the mobilization of the necessary humanitarian resources, including food aid.[36] The U.S. Agency for International Development established the Famine Early Warning Systems Network (FEWS NET) in 1985 for tracking and reporting food security conditions in the world's most food-insecure countries. As a food systems practitioner, I have found FEWS NET to be a rich, impartial resource for assessing acute food insecurity around the world.[37]

Society's ability to manage climate-related risks is enhanced by the provision of timely and decision-relevant climate knowledge and information—a suite of tools and activities known as *climate services*.[38] Development and deployment of climate services strengthen the ability of societies to adapt to climate change by building resilience to shocks and stresses. My colleagues at Columbia's International Research Institute for Climate and Society have led the development and dissemination of seasonal climate forecasts that draw on historical data to provide probabilistic outlooks for temperature and rainfall.[39] These forecasts are valuable to policy makers, program planners, and farming communities alike. For example, in expectation of a drier than normal season, a farmer can switch to a more drought-resilient crop or variety. Climate services can also enable users to take advantage of opportunities in more favorable conditions. By knowing that a favorable season is likely, farmers will be encouraged to invest in inputs like fertilizer and improved seed to increase their production. Climate services like seasonal forecasting are important assets for governments, development agencies, and businesses in planning and deploying resources to improve productivity, sustainability, and resilience of food systems.

Adaptation to climate change must be embedded in global, national, and local strategies toward climate action. I have noticed, however, that policy makers often use the term *adaptation* with a tone of accepting the inevitability of a warmer world with increased stresses and shocks: "We must adapt." *Total* reliance on adaptation would appear to be a risky strategy if the fears of a world 3°C or 4°C warmer are realized. As outlined in this chapter, there are myriad opportunities to adapt to a changing climate through improved resilience. Investing in resilience must serve as part of a two-pronged strategy for countering climate variability and change. The other prong is mitigation.

MITIGATION

Our planet will get warmer if we continue to emit GHGs at a faster rate than we trap GHGs, a trend made more complicated by the lag period that exists between reduced emission and its impact on temperature.[40] *Mitigation* of climate change is defined as "human intervention to reduce the sources or enhance the sinks of greenhouse gases (GHGs)."[41] The Paris Agreement—adopted by 196 parties at COP 21 in Paris on December 12, 2015—articulated an international commitment to restricting global temperature rise to "well below 2°C above pre-industrial levels and pursuing efforts to limit the temperature increase to 1.5°C above pre-industrial levels."[42] Parties to the Paris Agreement were required to make commitments, known as nationally determined contributions (NDCs), to indicate a country's contribution to global mitigation efforts and domestic adaptation initiatives. Current national pledges are significantly below that needed to achieve the 1.5°C target.[43]

The required investments in resilience to enhance adaptation will increase as the temperature rises. Maize crops will need to be more drought tolerant. Roads and canal systems will need to be redesigned to ensure their stability in more extreme weather events. Insurance premiums for farmers will need to increase to cover the cost of more crop failures. Transformational changes like abandoning large tracts of agricultural land and developing climate-protected alternatives, such as irrigation schemes and indoor farming, may entail significant economic, social, and environmental costs. For all these reasons, it is essential to invest in mitigation concurrently with investing in resilience.

As I noted earlier, food systems are a major source of GHG emissions. Changes in agriculture and other components of the food system can contribute to net GHG emissions reduction in two broad ways: by reducing current emissions and by capturing (*sequestering*) carbon. Reducing net GHG emissions can be achieved within different segments of the food system, as I outline here.[44]

Reduced deforestation. Agricultural land expansion continues to threaten forests, especially in tropical areas of South America and sub-Saharan Africa.[45] With 7 percent of net GHG emissions coming from land conversion, this strategy has high mitigation potential.[46] By reducing the clearing of forests, peatlands, and coastal wetlands for food production, we cut GHG emissions and perpetuate important ecological cobenefits, including building resilience. Consider the case of mangroves: a coastal forest ecosystem adapted to saline and brackish environments in the tropics and subtropics. By reducing mangrove deforestation for aquaculture development, we not only reduce carbon losses, we preserve their roles as a breeding ground of wild fisheries and as a physical buffer against coastal storm surges.

Changes in agricultural practices. Direct crop and livestock production activities—excluding land clearing and postfarm activities—contribute 15 percent to net GHG emissions, mainly through the more potent non-CO_2 GHG emissions.[47] The main opportunities for reduced emissions come from reduced enteric fermentation by ruminants through better feed and animal management; improved rice farming methods, primarily through improved water management; and better management of crop nutrients, both as fertilizers and as manures. More efficient use of water and nutrients can also make farming more profitable and reduce negative externalities such as groundwater depletion and water pollution.

Changes in consumer behavior. Consumer behavior can contribute to reduced emission in two ways: by reducing consumption of foods responsible for high levels of GHG emission and by reducing food waste. Reducing consumption of meat and dairy from ruminants on a global scale will ultimately reduce pressure to clear land and reduce enteric fermentation. Reduced food waste will lessen pressure on supply while also reducing CH_4 losses associated with landfill disposal. In addition, cutting food waste can benefit the consumer through food cost savings as well as reducing pressure on natural resources like water and energy (see chapter 11).

Carbon capture. In addition to reducing emissions, there is potential for the food system to remove CO_2. Planting trees has the greatest potential through new plantings and reforestation. Agroforestry—planting trees and shrubs on crop and pasture lands—can also capture carbon.[48] Converting a primary (old growth) forest to agroforestry, however, will likely reduce soil carbon. Conservation farming—a system of reduced tillage farming combined with soil cover and diversified cropping patterns (see chapter 3)—has been promoted as a way to capture carbon in soils.[49] The results vary and cannot be extrapolated to all conservation farming systems.[50] For both agroforestry and conservation farming, we need to consider important cobenefits and trade-offs in terms of income and environmental services. It is necessary to consider the baseline ecosystem state against which these interventions are assessed. For example, primary forests and degraded croplands represent two extreme baselines for assessment.

Overall, deploying these four measures—reduced deforestation, changes in agricultural practices, changes in consumer behavior, and carbon capture—could contribute about 30 percent, or 15 billion metric tons of CO_2 equivalent per year, of the global mitigation needed to meet the 1.5°C target by 2050.[51] All the above methods, however, are context-specific in terms of their impact on net GHG emissions. Estimates of their efficacy and impact vary greatly between different modeling and field studies. For example, Stephanie Roe and colleagues reported in *Nature Climate Change* that the total mitigation potential of the three supply-side (production-related) interventions—reduced deforestation, changes to agricultural practices, and carbon capture—amounted to 2 to 36.8 (median 10.6) gigatonnes of CO_2 ($GtCO_2$) equivalent per year in 2020–2050.[52] This extraordinarily wide range of estimates implies the need to measure the mitigation potential of specific actions in context-specific settings. Furthermore, as illustrated in each of the mitigation categories listed, there is potential for cobenefits and trade-offs that might enhance or detract from the contribution of an intervention in the face of uncertainty.

The livestock sector offers major opportunities for GHG mitigation. Mario Herrero and colleagues have estimated that changes to livestock production systems and meat consumption patterns could meet up to 50 percent of the global mitigation potential from agriculture, forestry, and land use.[53] They noted that high-income countries have high absolute emissions, but significantly lower emissions intensities (GHG emissions per unit of protein produced) than the developing world, due to higher productivity associated with better livestock diets, genetics, health, and management practices. There are opportunities for simultaneously increasing productivity and decreasing emissions intensity in the developing world. However, policies that deemphasize livestock production and moderate consumption of animal products should recognize the important nutrition and health contributions of livestock products in low-income settings (see chapter 12).

CLIMATE-SMART FOOD SYSTEMS

The message is clear, I hope. Climate has shaped agriculture and the broader food systems we rely on today. How we manage our food systems today and into the future will, in turn, influence our climate.

Most countries have acknowledged the importance of agriculture by including commitments in their NDCs. A review of NDCs undertaken by FAO in 2019 found that 96 percent of the 194 countries that had submitted NDCs included agriculture, land use, and forestry in their mitigation or adaptation contributions.[54]

To achieve universal food security, we will need to improve and sustain agricultural productivity while we adapt to changes occurring even today *and* contribute meaningfully to reduced GHG emissions. In practical terms, the question is: How will we achieve universal food security while simultaneously adapting to and mitigating climate change?

Over the past decade or so, an approach known as climate-smart agriculture (CSA) has gained prominence among academics and practitioners. FAO first introduced CSA at The Hague Conference on Agriculture, Food Security and Climate Change in 2010. FAO defined it as "agriculture that sustainably increases productivity, resilience (adaptation), reduces/removes GHGs (mitigation), and enhances achievement of national food security and development goals."[55] Through advocacy in member countries and through publication of case studies and sourcebooks, FAO has actively promoted CSA to "transform and reorient agricultural systems to effectively support development and ensure food security in a changing climate."[56] More than thirty countries specifically referenced CSA in the NDCs to the Paris Agreement.[57]

The CGIAR has adopted CSA in its research program on Climate Change, Agriculture, and Food Security (CCAFS). The CCAFS program research agenda included policies and priority setting to support CSA adoption and scale up climate-smart technologies. Working with CGIAR centers, national research partners, and nongovernmental organizations, CCAFS developed the Climate-Smart Village approach for testing CSA interventions at the village level. For example, in the degraded and climatically vulnerable Sahel region of West Africa, Climate-Smart Villages have introduced natural tree regeneration and species diversification, along with soil fertility improvement and water conservation. These interventions are designed to combat desertification and increase farmers' incomes, while sequestering carbon through trees and pastures.[58] The CCAFS program closed on December 31, 2021. Several CGIAR-sponsored programs are continuing to focus on climate adaptation and mitigation.[59]

The World Bank has embraced the CSA approach through its *Climate Change Action Plan*, committing to develop investment plans in at least 40 countries.[60] The World Bank plan focuses on introducing improved seeds and carbon-capture practices for crops; implementing high-efficiency and/or low-energy-use irrigation; increasing livestock productivity; promoting energy solutions for agribusinesses; and mainstreaming risk management. In 2020, 52 percent of World Bank financing in agriculture targeted climate adaptation and mitigation.[61] For example, in Bangladesh, a World Bank project aims to boost the resilience of livestock farmers and support climate mitigation through improvements in feeding strategies, animal health, breeding, manure and waste management, and low-emission technologies for milk chilling and transport.[62]

With high-profile organizations like FAO, the CGIAR, and the World Bank embracing CSA, the concept is now gaining increased acceptance as a framework for research, development, and investment. To learn more, I spent three days in October 2019 attending the Fifth Global Science Conference on Climate Smart Agriculture in—of all places—Bali, where forty-four years earlier, as a graduate student, I had been confronted with the menace of unpredictable weather. In fact, the conference venue was tucked away on a parched, west-facing slope of the Bukit Peninsula, just a few kilometers from the village of Kelan.

This conference has been held every two years since 2011 and serves as "the key global forum for scientific exchange to underpin CSA implementation." There were discussions about how to reshape supply chains to be more "climate smart," and this necessarily led to the role of businesses and finance. And there were important deliberations on the role of farmer organizations and gender. A key focus was "scale up": how to go beyond pilot projects to achieve large-scale impact.

This event assembled the best scientists and practitioners working on climate-smart agriculture—there were more than 410 participants from more than 200 institutions based in over 60 countries.

So what did I learn?

Adaptation. We have a large portfolio of technologies and policies with the *potential* to improve adaptation and resilience of farmers to a changing climate. The main challenge is to reach more farmers. There was less clarity about what other actors in the food system should do to take solutions to scale beyond the predictable calls to build partnerships, develop new business models, and show leadership.

Mitigation. We have an impressive portfolio of technologies and policies that can reduce net GHG emissions and thus mitigate the changing climate. There was a major emphasis on the importance of consumer behavior, especially in terms of diet and food waste. But again, I struggled to find clear direction on *how* to translate good intent into measurable change.

From CSA to a food systems approach. There was consensus that to bring about transformational change to combat climate change, we must take the agenda beyond CSA to take a *food systems* approach with an understanding of the interactions, synergies, and trade-offs throughout the food value chains.

In speaking with CSA experts at the Bali conference and on other occasions, one thing is clear: the objective of *simultaneously* improving productivity and food security, strengthening adaptation and resilience, and contributing to mitigation is aspirational and remains elusive. My conclusions are consistent with a 2014 review of CSA, undertaken by Leslie Lipper and colleagues, who concede that the CSA approach "does not imply that every practice applied in every location should generate 'triple wins,'" and indeed, "CSA requires consideration of all three objectives, from the local to the global scales and over short and long time horizons, to derive locally-acceptable solutions. The relative importance of each objective varies across locations and situations, as do potential synergies and trade-offs between objectives."[63]

Incentive systems (or lack thereof) help explain why improved productivity, enhanced resilience, and mitigation may not be achieved simultaneously. For example, a maize farmer in Malawi or Iowa has strong motivation to adopt practices that improve productivity and income, as well as those that lead to greater resilience in the face of climate change. In such cases, the benefits of adoption of these interventions will likely be directly captured by the farmers.

The beneficiaries of GHG mitigation efforts are spread across the entire planet. Actions of farmers in Africa and the United States can have positive (or negative) impacts on small island states in the Pacific. But unless we establish mechanisms to compensate (or penalize) those farmers, there is no obvious incentive for them to act beyond their own self-interest. Yes, there are some interventions that bring cobenefits of mitigation, along with increased resilience. But passively relying on cobenefits is unlikely to be sufficient to bring about the levels of mitigation required to stay within 1.5°C or 2°C of preindustrial levels. Distinct policies (such as subsidies) to advance GHG mitigation will be needed to augment the local adaptation benefits of CSA.

An understanding of how to make the transition from CSA to climate-smart *food* systems and indeed climate-smart *economies* is needed to ensure that all three CSA objectives are met at an aggregated global level.[64] We need to understand both the synergies and the trade-offs across the three objectives. CSA will likely continue as an agenda for identifying opportunities to transform agriculture in a changing climate. The Bali conference drew attention to the challenges of implementation, identifying key levers of change: governments (through their NDCs to the Paris Agreement), women and youth, and farmer and consumer organizations. But food-sector businesses can also be important agents of change, supported by public-sector policies and investments and shaped by broader public sentiments.

Climate has shaped the distribution and productivity of agriculture and food systems for millennia; stresses and shocks to the food system are not a new phenomenon. As we move from the relative calm of the Holocene to a more volatile Anthropocene, we can expect increasing disruptions across the food system. Climate-smart food systems should not be yet another development buzzword—hence my refusal to assign it a new acronym. Farmers have long realized the need to be *climate-smart* in deciding what, where, and when to plant. To get that wrong could have catastrophic consequences. But to meet the challenges of a warming world, we must look beyond agriculture. Climate-smart food systems are an acknowledgment that unprecedented climate variability and risks—brought on by global warming—require that we take actions across the *entire* food system to build sustainability and resilience in order to achieve universal food security.

7

HUMAN NUTRITION

As an agronomist, I find it a daunting yet an exciting and necessary challenge to venture into the field of human nutrition, a discipline known for its professional disagreements and its dearth of cross-disciplinary collaboration. Most of us are interested in nutrition, and many have developed opinions on what is good and bad based on personal experience, persuasive anecdotes, and popular media. But that is surely not good enough if our goal is to understand the full complexity of food systems and food security, and to design positive transformations. We need to act on the best evidence out there. To that end, I needed to find a colleague from this field who had the knowledge, skills, and patience to work with an old-school field practitioner more attuned to the nutrition of a rice plant than that of an infant or a pregnant woman. That colleague was Jessica Fanzo, a professor at Johns Hopkins University.

A midcareer nutritionist with a doctorate from the University of Arizona, Fanzo has worked as a lab researcher, a foundation grant giver, an award-winning graduate-level teacher, a United Nations adviser, and an international research leader. She was the first laureate of the Daniel and Nina Carasso Foundation's Premio Daniel Carasso in 2012 for her work on sustainable food and diets for long-term human health. Fanzo cochaired the *Global Nutrition Report* from 2017 to 2019 and has served as team leader for the High Level Panel of Experts on Food Security and Nutrition for the UN Committee on World Food Security. She was also a member of the EAT–*Lancet* Commission on healthy diets for sustainable food systems. In 2021 Fanzo published *Can Fixing Dinner Fix the Planet?*, calling on consumers, governments, and international agencies to change their food policies and practices or face dire consequences.[1]

To state the obvious, Fanzo knows a lot more about nutrition than I do, and many regard her as without equal when it comes to understanding and integrating the science, policy, and practice of human nutrition. Fanzo has served as my informal personal adviser on nutrition for more than fifteen years, including a year in Nairobi where we worked together with Columbia University's Earth Institute. We have subsequently collaborated on food systems projects in Myanmar, Timor-Leste, and Tajikistan.

As I struggled with the complexities of human nutrition, not least my own efforts to eat healthily, I went to Fanzo with an epiphany that *the only thing that really matters is the quantity and quality of what we eat and drink.* By and large, minor caveats aside, she confirmed that we should be striving for a world where every person consumes an *optimal diet*, calibrated to the level of physical exercise they undertake.[2] An optimal diet should comprise food items that can deliver the right amounts of nutrients we all need to survive and thrive.

THE FUNDAMENTALS

A focus on nutrition brings us to the most fundamental purpose of the food system: to nourish our bodies with an array of essential and useful nutrients. Not unlike the rice plants I am more familiar with, the human body requires appropriate levels of nutrients to live a healthy, productive, and long life. We require these nutrients in the right amounts—and at the right time—for growth and development, for building and maintenance of our immune systems, for reducing the risks of noncommunicable diseases (NCDs) like heart disease, type 2 diabetes, and cancer, and for recovering from illness and injury. When we get the amounts and balance right, we achieve a state of *good nutrition*.

Food is the main source of human nutrition except when we are in the womb or being breastfed. But even then, food intake by the mother (including prior to conception) directly affects the nutrition of the unborn and newborn child. Diet refers to the pattern of food and beverages consumed over time. A person's diet implies a sense of habitual nourishment, whereby a menu of foods and beverages is consumed with some degree of regularity. Nutrition and diets are therefore not one and the same, but they are closely correlated. Understanding diets provides important insights on nutrition. Understanding nutrition provides the foundation of a healthy diet.

Changing diets by choice or necessity can improve or worsen nutrition, with consequences on health, productivity, and longevity. In chapter 12 I will share some practical opportunities for improving nutrition through the process I like to think of as dietary convergence. This concept acknowledges that about half of the world's people are consuming either too little or too much, and that convergence on a *just right* diet should be our individual and collective goal.[3] But before exploring how to get the diet *right*, it is necessary to understand exactly what we mean by *good nutrition*, how to measure it, and what the consequences are of not achieving it. That is the purpose of this chapter.

The nutrition literature, the popular press, and the internet are replete with prescriptions on what constitutes good nutrition. We are bombarded with information on "the next superfood" and "the only diet that really works." As an outsider, it strikes me that the nutrition community is an argumentative lot, divided on science, policies, and practice. Fanzo agrees and believes that the diversity of positions and prescriptions has to do with the biological complexity of the human body and its needs, coupled with the diversity of foods and their nutritional content. These few pages will not resolve the many controversies and contested views on nutrition and diets. Instead, my goal is to establish some key concepts, definitions, and language for better understanding nutrition and its likely effects on human health and well-being.

Essential nutrients are defined as "chemical substances found in food that cannot be synthesized at all or in sufficient amounts in the body, and are necessary for life, growth, and tissue repair."[4] The essential nutrients are as follows:

- **water**, a nutrient in its own right; we will not survive long without water
- **carbohydrates**, not to be confused with energy, which comes from protein and lipids as well as from carbs
- **protein** and its constituent amino acids, nine of which are essential
- **lipids** (fat), often referred to as essential fatty acids
- **minerals**, that is, inorganic nutrients such as calcium, potassium, iron, zinc, and iodine
- **vitamins**, thirteen of which are normally considered essential
- **dietary fiber**, that is, plant parts that cannot be digested or absorbed; fiber is essential for a healthy diet but sometimes is not included in the list of essential nutrients

In addition, plants contain beneficial phytochemicals that supplement the needs of the human body by acting as antioxidants.[5] While antioxidants have been shown to have health benefits, they are generally not described as essential nutrients. Fruits and vegetables are the main sources of natural antioxidants, as well as being carriers of most of the essential nutrients.[6]

Some essential nutrients are required in very small amounts for normal growth and development. These minerals and vitamins are known as micronutrients, and their deficiency can cause severe and sometimes life-threatening conditions. Deficiencies in iron, vitamin A, and iodine are the most common, particularly in pregnant women and children. They more often occur simultaneously, rather than in isolation, and are associated with complex causal factors including food insecurity, infectious diseases, and poor access to clean water and sanitation.[7]

In 2019 an estimated 1.74 billion people had anemia, with iron and vitamin A deficiency being the major contributors.[8] As noted in discussing Golden Rice in chapter 5, the World Health Organization (WHO) estimates that 250,000 to 500,000 children become blind each year because of vitamin A deficiency.[9] As many as two billion people have inadequate iodine, a deficiency that can be easily addressed with fortified table salt.[10] Folate (a term covering multiple forms of the essential B vitamin) and zinc are also important micronutrients.

Just because these nutrients are considered essential does not mean they can be consumed in unlimited amounts. Some essential nutrients, if taken in excess, can lead to noncommunicable diseases. This applies to carbohydrates, especially added sugar and refined carbohydrates; some lipids, including saturated fats; and some inorganic nutrients, such as salt.

When the intake of essential nutrients falls below critical levels needed for normal body functions, a person is said to be in a state of undernutrition. The broader term of *malnutrition* covers undernutrition as well as intake of excessive nutrients that can lead to a variety of NCDs. The term *overnutrition* is often used to describe the latter. Fanzo, however, advises that this term is losing favor within the nutrition community. Instead, the growing preference of nutrition specialists is to use the somewhat clumsier phrase *overweight and obesity* or sometimes simply *obesity*, which itself can be misleading when excessive intake of a nutrient like sodium, iron, and some vitamins does not manifest as weight gain. A further challenge is experienced when authors increasingly use the term obesity to cover both overweight and the more extreme case of obesity, the difference between which I will explain later in this chapter.

My conclusion is that both undernutrition and overnutrition are, in equal measure, generalizations that do not fully capture the complexities of nutrient deficiencies and excesses and therefore can be used sparingly with that understanding. This is not the last time we will encounter pedantic arguments in the field of nutrition, which may in part explain the difficulties in reaching consensus on what constitutes good and bad nutrition practice.

UNDERSTANDING HUNGER

Hunger has long been understood as a condition that arises from food deprivation. We have all been there. In everyday use, hunger describes a feeling of discomfort or pain associated with not having eaten according to a normal pattern. Skip a meal or two and your body responds with signals to eat and return to a normal state. In the international development community, the term takes on a more specific technical definition. Hunger is synonymous with another term—*undernourishment*—and is measured in populations by the prevalence of undernourishment (PoU).

The UN Food and Agriculture Organization (FAO) defines hunger (and synonymously, undernourishment) in terms of dietary energy intake only. PoU is defined as "the number or fraction of people whose dietary energy intake is below the threshold Minimum Dietary Energy Requirement (MDER) for their age, sex and height."[11] This threshold represents the minimum amount of energy needed to conduct light physical activities for a minimum acceptable body weight to height ratio.[12] Dietary energy intake is expressed as kilocalories per person per day and is country-specific owing to national variations in age distribution. Carbohydrates, proteins, and lipids are the main sources of dietary energy.

The method used by FAO to calculate PoU is a little complicated and controversial. To determine the state of undernourishment objectively, it would seem most logical to survey a population to determine the amount of energy consumed by each individual and then relate that to a clinically determined reference threshold for energy adequacy for different age groupings. Those falling below a specified threshold would be considered hungry or undernourished. Unfortunately, the cost of regularly undertaking these surveys to ensure accuracy and validity is prohibitive. Few countries have nationally representative dietary intake surveys to draw on for monitoring purposes. We therefore need a proxy.

FAO has developed a more practical, cost-effective method that estimates the state of hunger in a country. The starting point for estimates of PoU is the Food Balance Sheet (FBS), using a three-year moving average. The FBS calculates the amount of food—strictly in terms of energy (kilocalories)—consumed in a country based on domestic production, imports, exports, non-food uses, postharvest losses, and carryover stocks. Then, using the data from periodic Household Consumption and Expenditure Surveys (HCES), FAO establishes a probability distribution that reflects interhousehold variance in energy consumption patterns within the population. This household information is then matched against the minimum dietary requirement for each age–sex group within the population.[13] From that information, we can then calculate what proportion of the population is likely to be energy deprived or hungry.

The PoU was the progress indicator used for Target 1.C (halving hunger) of Millennium Development Goal 1 and is now one of the Sustainable Development Goal indicators (2.1.1).[14] PoU is probably the most widely cited measure of hunger, food insecurity, and undernutrition. To illustrate the use of this terminology, *The State of Food Security and Nutrition in the World 2019*—the UN's flagship report on food and nutrition—reported that, in 2018, world hunger rose for the third consecutive year: "There were 821 million chronically undernourished people in the world last year [2018], up from 811 the previous year. One in nine people in the world now faces hunger. Africa is the region with the greatest prevalence of hunger in the world—the figure changes to one in five people hungry in Africa. Hunger is also on the rise in Western Asia."[15]

As described earlier, these numbers are calculated by FAO, country by country, and are aggregated regionally and globally. When we see or hear a statement like "more than 800 million are going to bed hungry," those numbers are based on PoU. It is an energy-based representation of one aspect of chronic food deprivation. Critics of the PoU methodology point to errors in measuring the FBS components, as well as the unreliability of household dietary recall data from the HCES.[16] For these reasons, we should not get too carried away in interpreting relatively small changes in this indicator, for example, the stated increase in the number of hungry people from 811 million to 821 million between 2017 and 2018. Of course, the thought of an additional 10 million hungry people in the world is disturbing, but only if it is real. Harsher criticisms point to a high risk of misclassification of hungry and nonhungry people using current methods.[17]

A more recent *The State of Food Security and Nutrition in the World 2020* shows that prior to the COVID-19 pandemic, almost 690 million people, or 8.9 percent of the global population, were hungry.[18] However, there is no reason to pause and celebrate a sharp decline in undernourishment. New data from China and several other countries required FAO to revise its hunger estimates going back to the year 2000, resulting in a significantly lower estimate of undernourished people worldwide. Despite these absolute numerical adjustments, FAO maintains that the trend reported in 2019 stands: since 2014 the number of hungry (energy-deprived) people worldwide has been slowly rising, which is nothing to celebrate.

Aside from measurement concerns, hunger, according to this definition, should not be automatically equated with undernutrition. Measuring energy deprivation does not consider deficiency of the essential nutrients described earlier, other than for energy. When there is a sustained deprivation of all foods—for example, as a result of extreme poverty, crop failure, or conflict—it is likely that people will suffer both hunger and at least some other form of undernutrition. Energy may be derived from staples like rice, wheat, and maize, which are high in carbohydrates but deficient in protein, lipids, minerals, and micronutrients. In such cases, a person might not be hungry based on the PoU but could be experiencing undernutrition through deficient intake of one or more of the essential non-energy-supplying nutrients.

A final point on the use of hunger is in relation to nutrition. While FAO and other development agencies use the term to delineate the adequacy of dietary energy, *hunger* is often used as a broader catch-all term that includes other forms of undernutrition—indeed, I have done that in phrasing the subtitle of this book. The UN Millennium Project Task Force on Hunger (2002–2005) was commissioned by then secretary-general Kofi Annan to develop a practical strategy for halving hunger by 2015. The Hunger Task Force rightly developed recommendations beyond dietary energy that included a range of nutrition interventions for mothers and infants.[19] The Irish Hunger Task Force (2006–2008) adopted a similar approach, recommending actions to reduce maternal and child undernutrition.[20]

More recently, the Sustainable Development Goals included a food-related goal—SDG 2, "End hunger, achieve food security and improved nutrition, and promote sustainable agriculture." In search of an overarching moniker to communicate SDG 2, the UN decided to use "Zero Hunger," drawing inspiration from Brazil's national hunger initiative—*Fome Zero*—introduced by President Luiz Inácio Lula da Silva in 2003 to eradicate hunger and extreme poverty in the country.[21]

The term hunger carries deep emotional significance. Hunger connects food deprivation with personal discomfort and pain, extending to starvation and death in the extreme. Hunger will remain in the development lexicon. But as we strive to understand and respond to the more nuanced nature of undernutrition, we need to be aware of the limitations of using the term hunger, and the potential for misrepresenting the problem that we are seeking to correct. Equating hunger with "going to bed hungry" may serve purposes of advocacy but is less useful in monitoring progress toward improving nutrition. The multidimensional challenges of undernutrition require a different language.

CHRONIC UNDERNUTRITION

Stunting—being too short for one's age—has emerged over the past decade as the most widely used indicator of chronic undernutrition in children. Based on WHO child growth standards, children are *moderately stunted* if they are two to three standard deviations below the median height for their age within a healthy reference population. If a child is more than three standard deviations below the median height for their age, they are classified as *severely stunted.*

Most indicators focus on children under the age of five years; their nutritional status serves as a proxy for the health and well-being of the larger population. Although the global prevalence of stunting has declined over the past two decades, about 150 million children under five years of age were stunted in 2020. That is an extraordinary 22 percent of the world's children in this age group. The highest rates of stunting are in South and Southeast Asia, sub-Saharan Africa, and Oceania. In each of these subregions, at least a quarter of children under five are stunted. Around 40 percent of children in South Asia are stunted, even though the stunting rate and the absolute numbers of stunted children have declined markedly since 2000. In all subregions of sub-Saharan Africa, the stunting rate has declined over the same period.[22] These statistics are yet to reflect the possible delayed impacts of the COVID-19 pandemic (to be discussed in chapter 14).

Stunting is the manifestation of impaired growth and development, especially during the period from conception to two years of age—the first thousand days.[23] Several factors can contribute to childhood stunting: poor maternal health and nutrition; inadequate infant and child feeding practices; and infections, including those related to unsafe water, poor sanitation, and poor hygiene practices. One way or another, each of these causes a shortage of essential nutrients for optimal growth and development.

The United Nations Children's Fund (UNICEF) conceptual framework for the determinants of undernutrition illustrates the complex hierarchical and interconnected causal pathways that lead to these poor maternal and child nutrition outcomes (figure 7.1). The most immediate causes of undernutrition are inadequate diet and disease. Underlying these are household food insecurity, inadequate care, inadequate health services, and an unhealthy environment, including unsafe water, poor sanitation, and inadequate hygiene practices. These factors, in turn, are shaped by deeper socioeconomic forces that include poverty and inequality, poor governance, lack of institutional and financial capacity, and cultural traditions. Developed in 1990 as part of UNICEF's nutrition strategy, the underlying problem tree can be reversed to identify a portfolio of potential solutions that lead to optimal maternal and child nutrition.[24] The UNICEF framework remains the most powerful and widely used conceptual tool for understanding the causes and consequences of undernutrition. In chapter 12 we will see how the UNICEF framework can serve as a basis for action that necessitates integrated investments across sectors and institutions.

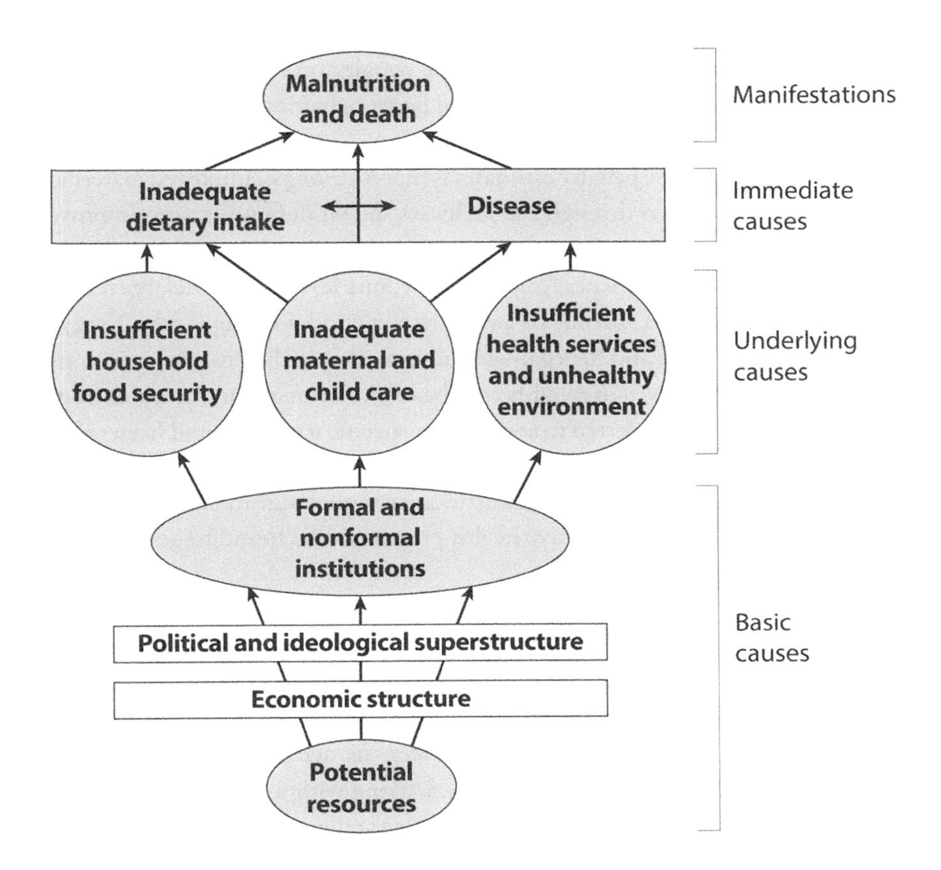

FIGURE 7.1. The UNICEF conceptual framework on the causes of malnutrition.

Source: United Nations Children's Fund. Redrawn with minor editorial changes from UNICEF, *Strategy for Improved Nutrition of Women and Children in Developing Countries*, June 1990.

Chronic undernutrition has several consequences.[25] Children who are stunted have impaired brain development, resulting in poorer cognitive and educational outcomes in later childhood and adolescence. These early developmental setbacks are largely irreversible and can have dire economic consequences later in life. Income earning potential is reduced. Stunting can also trigger a metabolic response by the body that results in overweight and obesity later in life.[26] When these effects are aggregated at the national level, stunting can place a major burden on a nation's economy in terms of health costs and lost economic potential. In an assessment paper for the Copenhagen Consensus on Human Challenges, Sue Horton and Richard Steckel estimated that stunting reduced annual gross domestic product (GDP) by as much as 12 percent in low-income countries during the twentieth century.[27] The World Bank estimates that the neglect of stunting has resulted in a per capita income penalty on the economy of about 7 percent. Africa and South Asia incur even larger penalties of 9 to 10 percent of GDP per capita.[28]

The corollary of these penalty estimates is that *investing* in improved maternal and childhood nutrition through the pathways shown in figure 7.1 can improve educational performance, employment prospects, and economic growth.

A recent review of experience from selected countries that showed large reductions in stunting over a fifteen- to twenty-year period—Ethiopia, the Kyrgyz Republic, Nepal, Peru, and Senegal—found that, despite heterogeneity in context, progress was achieved in roughly equal measure through interventions from the health sector (often referred to as nutrition-specific strategies) and from other sectors like education, water and sanitation, and agriculture (nutrition-sensitive strategies).[29] Investments in maternal nutrition, maternal education, maternal and child health care, and reproductive health practices were found to generate the greatest reductions in stunting.

In 2012 the World Health Assembly (WHO's governing forum with representation from all member countries) endorsed a comprehensive implementation plan on maternal, infant, and young child nutrition that specified six global nutrition targets for 2025. The first of the six targets is a 40 percent reduction in stunting between 2010 and 2025 for children under the age of five years; the UN adopted this as an interim target for SDG 2, along with a target of completely eliminating stunting by 2030. Despite a 50 percent reduction in the global stunting rate during the past two decades, the World Bank estimated that stunting will decrease by only 20 percent between 2010 and 2025 and will therefore fall well short of the World Health Assembly and SDG targets. All experts I have consulted confirm that ending stunting on a global scale by 2030 is not achievable. Yet we know the causes and the consequences. Why is it that progress is not on track to meet our global targets? How do we shape a new trajectory of progress that leads to an elimination of stunting? We will return to this question in chapter 12.

FIRST THOUSAND DAYS AND BEYOND

A series of papers published in the *Lancet* in 2008 focused on maternal and child undernutrition, identifying the need to focus on the period from conception to a child's second birthday—the first thousand days.[30] Over the past decade, an emphasis on the first thousand days has created a compelling focus and agenda for action.[31] There is little argument that the nutritional status of women and children during this critical window has lasting effects on health and well-being throughout life. In addition to supporting good maternal nutrition and health, there has been strong emphasis on exclusive breastfeeding (breast milk only for the first six months of life) and timely and adequate complementary feeding thereafter.[32] It is necessary, however, to recognize that nutritional needs do not stop at a thousand days.

Good nutrition is needed at all stages of the life cycle, including by older people, who often play an important role in childcare. It is evident that there is an intergenerational cycle of undernutrition (and overnutrition, as we will discuss later). For example, a woman with anemia is likely to have a baby with a reduced birth weight, thereby increasing the risk of stunting, wasting, and NCDs later in life.[33] Therefore, while the thousand days agenda brings focus and resources to a critical time frame for intervention, actions to reduce undernutrition, as well as overnutrition, require a whole-of-household, whole-of-community, and whole-of-population approach.

ACUTE UNDERNUTRITION

Wasting—a measure of acute undernutrition—occurs when a child loses weight and becomes excessively thin. WHO classifies the degree of wasting as moderate or severe, according to the WHO growth reference for weight-for-height.[34] WHO considers wasting to be a public health problem when 5 percent or more of the population are affected. In 2020 there were about forty-five million children—or almost 7 percent of the world's children under five years of age—with wasting, a third of whom exhibited severe wasting.[35]

As in the case of stunting, a combination of interacting factors results in rapid weight loss in children:[36]

- poor food security and inadequate diet in terms of quantity and quality
- poor childcare and feeding practices, including inadequate breastfeeding
- poor access to safe water, sanitation, and hygiene services
- poor access to health care services, especially the timely treatment of diarrheal disease

These factors are most common in settings of extreme poverty and are exacerbated in times of climate stress and conflict. This vicious cycle of food insecurity, infectious diseases, and inaccessibility of treatment presents a major challenge to households, communities, governments, and development and humanitarian agencies.

Children who are wasted have weakened immunity, are susceptible to developmental delays, and are at a heightened risk of death. It has been estimated that a severely wasted child is at least eleven times more likely to die prematurely than is a healthy child, and that as many as two million children die annually as a result of severe wasting.[37] Severe bouts of wasting can lead to the more permanent impacts associated with stunting. It is possible to be both wasted and stunted: thin-for-height and short-for-age. WHO, however, urges recognition that "wasting confers double the risk of mortality associated with stunting, and being both stunted and wasted confers an even higher risk."[38]

As with stunting, the highest prevalence of wasting is in South and Southeast Asia, sub-Saharan Africa, and Oceania. Based on the UNICEF/WHO/World Bank Group Joint Child Malnutrition Estimates for 2020, South Asia has by far the highest wasting level: 14 percent of children under five are moderately or severely wasted. This subregion represents more than half of the world's wasted children.[39]

Children who experience wasting require urgent feeding, treatment, and care for recovery and survival.[40] While wasting tends to be associated with disasters like severe droughts and conflict, WHO notes that most wasted children live outside the usual humanitarian response settings that are the focus of treatment programs. UNICEF notes that progress has been particularly slow in nonemergency contexts.[41] Less than 15 percent of wasted children receive treatment.[42]

One of the World Health Assembly targets is to reduce wasting to 5 percent by 2025. As with stunting, the world is unlikely to reach this target. But given the life and death implications of wasting, the UN has recognized the need to step up efforts. Protocols for identification and treatment are well established. Individuals with moderate acute malnutrition (MAM) are typically cared for as outpatients, and provided with supplementary foods when necessary. Hospitalization has been recommended for children with severe acute malnutrition (SAM), demonstrating severe weight loss and nutritional edema—swollen feet, face, and limbs. But hospitals and clinics are often inaccessible to poor and remote communities, and there are additional challenges to users associated with crowding, risks, and cost.[43]

Community Management of Acute Malnutrition presents an alternative integrated approach to managing SAM and MAM.[44] This approach has four components: "(1) community outreach and mobilization; (2) outpatient management of SAM without medical complications; (3) inpatient management of SAM with medical complications; and (4) services or programs to manage MAM, such a supplementary feeding program."[45] Ready-to-use therapeutic food (RUTF) has transformed the treatment of SAM.[46] RUTF is a high-energy, micronutrient-enhanced paste. RUTF has emerged as an effective and practical solution where cooking facilities and fuel are limited. It has a long shelf life and does not require clean drinking water for preparation.

While saving lives of children affected by wasting is the most immediate priority, the long-term solution requires multisector investments that address the underlying causes of wasting:

- more productive, diverse, and resilient agricultural systems that enhance food security and improve diets
- enhancements in education and health services to promote improved childcare and feeding practices
- improvements in water and sanitation infrastructure, coupled with complementary behavioral change investments, such as regular handwashing
- more effective, decentralized health surveillance systems to improve early diagnosis, treatment, and prevention of wasting and infections that contribute to wasting

These interventions—all of which are just as relevant to tackling the stunting challenge—must be accompanied by political awareness and commitment to ensure adequate funding, execution, and coordination of complex multi-institutional efforts to reduce wasting.

FOOD SECURITY AND GOOD NUTRITION

Stunting and wasting are the *consequences* of poor nutrition. For a variety of reasons just outlined, almost half of the world's population remains unsuccessful in providing the required amounts of essential nutrients for a healthy, productive, and long life. Dietary intake is not the only factor. A healthy diet is of little value if the body is racked by diarrhea and other diseases. Clean water, improved sanitation, and good hygiene practices (known as WASH) are critical complementary investments for reducing all forms of undernutrition. The work of UNICEF provides excellent guidance in this aspect of the nutrition challenge.[47]

There is a strong consensus that food security and dietary diversity are critical determinants of good nutrition. At the World Food Summit held in Rome on

November 13–17, 1996, assembled heads of state and government pledged to achieve food security for all. The Rome Declaration on World Food Security that emerged from this historic meeting provided us with a comprehensive and robust definition of food security—incorporating the notion of "sufficient, safe and nutritious food"—that remains widely accepted: "Food security exists when all people, at all times, have physical and economic access to sufficient, safe and nutritious food to meet their dietary needs and food preferences for an active and healthy life."[48]

These multiple dimensions of food security present a challenge in measuring progress toward its attainment. There is no single measure that captures the complexity of both the causes and effects of food insecurity. The PoU indicator has been outlined and critiqued earlier in the chapter. It serves as an indicator for SDG 2.1: "By 2030, end hunger and ensure access by all people, in particular the poor and people in vulnerable situations, including infants, to safe, nutritious and sufficient food all year round." The Food Insecurity Experience Scale (FIES) has recently emerged, however, as a complementary indicator of food access and is now used alongside the PoU as an indicator of SDG 2.1.

The FIES survey explores, through eight simple questions, the experiences of individuals or households in accessing food. The survey asks whether there was a time, during the previous twelve months, when, because of lack of money or other resources:

1. You were worried you would not have enough food to eat.
2. You were unable to eat healthy and nutritious food.
3. You ate only a few kinds of foods.
4. You had to skip a meal.
5. You ate less than you thought you should.
6. Your household ran out of food.
7. You were hungry but did not eat.
8. You went without eating for a whole day.[49]

An analysis of the "yes/no" responses enables determination of the food security status of a population on a scale that ranges from "food secure" to "severely food insecure." Used in combination with other indicators like PoU, prevalence of stunting, and prevalence of wasting, the FIES contributes to a more comprehensive understanding of the causes and consequences of food insecurity. According to FAO, the FIES is easy to use and can be easily integrated into other population surveys.[50]

After methodological development and more than three years of data collection in over 140 countries, FAO regards the FIES as a reliable and valuable contribution to global food security monitoring.[51] The FIES is used to determine the prevalence and severity of food insecurity, identity vulnerable subpopulations, assess the progress of food-security policies and programs, and identify risk factors and consequences of food insecurity over time and across different populations.[52] While the FIES includes questions on "healthy and nutritious food" (question 2) and "kinds of foods" (question 3), the limitation of "yes/no" responses means that neither the FIES nor PoU assesses dietary adequacy or diversity. A different kind of measure is needed if we are to obtain a more reliable assessment of nutritional intake.

DIETARY DIVERSITY

Nutritionists have long appreciated that a diverse diet is important for good health. If the plants and animals that provide different foods and comprise different food groups were identical, or at least similar, in their nutritional composition, dietary diversity would not matter, aside from aesthetic qualities of appearance, taste, and texture. But the foods we consume vary greatly in their nutritional content. For this reason, the dietary guidelines of most countries recommend eating a diverse range of foods within and across food groups.[53]

In low-income settings, there is often a strong dependency on staple foods that are high in carbohydrates. Bennett's law, as explained by C. Peter Timmer, Walter Falcon, and Scott Pearson four decades ago, states that "the proportion of calories that an individual derives from the basic starchy staples (mostly grains and root crops)—the starchy staple ratio—falls with rising income as the consumer diversifies the food consumption bundle to include higher-priced calories." They argue that Bennett's law reflects "the seemingly universal desire for variety in the diet and for high-quality protein and refined sugar."[54]

Marie Ruel defines *dietary diversity* as "the number of different foods or food groups consumed over a given reference period."[55] Household Dietary Diversity Score (HDDS) is a population-level indicator of household food access. The HDDS measures household consumption ("yes" or "no") of twelve different food groups with a twenty-four-hour recall period. There is no indication of the quantity of food consumed, which is an obvious limitation of the measure. Some studies use a more limited number of food groups and apply a minimum consumption quantity for the food group.[56] Variations on the HDDS approach include measurements at the individual level, such as the Minimum Dietary Diversity for Women and the Minimum Dietary Diversity for children six to twenty-three months old.[57]

While these measures of dietary diversity give us broad indications of the nutritional adequacy of diets in different populations and target groups, they provide limited guidance on the magnitude and causes of nutritional gaps for different segments of the population, for example, pregnant or lactating women, children under two years, and adolescent girls. To improve nutrition situation analysis with a view to formulating recommendations and priorities for action, the UN World Food Programme, with technical support from several partners, has led the development of a valuable methodology known as "Fill the Nutrient Gap" (FNG).[58]

The FNG analysis is an approach that requires extensive consultations and partnerships across multiple sectors working in the food and nutrition sector. It begins with a comprehensive review of secondary information on malnutrition characteristics, the enabling policy environment, availability of nutritious foods, household access to nutritious foods, nutrient intake, and local preferences and behaviors. The analysis includes an assessment of the affordability of nutrient sources using a "Cost of the Diet" tool that applies linear programming to estimate the minimum cost of locally available diets that meet the nutrient requirements of specific target groups. Drawing on household income and expenditure data, conclusions can be made on the affordability of "filling the nutrient gap" with and without a range of interventions.[59]

For a recent FNG analysis in Tajikistan, which I elaborate in chapter 12, the research team found that a nutritious diet from locally available foods was unaffordable for between 29 and 56 percent of the population, depending on the region of the country.[60] These unaffordability levels were almost halved with an optimized package of interventions that included fortified wheat, micronutrient supplements to adolescent girls and lactating women, supplementary feeding for children under two years old, and improved school meals. The FNG methodology is practical and participatory in developing an evidence base for policy makers to support dietary diversity leading to better nutrition outcomes, and warrants wider application.

THE ROAD TO OVERNUTRITION

It was March 1984. The final stop on my first trip to Africa was Tanzania. Julius Nyerere (1922–1999), father of the nation, was in the penultimate year of his rule.

In contrast to Malawi, where I had been the week before, Tanzania retained some distinctive elements of socialism from Nyerere's two decades as president: dimly lit streets, empty stores, and surly government drivers.

When Nyerere stepped down in November 1985, the *New York Times* reported that "the remarkable is happening in this East African nation of 20 million people—a sitting President has chosen to retire, declaring that he wished to pass the reins peacefully to a new generation." Moreover, as a farewell gift, the departing president "was feted by the Chamber of Merchants and given three dairy cows, a lion skin and 'assorted household items' by various women's organizations."[61]

In retrospect, I now see the practical value of the three cows. In March 1984 there was no butter to be found in the Embassy Hotel, one of Dar es Salaam's premier lodgings at that time. The country was in economic ruin. Agriculture, once the mainstay of the economy, had steadily declined with *ujamaa* socialism, causing a foreign exchange shortage and a crisis in manufacturing.[62] The private sector was weak and fragmented. Gross national product (GNP) per capita was around US$200. The roads, with unsurprisingly few vehicles, were by far the worst I had seen in any African capital, albeit a selection limited to Lusaka, Harare, and Lilongwe at that point.

Three decades later, in a taxi from Julius Nyerere International Airport to a downtown hotel, three changes stood out to me. First, the roads were wide and pothole-free. Second, the traffic was horrendous, the country's population having already passed fifty million, more than doubling since 1984. And third, there were hundreds of advertising billboards down the median strip and scores of larger promotional structures on each side of the highway. The drivers were not at all surly, perhaps distracted by the plethora of roadside billboards, half of which were promoting sugar-sweetened beverages.

As we stopped in the traffic, vendors were moving vehicle to vehicle selling high-carb packaged snacks. A quick check of the statistics on my mobile phone, seamlessly operating in roaming mode, revealed that adult overweight (including obesity) prevalence in Tanzania had risen sharply since the turn of the century, with the rate in women being roughly double that of men.[63] The Tanzania Demographic and Health Survey showed that the prevalence of overweight reproductive-age women increased from 11 percent in 1991–1992 to 28 percent in 2015–2016, while the prevalence of underweight did not improve over that same period.[64] I was witnessing what is called the *nutrition transition*: a concept, first described by Barry Popkin of the University of North Carolina, whereby changes in dietary intake and physical activity patterns were leading to obesity and other nutrition-related noncommunicable diseases in low-income countries that were previously associated mainly with undernutrition.[65]

Overweight and *obesity* are terms nutritionists use to describe degrees of excessive fat accumulation as measured by body mass index (BMI). We calculate BMI by dividing the weight of a person in kilograms by the square of their height in meters: kg/m^2. WHO describes BMI as a useful population-level index for overweight and obese adults for both sexes. Overweight is used where the BMI is greater than or equal to 25. The term obesity applies to overweight adults with a BMI greater than or equal to 30.

These measures should be contextualized and adapted to specific body types. Fit and muscular athletes like American football quarterback Tom Brady, whose BMI is around 27, would hardly be considered overweight and in danger of tipping into the obese range. Similarly, BMI may underestimate risks for older people who have lost muscle mass.[66] WHO acknowledges that BMI is a "rough guide" and should be used accordingly.[67] In measuring overweight and obesity in children and adolescents, BMI is calculated as for adults but applied in relation to WHO growth references for different ages.

The prevalence of overweight and obesity is rising everywhere. Fundamentally, we are observing an energy imbalance between calories consumed and calories used. Those sugary drinks and high-fat, high-carb processed foods being marketed along the airport road in Dar es Salaam are indicative of a global crisis. Urbanization has changed eating patterns, diets, and food environments. When coupled with reduced physical activity, associated with new job types and transportation modes, it is unsurprising to find that no country has been able to reverse the increasing trend in overweight and obesity in adults, and very few have made progress in children and adolescents. Calories in; calories out. Whether you live in Dar es Salaam or New York City, it is as simple as that.

Being overweight or, worse still, obese has potentially serious health consequences: the risk of NCDs increases with increases in BMI.[68] Cardiovascular diseases and type 2 diabetes are on the rise, especially in low- and middle-income countries. In Tanzania, diabetes prevalence has passed 6 percent for both men and women.[69] Other NCDs that accompany raised BMI include musculoskeletal disorders and some cancers.[70] Childhood obesity is associated with breathing difficulties, insulin resistance, psychological effects, among other disabilities, and premature death.

The 2021 Global Nutrition Report revealed that 2.2 billion adults aged eighteen years and older were overweight in 2019, and of these, more than 770 million were obese. In 2016, 340 million children and adolescents aged five to nineteen were overweight or obese. Another 39 million children under five years of age were overweight.[71] There is little doubt, based on current trends, that overweight (including obesity) prevalence across all age groups

FIGURE 7.2 Sugary drinks for sale in Shaartuz, Tajikistan (2018).

Credit: Glenn Denning.

will continue to grow through the 2020s, probably passing three billion by mid-decade.[72]

Long considered a problem of the rich world, more than 70 percent of overweight (including obese) adults live in low- and middle-income countries. There will be negative impacts on productivity and economic development. Health services that have struggled to reverse the prevalence and mortality of infectious diseases will be stretched to meet the growing demands from NCDs. A World Bank report in 2020, *Obesity: Health and Economic Consequences of an Impending Global Challenge*, estimated that overweight and obesity will cost more than US$7 trillion in developing countries over the next fifteen years. These costs are not only the result of early death. Other consequences include increased health care costs, early retirement, and reduced productivity.[73] The U.S. Centers for Disease Control and Prevention has included obesity and diabetes as important preexisting conditions that heighten the risk of morbidity and mortality from COVID-19.[74]

The World Bank describes overweight and obesity as a "ticking time bomb" with "huge potential economic and health impacts, especially for the poor." Estimates vary greatly because of different settings and methodologies. Costs in the United States range from US$89 billion to US$212 billion annually (in 2010 U.S. dollars); those from China are estimated at almost 9 percent of GNP by 2025, while Brazil projects a doubling of obesity-related health care costs reaching US$10 billion annually by 2050.[75]

Even the World Bank's ticking time bomb metaphor falls short in expressing the urgency to act. Overnutrition "bombs" are already exploding every day in every country across the planet. The overnutrition pandemic represents a relentless attack on the health and well-being of populations and economies everywhere. Although there are few successes to draw on, an evidence base is slowly emerging as nations recognize and begin to address the scale of the problem. Policies and interventions with the potential to succeed include the following:

- taxes and subsidies to encourage healthy eating
- regulations on marketing and advertising
- education sector policies and programs
- transport and urban design interventions
- food environment design interventions
- targeted nutrition interventions
- promotion of physical activity

The good news is that the international development community is finally taking overnutrition seriously, to a degree. The SDGs calls on nations to "end all forms of malnutrition" (Target 2.2)—an emotive headline *call to action*—with an indicator that only considers children under five years of age in relation to the prevalence of overweight (and wasting) (indicator 2.2.2).[76] Under SDG 3, Good Health and Wellbeing, is a target to "reduce by one third premature mortality from non-communicable diseases," though, incomprehensibly, there is no target or indicator within the SDGs on adult overweight or obesity.

In early 2015, on behalf of the UN Sustainable Development Solutions Network (SDSN), I canvased some of the world's top nutrition experts for key indicators to include for measuring SDG progress. There was full agreement among the nutrition gurus that overweight and obesity, disaggregated by gender and age, should be among them; it was therefore listed among SDSN's recommended one hundred indicators across the seventeen SDGs. But for reasons that remain a mystery to me, Jessica Fanzo, and others, that indicator was not among the 169 indicators ultimately approved by the UN.

THE DOUBLE BURDEN

In this chapter we have looked at the two extremes of malnutrition: undernutrition (hunger, stunting, and wasting) and overnutrition (overweight and obesity).

The imbalance of treatment of these topics within the chapter reflects, in part, the historical emphasis of the international development community, of which I am a member. There is no doubt that we have focused mainly on undernutrition over the past sixty years, especially in low-income settings. We have seen undernutrition and overnutrition as two distinct sets of issues. The former was often related to poverty and conflict, a so-called third world problem; the latter was associated with wealth and overconsumption, more of a first world problem. These different forms of malnutrition have been viewed as discrete problems and have been addressed in silos by different communities of science and practice, and managed through different programs, governance structures, and funding.[77]

These generalizations are now rightly under fire in both the nutrition and development communities. The data make it clear that undernutrition and overnutrition coexist across a range of levels: within countries, communities, households, and even at the level of an individual.[78] This phenomenon—a result of the global nutrition transition—is known as the double burden of malnutrition. As a consequence of "this new nutrition reality," authors of one of the papers in the *Lancet*'s 2020 WHO Series on the Double Burden of Malnutrition argue: "We can no longer characterise countries as low-income and undernourished or high-income and only concerned with obesity."[79]

At the household level, there may be an overweight mother (or father) and a child with stunting or wasting. At the individual level, that same overweight mother may be experiencing vitamin and mineral deficiencies. Also, stunting from childhood undernutrition can result in overweight and obesity later in life, increasing the risk of noncommunicable diseases.[80]

An emerging alternative approach to improving human nutrition—known as *double-duty actions*—recognizes the coexistence and interrelationships between undernutrition and overnutrition. Double-duty actions are based on the rationale that all forms of malnutrition share common drivers, which include early-life nutrition, diet diversity, food environments, and socioeconomic factors. As a result, many of the same policies and interventions can bring positive outcomes across the malnutrition spectrum. In chapter 12 we will examine these double-duty actions to address both undernutrition and overnutrition in greater detail in the context of food system transformation to achieve healthy and sustainable diets.

8

FOOD SYSTEMS TRANSFORMATION

Food systems comprise all the actors, activities, and forces influencing the production, aggregation, processing, distribution, consumption, and disposal of food products.[1] Food systems are inherently economic systems in which decisions are made by a multitude of individuals and organizations, entities we tend to call "actors." These actors allocate resources to a range of activities that are undertaken within biophysical, socioeconomic, political, and cultural constraints. Food systems fundamentally connect producers and consumers with outcomes affecting nutrition and health, livelihoods and economies, and the environment. The nature and quality of decision-making by food system actors will ultimately determine our success in achieving a food-secure world.

The term *global food system* is often used to highlight the international interconnectedness of food systems around the world. The global food system is a dynamic and complex aggregation of thousands of constituent subsystems that reflect interacting factors of geography, history, culture, economics, policies, and politics.[2] We measure the outcomes of our global food system in terms of food security and environmental impact. In earlier chapters I disaggregated agricultural production into its biophysical components, namely, soil and land (chapter 3), water resources (chapter 4), and seeds (chapter 5). These are the fundamental building blocks of diverse food systems that are further shaped by climate (chapter 6) and humanity's demands for nutritious food (chapter 7). It is my hope and assumption that chapters 3 to 7 have provided readers with the fundamental knowledge and language for understanding food systems and the drivers and impacts of food systems transformation.

In this chapter I will first track the evolution of food systems of societies, starting with those that hunted and gathered food in essentially natural environments. Those societies evolved and transformed to market-mediated, often globalized, supply chains that have come to dominate the global food system. I will explain the ubiquitous process of structural transformation that has been a driving force of change in food systems over the past two centuries. Understanding and managing structural transformation are essential for achieving food systems transformation. Finally, I will address two critical and strategic questions: What are the desired attributes of food systems? And what does it really mean and take to achieve universal food security?

BEYOND SUBSISTENCE

Starting around twelve thousand years ago, the warmer temperatures of the Holocene geological epoch enabled the transition from small, nomadic bands of hunter-gatherers to settled agricultural societies growing domesticated plants and animals.[3] Initially, these settled farmers, like their nomadic predecessors, focused on survival, often on a season-to-season basis. This mode of production is known as *subsistence agriculture*. Subsistence farmers manipulated their domesticated species within the constraints set by the natural environment to achieve food self-sufficiency at the household or community level. The primary risk for subsistence farmers was poor yield. A flood, a drought, a frost, or a heatwave could reduce food supply below the household requirements. Pest and disease outbreaks, sometimes weather related, could also cause catastrophic outcomes, including starvation, for subsistence households.

One form of subsistence agriculture is known as *shifting agriculture*, which is likely one of the oldest farming systems.[4] Traditional shifting cultivation systems (also known as swidden cultivation) were practiced throughout most of the world, including in the early farming settlements in Europe and the United States. These days, shifting cultivation is found in pockets throughout the humid tropics and usually consists of forest clearing, burning the cut vegetation, a short cycle of one to three crops, and a fallow of secondary forest regrowth of up to twenty years. This production system is usually referred to as *slash-and-burn farming*, a system we encountered in chapter 3 in describing land degradation in Madagascar.[5]

I first observed this practice in the mountains of Zamboanga del Sur in western Mindanao, Philippines, while working there as an agronomist between 1977

FIGURE 8.1 Slash-and-burn farming in northern Lao PDR (1990).

Credit: Glenn Denning.

and 1980. Each year, early in the dry season, mostly indigenous farmers cleared patches of primary or secondary forests using machetes, often coming in after loggers had removed the most valuable timber trees. Toward the end of the dry months, the farmers set fire to the desiccated vegetation, the ashes were spread over the sloping, unterraced fields, and, with the first rains (usually in May), seeds of upland rice (also called *dryland rice* due to the absence of bunds to contain water) were sown in holes made with pointed planting sticks.

The first crop after burning usually flourished on the nutrients released from the ash and the soil moisture from abundant rainfall. But after one to three seasons, the soil fertility declined through crop removal and erosion, and the native or sometimes invasive exotic vegetation returned to compete aggressively with the crops. These fields were then abandoned, and the farmers would move to another location. In systems like this, farmers return to the original plots once the vegetation has regrown after fifteen or more years. With increasing population growth, however, the cycle is shortened, and yields are lower, even for the first crop grown after the fallow period.

Shifting cultivation has largely disappeared from Mindanao, and only a few isolated pockets remain across Asia.[6] Although considered sustainable at low population densities, shifting cultivation inevitably is replaced by unsustainable slash-and-burn farming with short fallow cycles that fail to restore nutrients to anything like the former levels. In Mindanao, these areas were often abandoned to low-productivity grasses and shrubs. What may have been sustainable under low population pressure had become degraded and unproductive.

During my three years in Mindanao, extensive areas of hilly land were deforested in this way. The dark green tree line retreated up the slopes, replaced by yellow cogongrass (*Imperata cylindrica*) that was periodically burned to extract some nutrients for a single crop or some green shoots for grazing. Subsistence farmers barely survived on their upland rice, along with some root crops, vegetables, and small livestock and poultry. They were generally far from markets and government support services. Coupled with weak enactment of land reforms, an increase in illegal logging, and an aggressive military presence, this region of immense potential became an arena of ethnic and social conflict. The collapse of shifting cultivation and a downward spiral of productivity have held back the development of large parts of Mindanao to this day.

Three longtime colleagues of mine—Pedro Sanchez (the legendary soil scientist whom you met in chapter 3), Tom Tomich (a former colleague at the World Agroforestry Centre and now distinguished professor of sustainability science and policy at the University of California, Davis), and Cheryl Palm (a former Columbia Earth Institute colleague, now research professor with the University of Florida)—have spent much of their professional lives working with hundreds of researchers and policy makers to improve our understanding of slash-and-burn systems and to develop sustainable alternatives.[7] Sanchez details the following range of different alternative systems, from the least to the most intensive, that have proven to be promising.[8]

Managed forests. Four broad forest management systems generate income for local people while minimizing environmental damage:

- nonextractive forest reserves where local communities are responsible for protection and receive benefits from enterprises, such as ecotourism or carbon credits (e.g., Sabang mangrove forests in Palawan, Philippines)[9]
- selective logging with strict adherence to permits and coupled with food production opportunities in forest margins (e.g., riverine communities along the Amazon)
- extractive reserves where local people extract nontimber forest products (e.g., Brazil nut extraction in the Amazon)[10]
- community-managed forests whereby restricted low-intensity logging is permitted alongside crops and pastures within a single land unit (e.g., community forest enterprises in Mexico)[11]

Complex agroforests. These are mainly indigenous systems with a diversity of tree and crop species. Complex agroforests are found in the tropics of Asia, Africa, and the Americas.[12] Eventually the annual crop species die out, and farmers harvest a variety of income-generating products such as rubber, cacao, coffee, fruits, and medicinal plants. These forests can also allow for regeneration of high-value indigenous species. Complex agroforests can mimic the diversity of natural ecosystems and thereby offer sustainability and resilience in the face of fluctuating weather and commodity prices.

Simple agroforests. These systems have less diversity than the complex agroforests and normally involve a dominant tree crop with one or more crops growing in the understory. My first exposure to simple agroforests was seeing maize and cassava growing under coconuts on the island of Bali. As part of my master's research, I made the system more complex by undersowing a pasture legume (*Stylosanthes guianensis*) in the maize and cassava to provide fodder for the indigenous Bali cattle. Sadly for me, but perhaps not for the local community, this simple agroforest system has been replaced forty years on by tourist accommodation and related commercial establishments. Other examples of simple agroforests are cacao and oil palm. Simple agroforests demonstrate less sustainability and resilience than complex agroforests.

Improved pastures. Most notably in the humid tropics of Latin America, forests have been cleared with the purpose of establishing pastures, often preceded by one or two food crops. More than 40 million hectares of the Amazon have been cleared for pastures for beef production. Much of this has been subsequently degraded with the loss of soil nutrients and weed infestation. In recent years, improved pastures with well-adapted species and adequate nutrient replenishment, complemented by infrastructure investments and land tenure policies, have improved the profitability and sustainability of these systems.[13]

Continuous crop production. The most widely practiced farming system following deforestation is continuous cropping. Virtually all land growing crops today was once occupied by forests or grasslands. Depending on the soil and rainfall conditions, farmers have devised and executed systems of crop production to meet their food security (subsistence) and generate income by selling surpluses. On some flat and rolling hillsides, lands have been cultivated continuously for hundreds of years, though, as we saw in chapters 3 and 4, significant swathes of agricultural land have been abandoned through poor management of soils and water.

In the case of rice, terracing has been used to retain water and reduce erosion. Rice terraces are among the most sustainable food production systems in the world.[14] But as I saw in the World Heritage–listed rice terraces of Jatiluwih, Bali, these labor-intensive systems are now under threat. Bali's booming tourism economy has provided alternative employment and higher wages that are drawing away workers from these once productive and profitable fields. Agricultural tourism may be the only way the rice systems of Jatiluwih will survive. In the wake of COVID-19, however, and given the possibility of future pandemics, a high dependence on tourism places these farmers in a vulnerable position. We will return to the impact of and responses to COVID-19 in chapter 14.

Urban agriculture. At the extreme end of the evolution from slash-and-burn farming is urban agriculture, where farming is practiced in urban settings. An even more intensive variant of urban agriculture is indoor farming—where we remove most of the risks of the outside environment. Most recently, we have seen the rise of *vertical farming*. For the latter, producers have created an artificial, high-rise environment with precision delivery of nutrients and light and some level of independence from climate change.[15] Indoor farming, including vertical farming, is most successful when attuned to niche markets of consumers willing to pay for its high-value products.

The common thread of all these alternatives to traditional shifting cultivation systems, and for the transition from subsistence to increasingly intensive commercial farming systems, is connectivity to markets. As articulated by Professor C. Peter Timmer—a Harvard economist, whose work I have drawn on extensively in this chapter—a market economy is "the only kind of economy with a successful track record of raising productivity, and hence living standards, over many generations." Markets connect producers with consumers, a topic I will explore further in chapter 10. Timmer argues that markets reflect scarcity and abundance, sending price signals that inform investment decisions:

> Markets serve as the arena for allocating society's scarce resources to meet the virtually unlimited needs and desires of consumers. This allocation process, when joined to reasonably efficient price formation, is the reason market economies have outperformed other forms of economic organization in the long haul. Efficiency in resource allocations is essential to raising economic output in a sustainable fashion and thus to reducing poverty and hunger.[16]

This market-positive view hangs somewhat precariously on the phrase "in a sustainable fashion." The clearing of Mindanao's rainforests was a response to international price signals for tropical timber. The ongoing conversion of the Amazon forest to pastures is responding to price signals for livestock products in China and other emerging economies. Markets are a powerful driving force affecting investment decisions of actors across the entire food value chain. Market prices, however, do not always reflect the full social or environmental costs or benefits of a product in the short or long term. It is a point I like to raise when interviewing urban farming practitioners and enthusiasts: Are we including all the costs?

Timmer acknowledges that "market processes sometimes fail to get this scarcity signal right (at least for long-run decision making)," though whatever the underlying cause of an imperfect price signal, "food scarcity and high prices for food mean significantly higher food insecurity and hunger."[17] Understanding this important caveat to the role of markets gives reason to critique and augment market-based approaches to achieving food systems transformation.

But first, it is important to understand a crucial economic phenomenon that has driven food systems change around the world for the past two centuries: structural transformation.

STRUCTURAL TRANSFORMATION

Memo to Willie Nelson: "Trying to stop the structural transformation does not work, at least for the poor. Investing in the capacity of the poor to cope with change and to participate in its benefits through better education and health does seem to work."[18]

These are the words of Professor Timmer in 2007, addressing the many individuals and organizations who seek to halt, reverse, or reject structural transformation: the process whereby economic activity is reallocated across three broad sectors: agriculture, industry, and services. The outcome of successful structural transformation in almost all countries is a reduction in the relative contribution of agriculture to the overall economy, accompanied by increases in the contributions of nonagricultural sectors—industry and services.

According to Timmer, the historical record is crystal clear: improvements in agriculture are needed to stimulate growth in the nonagricultural sectors, leading to overall growth in the economy. The paradox is that this process inevitably leads to agriculture's reduced overall role in the economy, including a decline in the number of farms and farmers, a point many supporters of farming communities find difficult to accept.[19]

American country music legend Willie Nelson fights for farmers. In 1985, along with fellow musicians Neil Young and John Mellencamp, Nelson organized the first Farm Aid concert to raise awareness and mobilize funds to keep farm families on the land, saying: "As farmers continue to face incredible threats to their survival, our grant decisions were guided by the need to first and foremost strengthen organizations that provide essential resources to keep family farmers on the land."[20]

The number of farms in the United States peaked at 6.8 million in 1935, fell sharply until the early 1970s, and thereafter declined more slowly to about 2 million farms in 2020.[21] Between 1900 and 2000 the agricultural workforce dropped from about 40 percent to 3.2 percent (figure 8.2).[22] By 2020 direct on-farm employment accounted for just 2.2 million jobs, or 1.5 percent of the U.S. workforce.[23] The rural (nonurban) population appears to have stabilized at about 14 percent of the U.S. population.

Nelson and friends—expressing a view shared by a significant proportion of Columbia students taking my Global Food Systems course over the years—were responding to the perceived decline in family farming over the past century, in some respects a textbook case of structural transformation, illustrated by the following changes that have taken place in the U.S. food system:[24]

- **Productivity improvement has driven farm output growth.** Technological developments have meant that, even as the amount of land and labor used in farming declined, total farm output nearly tripled between 1948 and 2019.
- **Agriculture is today a relatively minor contributor to the overall U.S. economy, though the broader category of food industries is more significant.** Agriculture, food, and related industries (like textiles and apparel) contributed US$1.1 trillion to U.S. GDP in 2019, a 5.2 percent share. Of this amount, farm output contributed US$136.1 billion—about 0.6 percent of GDP.
- **Food industries, though not agriculture, are an important employer.** In 2020 there were 19.7 million full- and part-time jobs related to the agricultural and food sectors—10.3 percent of total U.S. employment.
- **Agriculture and food exports are booming.** The United States is the world's largest agricultural exporter, with total export sales of farm and food products of US$146 billion in 2020.
- **Most farms remain family operated.** Family farms (owned by the operator or individuals related to the operator) of various types accounted

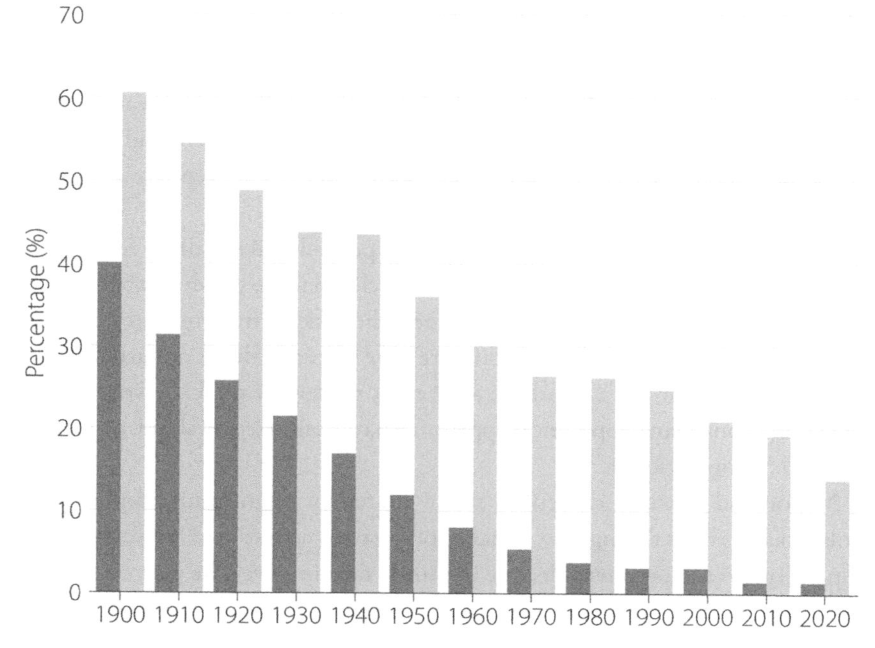

FIGURE 8.2 Decline in farm workforce and rural population in the United States, 1900–2020. The representation was inspired by Carolyn Dimitri, Anne Effland, and Neilson Conklin, *The 20th Century Transformation of US Agriculture and Farm Policy*, Economic Information Bulletin no. 3 (Washington, D.C.: U.S. Department of Agriculture, June 2005).

Sources: (1) Farm workforce data for 1900 to 1960, Stanley Lebergott, "Labor Force and Employment, 1800–1960," 119, National Bureau of Economic Research, https://www.nber.org/system/files/chapters/c1567/c1567.pdf NBER, 1966; farm workforce data for 1970 to 2000, U.S. Department of Agriculture Economic Research Service, "Farm Labor," https://www.ers.usda.gov/topics/farm-economy/farm-labor/#size; farm workforce data for 2010 and 2020, U.S. Bureau of Labor Statistics, "Employment by Major Industry Sector," https://www.bls.gov/emp/tables/employment-by-major-industry-sector.htm; (2) Rural (nonurban) population for 1900 to 2000: U.S. Census data prepared by State Library of Iowa, State Data Center Program, https://www.iowadatacenter.org/datatables/UnitedStates/urusstpop19002000.pdf; rural (nonurban) population for 2010, U.S. Census Bureau, "2010 Census Urban and Rural Classification and Urban Area Criteria," https://www.census.gov/programs-surveys/geography/guidance/geo-areas/urban-rural/2010-urban-rural.html; rural (nonurban) population for 2020, estimated from July 2019 by USDA's Economic Research Service (ERS) research cited in John Cromartie, "Modest Improvement in Nonmetro Population Change During the Decade Masks Larger Geographic Shifts," Amber Waves, https://www.ers.usda.gov/amber-waves/2020/july/modest-improvement-in-nonmetro-population-change-during-the-decade-masks-larger-geographic-shifts.

for nearly 98 percent of U.S. farms in 2020, with an average farm size of about 180 hectares, little changed since the early 1970s.

- **Agricultural production comes mainly from large family farms.** Large-scale family farms (having gross cash incomes of US$1 million or more) constitute just 3 percent of farms but 44 percent of the value of production.

These numbers describe a food system where the structure of U.S. agriculture has shifted dramatically over the past century. Technology has driven productivity improvements with increasing contributions from large-scale farms, though even small family farms continue to exist as lifestyle choices with greater reliance on nonfarm incomes. Agriculture contributes in a major way to related food industries and exports of both agricultural commodities and food. But its overall share in the economy has waned.

Agricultural modernization throughout the world, and throughout history, has often been characterized by the shift from subsistence to commercial farming. As subsistence farming households expand their vision beyond household food security and survival, they increase transactions and expand their reach and impact within the economy. Through agricultural productivity improvements, farmers generate food surpluses beyond their household needs. Those surpluses are bartered or sold in return for different foods, manufactured goods, and services. Ancient societies from the Sumerians to the Romans to the Incas prospered on the back of surpluses from settled farm communities, wherein farm products exceeded the consumption requirements of the producers. This was the dawn of economic civilization, opening opportunities for commercialization and the specialization of farms and nonfarm activities. It was no longer necessary for all households to be directly involved in food supply. Agriculture was no longer the sole economic activity of societies. The structural transformation of societies and economies was underway.

So should Willie Nelson, his buddies, and many of my students be fighting to keep farmers on the land? Should they instead accept the inevitable and find ways instead "to cope with change and to participate in its benefits," as Timmer suggests? I am reminded of a response by Australian treasurer Peter Costello in 2001 when challenged about his defense of globalization: "Globalization is not a value, it is a process. And ranting against globalization is like ranting against the telephone. You can use the telephone for good or for ill. So too the wider process of which the telephone is part, it can be a force for good or a force for ill. And what is more, you will not reverse the process."[25]

In similar vein, structural transformation is not a value, it is a *process* that is usually the consequence of both agricultural productivity improvement and market expansion. Structural transformation can bring positive and negative consequences to different actors in the food system. The balance and duration of positive and negative outcomes will be determined in part by policy interventions and investments. To understand better how to manage structural transformation for the benefit of society, including for the attainment of food security, we must dive deeper into this important topic.

As Timmer has observed, virtually all countries have experienced—or are in the process of—a structural transformation, which is characterized by:

- a shift from subsistence to commercial agriculture
- declining relative importance of agriculture in terms of employment and contribution to national economic output
- rising relative importance of industry and services in terms of employment and national economic output
- increasing urbanization associated with the growth of industry and services and the migration of workers from rural to urban areas
- increasing overall economic growth as industry and services increase in importance relative to agriculture
- reduced population growth rates related to urbanization, increased incomes, education investments, and women's empowerment[26]

Structural transformation is a characteristic of long-term economic growth and has been the main pathway out of poverty in almost all countries. Timmer describes structural transformation as "the defining characteristic of the development process" and argues that "no country has been able to sustain a rapid transition out of poverty without raising productivity of its agriculture sector."[27] The economic consequences of raising agricultural productivity of land and/or labor are as follows:

- **Labor requirements for agriculture are usually reduced**, especially with mechanization and land consolidation, resulting in a pool of labor that can be deployed in nonfarm sectors. With appropriate infrastructure and policy support, complemented with expanding export and domestic markets, these displaced workers can find employment in labor-intensive industrial and service sectors.
- **Due to the relatively inelastic demand for food, the demand for industrial products and services in the rural and urban areas increases with**

increasing farmer incomes, in turn, generating more employment and income in nonfarming rural and urban areas.

- **As more food is produced per capita through improved agricultural productivity, food prices stabilize and may decline, meaning industrial and service workers will also spend relatively less of their earnings on food.** This makes the effective wages in the industrial and service sectors more attractive to workers. With stable (i.e., nonrising) food prices, nonagricultural sectors can improve their competitiveness in export markets.
- **Increased real incomes of urban and nonfarm rural workers, *in principle*, could lead to improved nutrition and better health**, as they shift to more nutritious, diversified diets and spend relatively less on starchy staples (Bennett's law). However, the recently observed dietary transition in most developing countries to highly processed, high-sugar, high-fat foods, especially in urban areas, may counter any positive nutritional effects (see chapter 7).
- **More and lower-cost raw materials are produced for use in food processing and other industrial and service uses.** Food-processing industries expand with economic growth and urbanization.
- **Increased incomes of farmers can lead to increased investments in their children's education**, enabling greater mobility to higher-paying industrial and service jobs in both rural and urban areas.
- **Where increased agricultural productivity leads to increased exports, foreign exchange can be generated** to import capital goods that can, in turn, raise productivity of the industrial, service, and agriculture sectors.
- **Political stability (or absence of civil unrest), associated with stable access to food, can lead to a more secure environment for investment and employment.**
- **Increased economic power in rural areas can lead to a decrease in urban bias** and even regime change in more extreme circumstances.

From the multiple direct and indirect effects just outlined, born out by the historical records of dozens of countries around the world, it is evident that increased agricultural productivity facilitates improvements in the productivity of the nonfarm sectors. The latter may then induce further productivity improvements in agriculture by shifting demand to higher-value foods and increasing opportunities for farm diversification. Remittances from urban industry and service workers may also be reinvested in agriculture to raise productivity.

The paradox that improving agricultural productivity leads to its declining relative importance deserves further explanation. With productivity improving, both the agriculture and the nonagriculture sectors will likely grow in absolute terms. The industrial and service sectors, however, grow at a faster rate. Why is that?

In simple terms, there is a biological limit to how much food we can consume. The German statistician Ernst Engel (1821–1896) first described this concept. Engel's law states that, as incomes rise, the share of food in budget expenditures falls.[28] In contrast, our ability to consume industrial goods and services is practically without limit, opening endless opportunities for employment in the nonagricultural sectors. Therefore, even as the agriculture sector grows, it becomes less important than industry and services in terms of its *relative* contribution to the economy.

While Timmer and others make a good case for investing in agriculture as an "engine of growth," this approach should not imply that we need *only* invest in agriculture to achieve economic growth and poverty reduction. Indeed, in their highly influential paper "The Role of Agriculture in Economic Development," Bruce Johnston and John Mellor drew attention in 1961 to "the false dichotomy of agricultural vs. industrial development."[29] A successful structural transformation requires balanced public and private investment across agriculture, industrial, and service sectors. Infrastructure and policies to support the nonfarm rural economy and the urban economy—for example, urban electrification and public transport systems—help ensure a productive nonagricultural sector that absorbs labor and capital released from farming.[30] The growth of labor-intensive manufacturing for domestic and export markets has been an important complement to agricultural productivity improvement.

Investing in agriculture must complement urban and nonagricultural investments, and the growth of industrial and service sectors. The World Bank concluded that China's rapid growth in agriculture through the household responsibility system, liberalization of markets, and rapid technological change was largely responsible for the country's rapid decline in rural poverty in the last twenty years of the twentieth century.[31] Drawing on the historical record of seven countries across two continents—from the experience of England during the Corn Laws (late seventeenth to early nineteenth centuries) to Indonesia's promotion of rice production from the 1960s to the 1990s, Timmer makes a strong case for raising agricultural productivity as a prerequisite for successful industrialization:

> No country has managed to get rich without generating significantly higher agricultural productivity than existed when the country was poor (Timmer 2002). There are two-way cause-and-effect relationships in this process, of course, but the historical evidence argues (controversially for some) for a fundamental (and

probably prior) role for higher agricultural productivity as a driver of economic growth in the rest of the economy. A failure to modernize agriculture almost inevitably leads to a failed industrialization effort.[32]

Despite the universality of structural transformation as a feature of economic growth, there is a wide diversity of experiences across regions and among countries within regions. The prospects and policy options depend on many factors, such as historical context, agro-ecological conditions, market access (e.g., whether landlocked), natural resources endowments (e.g., land and water availability), institutional capacities, and human capital. Many—though not all—Asian economies have invested successfully in agricultural productivity improvement—the Asian Green Revolution—and have supported the growth of labor-intensive, export-oriented manufacturing. As a consequence, we have observed structural transformation and sustained economic growth in several Asian countries, illustrated by the experience of China, Vietnam, and Bangladesh in East Asia, Southeast Asia, and South Asia, respectively. In Latin America, structural transformation has been observed across the region, with agriculture declining and services growing, albeit with persistent urban poverty and environmental degradation. The process in Africa has been slower and less consistent.

In the *World Development Report 2008*, the World Bank showcased the powerful role of agriculture in reducing poverty in all regions of the developing world. The World Bank called for agricultural productivity improvements, especially for smallholder farmers in those countries where agriculture remains a large part of GDP and where most of the poor live in rural areas. It noted that in much of sub-Saharan Africa, agriculture had the potential to spur economic growth, reduce poverty, and improve food security, while stimulating broader economic growth.

Recent progress in food production in Africa, as outlined in chapter 2, suggests that investments in agriculture could support a positive structural transformation for all the reasons outlined earlier in this chapter. To that end, greater public investment is needed in infrastructure in both rural and urban areas, and complementary policies to improve market connectivity (see chapter 10). A coherent whole-of-economy approach will be needed, even in those countries that remain primarily agricultural. In the absence of such an approach, farmers will be discouraged by poor or erratic returns on their investment, while urban populations will be plagued with high inequality and social discord.

The WDR 2008 recognized that agriculture could also play an important role in two other types of economies: "transforming countries" and "urbanized countries." In transforming and urbanized countries, agriculture is no longer a major source of economic growth, contributing less than 10 percent to GDP growth.

However, poverty in transforming countries—such as China, India, Indonesia, and Morocco—remains overwhelmingly rural. The focus in these countries needs to be on reducing rural–urban income disparities. Rapidly developing urban markets provide opportunities for high-value products such as horticulture and livestock industries. Investments in infrastructure to better connect farmers to markets are a priority for governments seeking to reduce rural–urban inequality. Nonfarm rural industries provide employment opportunities as farms become more capital intensive, as we have observed in the United States.

In urbanized countries, agriculture contributes even less to GDP (less than 5 percent), and poverty is predominantly urban. Most countries in Latin America and many in Europe and Central Asia fall into this category. In these cases, however, agribusiness and the food industry and services account for as much as one-third of GDP. Corporate farming is growing in urbanized countries. Supermarkets have come to dominate the retail food sector in urbanized countries, though they are rapidly growing in importance in transitioning and even agriculture-dominant countries.[33] In such cases, the emphasis is again on improving access of smallholder farmers to urban and export markets. High-income urban consumers are demanding high-quality food, including a growing demand for certified organic products.

As countries move through these phases of development, traditional subsistence agriculture is progressively replaced by commercial agriculture. In subsistence farming, by definition, the producers are also the consumers. Once farmers are no longer producing exclusively for themselves, they tend to specialize. Farms become businesses, and farmers become business operators. Consumers express preferences and send price signals reflecting their demands. As their incomes increase, consumers diversify their diets. Agribusinesses emerge to provide farmers with inputs like seed and fertilizer. Moneylenders and banks provide credit so that farmers and nonfarm private actors can invest. Farmers face new risks associated with prices of both their inputs and their outputs. Governments intervene to shift decisions and outcomes to meet their objectives, which may or may not be in the public interest. These actors are all participants in structural transformation.

TRANSFORMATION FOR FOOD SECURITY

"Food security exists when all people, at all times, have physical and economic access to sufficient, safe and nutritious food to meet their dietary needs and food preferences for an active and healthy life."[34] This Rome Declaration on World

Food Security of 1996 provides an important but insufficient snapshot of what constitutes a food-secure world.

Universal food security requires changes in how and where we produce, distribute, and consume food within a dynamic global food system. The supply of food must be sufficient and accessible to *all* people at *all* times. And it must be nutritious and safe. Both quantity and quality are emphasized.

The expression "at all times" implies temporal stability and sustainability of food security. Therefore our policies, actions, and investments must consider a range of time frames, from intradiurnal (e.g., consider the impact of daily periods of child hunger on school performance) to multigenerational (e.g., consider the impact of resource depletion and degradation on future food supply). Directly applying the language of the landmark Brundtland Report's definition of sustainable development, sustainable food security "meets the [food] needs of the present without compromising the ability of future generations to meet their own [food] needs." Hungry and malnourished children and degraded soils each undermine "the ability of future generations to meet their own needs."[35]

The Rome Declaration articulates a vision whereby food security should be achieved at multiple levels: individual, household, national, regional, and global. Of course, individual food security is ultimately what matters. But that can be achieved only through policies, actions, and investments at those higher levels of aggregation.

With greater health and environmental awareness since the 1990s, we might now reconsider the blanket inclusion of meeting *food preferences* in our definition of food security. As described in chapter 7 and further elaborated in chapter 12, our food preferences might be bad for our health and for the environment. Preferences for sugary drinks, daily servings of red meat, and shark fin soup will need to be addressed. For now, we might insert the qualifier "healthy and environmentally sound" or simply "well-informed" ahead of the phrase "food preferences." In any case, the Rome Declaration remains a powerful expression of ambition that can help shape our transformation strategy.

Building on the Rome Declaration definition, I see **seven critical attributes** of a *transformed* global food system. Each is important, although, as I have learned from presenting these ideas to diverse audiences, none is free from caveats and controversy.

To achieve genuine universal food security, in alignment with the Rome Declaration definition, food systems must be:

1. **Productive.** Supply must meet demand. A productive food system meets the "dietary needs and [healthy and environmentally sound] food preferences" through efficient use of land, water, and energy. Research and extension are

important sources of innovation for improving food system productivity. Too often, productive is misinterpreted as simply *more* production. In the next chapter I will present the concept of "sustainable intensification" as a means of improving productivity *and* food supply without compromising the environment.

2. **Profitable.** Profits drive decision-making throughout the food value chain. Hobby farmers aside, most farmers make decisions based on an expected return on their investment, be it their land, their labor, or their financial resources, tempered by risk considerations. They rarely make decisions based on the greater good of society or the health concerns of urban consumers. Ask an Iowan soybean farmer or a Malawian tobacco producer what drives their investment decisions. If we want producers or retailers to incorporate nonfinancial social benefits, society must pay the difference, perhaps through subsidies. Alternatively, governments must act to change behavior through education and policies throughout the food system.

3. **Inclusive.** Food security for "all people" requires a careful disaggregation of producers, consumers, and others working in the food system. We must ask: Who are we leaving behind? Women? Children? Older people? Minorities? Smallholders? Migrant laborers? Landlocked countries? The food system should be equitable. My mentor M. S. Swaminathan has long championed a profoundly ethical approach to agricultural development, urging researchers, policy makers, and practitioners alike to take actions that are "pro-poor" and "pro-women." In chapter 13 I will introduce the role of social protection, safety nets, and other actions explicitly designed to include those who are left behind.

4. **Healthy.** Adequate, safe, and nutritious food is essential for a healthy and productive life. Our food systems must be transformed to meet this objective. Understanding human nutrition and applying that knowledge to support dietary change are among the great challenges in transforming the food system. Consumers make decisions based on multiple factors, but principally price and taste. In wealthy, highly educated settings, we are seeing a shift toward healthy diets. But even as awareness of healthy diets is increasing, individual decision-making falls short in delivering the needed transformation. This will be discussed in chapter 12.

5. **Sustainable.** There are myriad definitions of sustainable food systems. One could argue that all the other six characteristics in this list of key attributes must be present for the food system to be truly "sustainable," consistent with the three pillars of sustainable development: economic, social, and environmental. In this context, however, I want to highlight the importance of *environmental* sustainability. We must examine our systems of production, distribution, and consumption and assess whether they can be sustained ten, twenty, fifty, or one

hundred years from now. In reality, no system is perfectly sustainable, so long as population grows. Our challenge is to improve and extend sustainability to acceptable levels that do not breach our biophysical capacity to meet global demand indefinitely.

6. **Resilient.** Food system resilience describes the ability of the system and subsystems to withstand, adapt to, and recover from disturbances, both sudden shocks (e.g., hurricanes, earthquakes, and food price spikes) and longer-term stressors (e.g., population growth, climate change, and land degradation). Resilience is described as a capacity to continue to function despite such disturbances, characterized by the three Rs—resist, recover, and reorient—elaborated in chapter 6 within the context of climate change. Resilience and sustainability are complementary. Resilience is necessary for sustainability of the food system.

7. **Ethical.** Controversy and vigorous argument should not deter consideration of ethics in assessing food systems. Deliberations about human rights (such as the right to food), animal welfare, child labor, workers' rights, and fair trade, among other moral questions, are necessary as we assess and seek to shape the direction of food systems. Inclusive food systems address the ethical dilemma of who is being left behind. Resilience addresses those who are most vulnerable to shocks and stress. Sustainability addresses the ethical dilemma of future generations, whose voices cannot be heard.

We are seeking a global food system that is productive, profitable, inclusive, healthy, sustainable, resilient, and ethical. That is a big ask when each critical attribute is not a simple dichotomous variable. For all seven attributes, there are degrees of attainment, so we should identify thresholds to avoid and benchmarks for improvement. Objective measurement of system performance is challenging, more so when it comes to sustainability, resilience, and ethics. We also need to recognize that there are trade-offs among attributes—for example, increased profitability may result in food systems practices that are less sustainable and less healthy. But, however imperfect the answers appear, these are the questions to ask when planning investments for a food-secure world. To that end, I propose a transformation strategy that comprises what I am calling the *Big Five investment areas* (or the Big Five for short).

Why am I using the term *investment*? Why not programs, actions, pillars, objectives, outcomes, etc.? In one sense, it hardly matters, so long as the work gets done. But I have chosen the term investment to signify that the transformation work will get done only if we *invest* as individuals and institutions in terms of time and money. A food-secure world that meets the needs of ten billion people and more is *The Future We Want*, to borrow from the title of the Rio+20 outcome document.[36] We cannot just *want* a food-secure world, we must commit to investing our time and money to get there.

In the remainder of this book, I will argue that we must advance in these five strategic investment areas—the Big Five—while concurrently improving the performance of a number of related cross-cutting issues that affect the outcomes we are seeking (figure 8.3). The latter include women's empowerment; education; health (beyond nutrition); water, sanitation, and hygiene (collectively known as WASH); and good governance. All these are *essential* for food security and broader development. Each of these cross-cutting topics deserves a book of its own.

The Big Five investment areas—sustainable intensification, market infrastructure, postharvest stewardship, healthy diets, and social protection—are introduced here and explored in more detail in part III:

1. **Sustainable intensification.** The premise here—representing a broad consensus of evidence-based studies on the topic—is that we will need more food supply, even if we invest aggressively and strategically in the other four areas. At the very least, production in some locations will need to offset the production declines in others due to land degradation, climate impacts, and other factors. The most broadly accepted definition of sustainable intensification is a process or system where "yields are increased without adverse environmental impact and without the cultivation of more land."[37] Hardly a bulletproof definition; some argue that sustainable intensification is an oxymoron, while others seek a more radical sustainability-first transformation.[38] But this is nonetheless a good starting point for discussion about *how* to increase production in environmentally conscious ways. Sustainable intensification is the focus of chapter 9.
2. **Market infrastructure.** We have all heard the calls for better distribution of the food we currently produce, whether or not it is produced sustainably. You will hear, "There is enough food produced to meet the caloric needs of everyone on the planet. Food is rotting or discarded while people are hungry." If only, it was better distributed, right? As we learned earlier, food availability must be coupled with food access. Much more can be done to improve the functioning of markets. In simple terms, our goal should be to move food from where it is produced (sustainably) to where it is consumed (sustainably). Making markets work has been a theme of investment for decades. There are many historical cases of farmers being able to produce surpluses but being unable to sell their crops profitably. Physical infrastructure investment—in roads (and other transport conduits), electrification, and information technology—is key in connecting producers

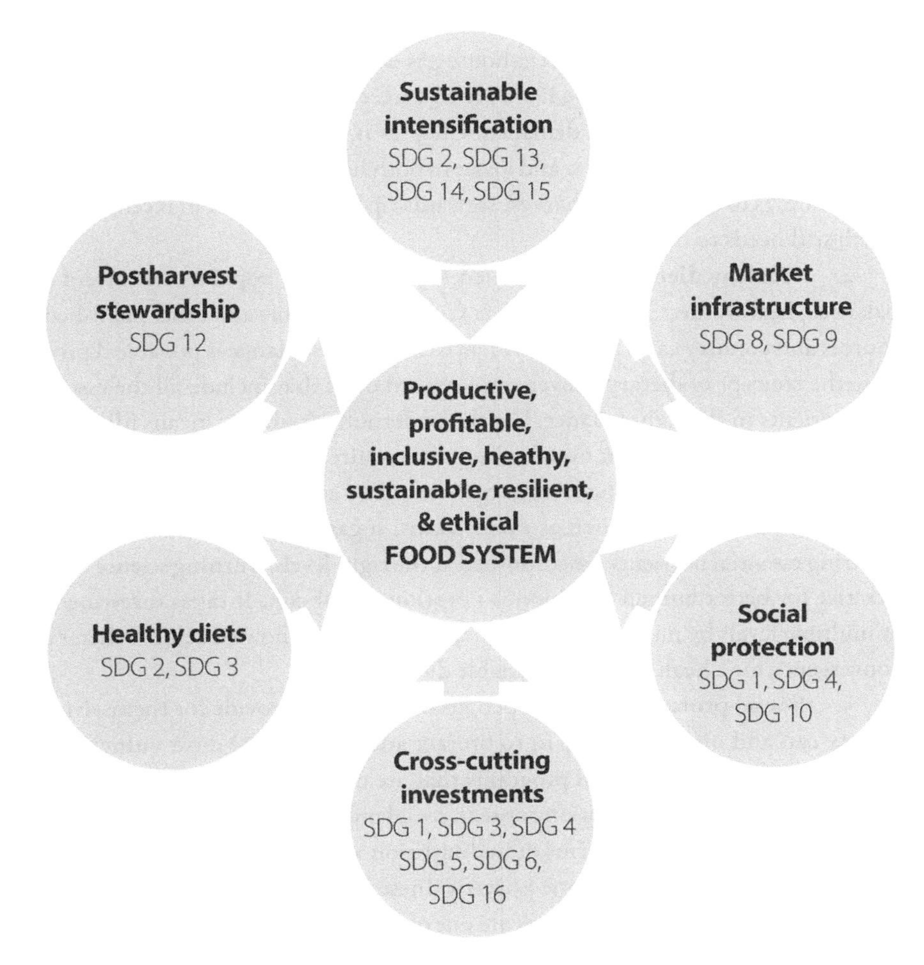

FIGURE 8.3 Key investment areas and related Sustainable Development Goals for transforming the global food system.

to consumers. Supportive policy environments, enabling movement of commodities within and between countries, are also needed. Chapter 10 explores this area of investment.

3. **Postharvest stewardship.** One-third of all food produced is lost or wasted due to poor practices in harvesting, storage, and transportation, and as market and consumer waste.[39] This represents lost opportunities to improve profits for farmers and reduce the cost of food to consumers. In addition, food loss and waste embody the resources (land, water, fertilizer, energy, and labor) that

were used to produce the unconsumed food and the environmental costs of deforestation, biodiversity loss, and greenhouse gas emissions. This level of wastage is a tragedy that cannot continue if we are to succeed in the challenge of sustainably meeting our future food demands. Chapter 11 examines the nature, causes, and consequences of food waste and losses in both high- and low-income settings. While efforts to improve postharvest stewardship have increased in recent years, much still needs to be done.

4. **Healthy diets.** What is needed to move diverse populations toward a balanced, healthy diet? The EAT–*Lancet* Commission report and other published sources and country cases provide evidence that dietary change is possible. I propose the concept of dietary convergence toward diets that include all the essential nutrients in the right balance. For the undernourished, this means filling in the nutritional gaps without overshooting the requirements with unhealthy outcomes. For the overnourished, convergence will be achieved through reducing unhealthy consumption (such as excess carbs, sugar, saturated fat, salt) while ensuring essential nutrients are consumed at the right levels. Turning science into practice for better human nutrition is notoriously difficult. It takes investment at multiple levels by multiple actors. Chapter 12 outlines how to achieve dietary convergence for a healthy and sustainable diet.

5. **Social protection.** When people are unable to provide for themselves, society can and often does step in to protect and support the most vulnerable through systems, policies, and programs that are collectively described as *social protection*. In a food and nutrition context, social protection helps meet the needs of people who face various forms of malnutrition as a result of conflict, natural disasters, poor health, or extreme poverty. These are people for whom the markets do not work, and the right diets are out of reach, physically or economically. A food-secure world requires attention being paid to such people, and social protection must be part of the solution. In chapter 13 a number of practical approaches to investment are presented and discussed, including food and cash transfers, school meal programs, and food voucher programs.

Each of the Big Five investment areas *can* be designed to advance each of the seven critical attributes for food systems transformation (figure 8.4). At first glance, it may be surprising to see that the Big Five investments score ticks across all seven attributes. Can sustainable intensification be inclusive? Can heathy diets be productive? Can social protection be profitable? The answer to all is yes. As we will see in part III, through creative use of technologies and policies, we can sustainably intensify the most marginalized farmers in rainfed areas, improve

BIG FIVE INVESTMENT AREAS	SEVEN ATTRIBUTES FOR FOOD SYSTEMS TRANSFORMATION						
	Productive	Profitable	Inclusive	Healthy	Sustainable	Resilient	Ethical
Sustainable intensification	✓	✓	✓	✓	✓	✓	✓
Market infrastructure	✓	✓	✓	✓	✓	✓	✓
Postharvest stewardship	✓	✓	✓	✓	✓	✓	✓
Healthy diets	✓	✓	✓	✓	✓	✓	✓
Social protection	✓	✓	✓	✓	✓	✓	✓

A tick (✓) indicates a positive or neutral impact.

FIGURE 8.4 Matrix of contributions of key investment areas to critical attributes of a transformed global food system.

productivity of more nutritious foods and increase the nutritional value of productive foods, and design productive, profitable, and sustainable social protection schemes.

No single area of investment will lead us miraculously to universal food security. Focusing on production and food availability through sustainable intensification, market infrastructure, and postharvest stewardship will not ensure food security for all, unless we concurrently and strategically invest in healthy diets and social protection. Universal food security requires a comprehensive, multipronged, context-informed investment approach, which is the focus of part III.

PART III

Strategy

9

SUSTAINABLE INTENSIFICATION

In July 2005 Professor Jeffrey Sachs, United Nations Resident Coordinator Michael Keating, and I met with President Bingu wa Mutharika at the lush, colonial-era State House in Zomba, southern Malawi, to discuss the nation's progress toward the Millennium Development Goals. Sachs was then UN Secretary-General Kofi Annan's special adviser on the MDGs. I headed the Millennium Development Goals Centre for East and Southern Africa in Nairobi, established by Sachs a year earlier to advise countries on best practices for achieving the goals.

Malawi was then one of the poorest countries on Earth, with a per capita gross national income of US$300 and a human development index ranking of 164 of 177 countries listed in the *Human Development Report 2005*, produced by the United Nations Development Programme (UNDP).[1] Income measures of poverty indicated that 52 percent of the population were living in extreme poverty and 22 percent were considered ultrapoor.[2] More than half of Malawi's under-fives were stunted through malnutrition.[3] Malawi was far from achieving the MDGs. Achievement of MDG 1, "Eradicate extreme poverty and hunger" (by 2015), was generously described in the *Malawi—Poverty and Vulnerability Assessment* (2007) as "unlikely."[4]

The focus of our discussion in Zomba quickly turned to the state of agriculture and food security in the country. Malawi had just experienced the worst maize harvest in a decade, with production falling almost a million metric tons short of the national consumption requirements. The rains failed during the critical flowering period for the national staple food, maize. Most of Malawi's more than two million smallholder households were anticipated to run out of maize well before the next harvest in April to June 2006. The UN identified over five million Malawians as vulnerable to famine and likely to require food aid within the following year.[5]

Mutharika explained his ambitious scheme to ensure the devastating harvest of 2005 would not be repeated in 2006. The plan called for increasing smallholder access to fertilizer and seed on a massive scale through a nationwide Farm Input Subsidy Program (FISP). Several years earlier, a Starter Pack program had demonstrated promisingly that the technology existed to lift production and that farmers would respond to increased availability of inputs.[6] Unfortunately, the donors shifted priorities—not for the first time in Malawi—and the Starter Pack program abruptly ended. This time, faced with a possible famine, the president instructed his government to ensure that all smallholders had access to enough fertilizer and seeds to meet and even exceed the food needs of a typical household farming on about half a hectare of nutrient-depleted land. He also committed to sustaining the program for as long it would take to end hunger in Malawi.

The operational details and early impact of the inputs subsidy program are well documented.[7] In brief, the government of Malawi allocated coupons to farmers to buy discounted fertilizer and seeds, representing an overall subsidy of 63 percent. Coupons were allocated across regions and then distributed to districts and traditional authorities (subdistrict government entities), who allocated them to village development committees, who identified the recipients. All the subsidized fertilizer and seeds were distributed through government agencies during the first year of operations.

Good rains favored production, enabling a strong response to fertilizer and improved (mostly hybrid maize) seed. National maize production in 2006 was more than double the 2005 harvest, resulting in a surplus of 510,000 metric tons above the national maize requirement. Maize yields averaged 1.6 metric tons per hectare (t/ha), double the 0.8 t/ha of the drought-affected 2004–2005 season, and well above the 1.1 t/ha obtained over the three years prior to the drought year. Incremental maize production attributed to the fertilizer subsidy was estimated at 300,000 to 400,000 t. The total cost to the government was about US$50 million.

For the following season, Mutharika implemented the input subsidy program on a similar scale. A total of 3.5 million coupons for maize fertilizer were distributed, targeted to maize-growing households. The harvest in 2007 was estimated at 3.44 million t, an all-time national record for Malawi, representing a surplus of about 1.34 million t of maize grain above national requirements. The incremental effect of the fertilizer subsidy on maize production was estimated at 670,000 t. By late 2007 Malawi had exported over 300,000 t of maize to Zimbabwe, not only generating income for its smallholder farmers but contributing to regional food security.

International donors had actively discouraged fertilizer subsidies for two decades. That was the direct effect and legacy of structural adjustment, a set of policy measures usually involving reduced government spending and advocacy of market-based solutions to development problems. The perceived high cost and trade-offs with other development investments, the absence of time-bound "exit strategies," and difficulty targeting the poorest of the poor had discouraged most donors from supporting input subsidy programs. Antisubsidy bias was well entrenched in Malawi in 2005, and there was little encouragement from donors for the government to implement the subsidy program. Indeed, Sachs and I were on the receiving end of harsh criticism from Malawi's most influential donors, inflamed by our public praise of the efforts of Mutharika and his government, exemplified by the following extract from our op-ed in the *Financial Times* in May 2007:

> Millions of people have averted hunger and its cruel manifestations. The focus has shifted from food aid to food exports; and from subsistence agriculture to longer-term rural economic transformation. This year the government's subsidy programme will promote crop diversification into higher-value cash crops to grow alongside their staple food production. The impact has stunned the sceptics and the doomsayers. It seems that an African green revolution is possible after all.[8]

I was then—and I stand today—in awe of the remarkable political will and sheer grit that Mutharika showed in the face of resistance and obstruction from the donor community, in particular, from Malawi's two leading donors, the World Bank and UK Department for International Development (now the Foreign, Commonwealth and Development Office). Despite their objections and unwillingness to provide financial support, the president pushed ahead and implemented the subsidy program using national budget resources. On June 4, 2008, at the World Economic Forum on Africa in Cape Town, Mutharika reflected on the first three years of the subsidy program and shared his frustration at having to request food aid year after year: "The president was supposed to do a ritual every year, going down on your knees and saying 'Please help us, give us food.' Until 2004, when I said 'Enough is enough. I am not going on my knees to beg for food. Let us grow the food ourselves.' And indeed we have."[9]

Since 2005, Malawi's smallholders have responded to the input subsidy program, generating surpluses over national maize requirements in all but three

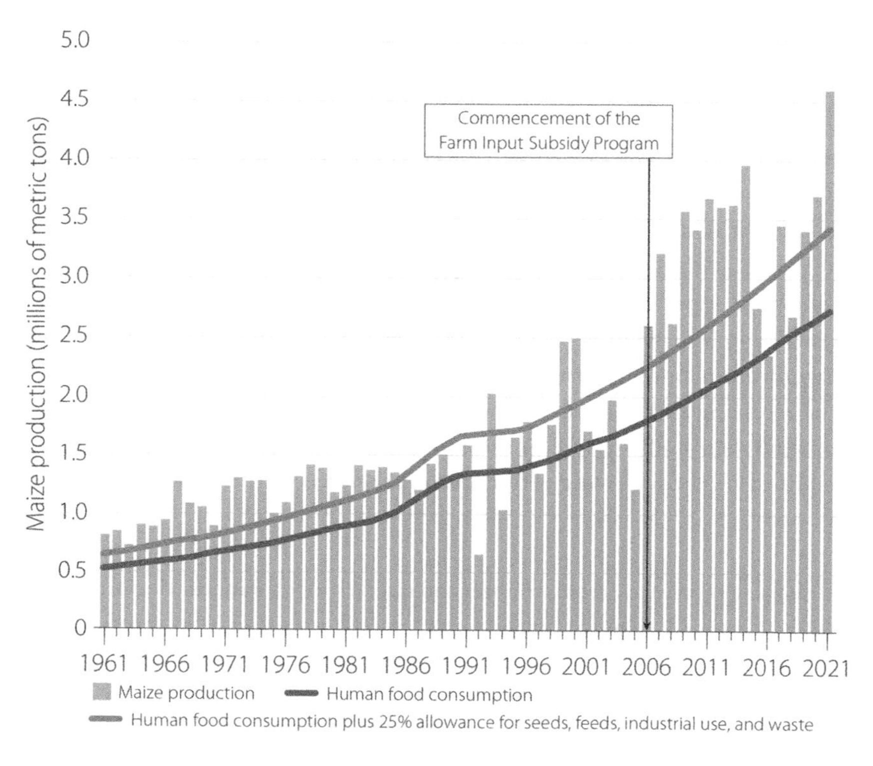

FIGURE 9.1 Maize production and consumption requirements in Malawi, 1961–2021.

Sources: FAOSTAT, http://www.fao.org/faostat/en/#data/QCL for 1961–2020; FAO, GIEWS Country Brief, Malawi for 2021 (Rome: FAO, February 8, 2022), https://reliefweb.int/report/malawi/giews-country-brief-malawi-25-march-2021 for 2020 and 2021 (estimate).

years—2015, 2016, and 2018 (figure 9.1). In the fifteen years following the 2005 drought, annual national production averaged 3.25 million t, with an average yield of 2.0 t/ha; in the three years prior to 2005, the corresponding figures are 1.72 million t and 1.1 t/ha.[10] Even in years affected by drought and floods, the country usually achieved an aggregate national surplus, which, together with strategic reserves from earlier years, ensured that food prices remained stable and the extent of hunger was limited. Unlike many African nations, Malawi was spared the destructive riots that accompanied the food price spikes in 2007–2008 and 2010–2011. The government has persisted with the program following the death of President Mutharika in 2012.

The demise of the subsidy program has been long sought and predicted by its detractors. The results on the ground, however, suggest a remarkable policy success in the face of national food insecurity. In February 2022 the UN's Food and Agriculture Organization estimated the maize harvest of 2021 to reach 4.6 million t, an astonishing 2.7 times the average annual production for the three years preceding (but not including) the devastating 2005 harvest.[11] FAO attributed the 2021 result to good weather conditions and the rollout of the government's Affordable Inputs Program (the new official name of FISP), which provided 3.8 million smallholder farmers with access to subsidized seeds and fertilizer.[12]

There have been many criticisms of the program, presenting the Farm Input Subsidy Program as an ongoing narrative of policy failure. My response to each is outlined here:

The subsidy program was designed to win political favor and reelection. Well, yes, that is true. Mutharika was reelected in a landslide in 2009, no doubt helped by his commitment to farm input subsidies. I had never before observed a country where politicians vigorously argued about fertilizer subsidies late into the night. And it was not about whether to subsidize, it was concerning *how much* to subsidize, and whether the subsidies should be extended to other crops, like tobacco. Riveted to the televised debate one evening in Lilongwe (Malawi's capital), I saw shoes flung and scuffles break out on the floor of Parliament.

The subsidy program was expensive. It is correct that the program was using up most of the Ministry of Agriculture budget. This was mainly because donors would not provide fiscal support to one of the poorest countries in the world. The budgetary allocation, which represented less than 7 percent of the 2005–2006 national budget (corresponding to a paltry US$5 per person per year for farm inputs), seems a remarkably small price to pay for national food self-sufficiency and widespread household food security. Donor aid to Malawi in 2005 was US$578 million, or about US$44 per person.

The subsidy program was not targeted to the poorest of the poor. For starters, most Malawian maize farming families were living in extreme poverty. There were very few large and wealthy farmers who stood to benefit. Under the input subsidy program, beneficiaries had access to the same quantity of inputs, irrespective of farm size: the smaller the farm, the greater the effective input level on an area basis. This is not to imply that fertilizer subsidies benefit all poor households. Where households have little or no land or are constrained by labor, alternative approaches such as social cash transfer programs are needed.

The subsidy program was not "sustainable." This was perhaps the most common refrain from opponents of the subsidy program. There was no question that the input subsidy program placed a heavy burden on a tight and already inadequate national budget. Donors, many of whom—without any apparent sense of hypocrisy—were heavily subsidizing their own farmers, were insisting on an early "exit strategy" that would somehow enable farmers to move seamlessly from what I call sub-subsistence—meaning they produce less than enough to feed their families even in a good season—to profitable surpluses that would not require subsidies.

In 2009 my colleagues and I at the UN Millennium Project argued that Malawi's subsidy program should be embedded as part of a broader, long-term economic transformation:

> Malawi (and African countries like it) will require a firm 10- to 15-year multi-stakeholder commitment to support rural economic transformation from sub-subsistence farmers into diversified, small-scale entrepreneurs, including major complementary investments in health, nutrition, education, family planning, infrastructure, water, and sanitation. The input subsidy program must continue for several more years to meet the immediate needs for national and household food security, while longer-term investments in economic and social infrastructure are made.[13]

For more than fifteen years, the case of input subsidies in Malawi has provided a focus for heated debate over strategies for achieving food security in Africa. There is no question that the input subsidy program enabled agricultural intensification and contributed to increased food availability in Malawi. Small-scale farmers in Malawi have two basic resources to draw on for their food security within a tough rainfed environment: their land and their labor. Fertilizer and improved seeds have made it possible to extract more from their land and labor. Productivity of both resources increased. Despite climate variability and likely climate change effects, and the high vulnerability of Malawi's predominantly rainfed farming systems, the country has experienced more than fifteen years of relative stability in terms of national food security.

Malawi's quest for food security may not meet the armchair expert's definition of sustainable intensification. Anyone who had spent any time with farmers in Malawian villages would roll their eyes at the assertion that easy "exit strategies" existed that would enable transition to a profitable and sustainable future. What was clearly neither sustainable nor dignified was a nation whose president would go "down on [his] knees to beg for food" year after year. To paraphrase Norman Borlaug, what Mutharika and his successors have done is help Malawians buy more time. Malawi's Green Revolution is work in progress, deserving of praise and continued support.

DO WE PRODUCE ENOUGH FOOD?

The concept of intensification, sustainable or not, is premised on the need to produce more food to meet a growing demand. In the context of universal food security, food *availability* (or supply) is one of the four pillars of food security, the others being access, utilization, and stability.[14] Physical availability of food is a function of production and distribution to points of potential *access* by consumers. And it was Malthusian concerns about availability that prompted the Green Revolution responses that were discussed in chapter 2. Production was of paramount concern. Without sufficient production, there is little point debating about who has access.

We can begin our discussion of food availability by asking the most fundamental question: Will more food be needed in 2050? The literature and popular media abound with calls to "double production" or at least increase supply by 60 or 70 percent to meet the needs of population growth, increased incomes, and dietary change. An alternative view, more often than not inducing nods of approval in nonpartisan, lay audiences, is that we *already* produce enough food; we just need to do a better job of redistributing the food and cutting our food waste.

As population grows, logically, more food will be required. As incomes rise, more food is required up to a point where consumers' needs and preferences are satisfied. But needs and preferences change over time. As introduced in chapter 7, Bennett's law states that the share of starchy staples (e.g., rice, wheat, maize) in a diet declines with increased income and the desire for greater dietary diversity.[15] This phenomenon describes a flattening of demand for basic food staples beyond threshold income levels. Countering a slowing per capita demand, however, is the increased consumption of animal products. Since some livestock systems require conversion from grain to animal products, even more production is needed. These scenarios—most evident now in Asia, where incomes have risen rapidly in the past forty years—strongly indicate that, on balance, production of both starchy staples *and* high-value foods such as livestock, seafood, vegetables, and fruits will need to increase to meet demand.

There are clear possibilities for moderating the overall demand for food. About one-third of all food produced is estimated to be lost or wasted.[16] Therefore if food loss and waste could be reduced along value chains (see chapter 11), there would be less pressure to increase supply. Similarly, a shift toward plant-based diets and away from livestock, especially grain-fed animal systems (see chapter 12), would reduce pressure on supply of at least some grains.

So how do we make sense of these different trends and reach an evidence-based position on the need to produce more food?

At different times there have been strong statements on the need to increase food production, generally in the context of "calls to action." In 2012 FAO projected an aggregate increase in demand of 60 percent by 2050 using a baseline of production for the period 2005 to 2007.[17] An earlier FAO report prepared for a High-Level Expert Forum on "How to Feed the World in 2050," held in 2009, projected the need for a 70 percent global increase (100 percent increase in developing countries) over the same time frame and was commonly repeated in popular literature.[18] Another widely cited source forecasted that global demand for crop calories would increase by 100 percent and global demand for crop protein would increase by 110 percent over the same period, resulting in statements that "we need to double food production to meet the growing population."[19] More recently, critics have called for a revision of the estimate to 25 to 70 percent above current (then 2017) levels, primarily reflecting production gains since 2005.[20]

In all such estimates, there are caveats related to (1) uncertainty surrounding future population trends, the speed of dietary change, progress on food loss and waste reduction, and the impacts of climate change on current agricultural production; (2) variation among different commodities that are otherwise aggregated as "food"; and (3) geographical and economic diversity in demand and the failure of market mechanisms and social protection to redistribute what might otherwise be an adequate aggregate supply.

There is no clear consensus on the amount of additional food required by 2050. But it is important to note that none of the science-based estimates of future food needs suggests that we can satisfy demand with current production levels. Even if we could, the observed decline in productivity in some regions, through resource degradation, would mean that some compensatory increases would be needed to offset any shortfalls. That said, net availability of food more likely than not must increase, at least until we see real evidence of tangible progress in redistributing surpluses through markets and social protection mechanisms, in dietary shifts toward plant-based diets, and in cutting food losses and waste. It would seem prudent to take this as a premise for increasing food availability, whether the needed increase is 25 percent or 100 percent.

In the absence of both perfectly functioning markets and perfectly functioning social protection systems, we cannot hide behind the delusion of universal food security based on the calculation of aggregate global food supply divided by the world's population. That would be akin to arguing against new local water supply systems in drought-prone countries because, on average, there is enough fresh water in the world for everyone's needs. At the very least, it will be necessary to increase local food production in specific countries or regions where production deficits are constraining food and nutritional security and are not easily resolved by markets and social protection systems.

On the balance of evidence, therefore, I am drawn to conclude that we will need to produce more food, both globally and, in specific situations, locally. It is impossible to predict the additional requirements needed given the uncertainties about our success in changing diets and cutting food loss and waste. We can intensify production on existing agricultural land, or we can extend production by expanding into new or previously abandoned areas of land. Both options have important implications for sustainability.

TO INTENSIFY OR NOT TO INTENSIFY?

In agricultural context, *intensification* means getting more out of the resources that the farmer is utilizing. For food production, we usually apply the concept to land resources. For a food crop like maize, increasing yield from 1.1 t/ha to 2.0 t/ha, as occurred in Malawi through input subsidies, represents an intensification.[21] Multiple cropping and intercropping are other forms of intensification that can be enabled by use of irrigation or deployment of shorter-duration crop varieties. In both cases, the farmer changes practices with the objective of obtaining more food from the land. The Asian Green Revolution illustrates a focus on intensification. For example, between 1966 and 1981 rice production in Indonesia increased by 140 percent. This was achieved through a 97 percent increase in rice yield and a much smaller 22 percent increase in the area harvested.[22]

This approach to increasing production contrasts with *extensification*, the process of increasing production by expanding the land area planted. A farmer could increase crop production by growing more per unit of area (intensification), by expanding the area planted (extensification), or by some combination of the two. Extensification requires a change in land use, for example, converting from grazing or fallow or abandoned land to crops, or through deforestation to bring new land into agricultural production. Oil palm production in Indonesia increased from 11 million t in 1990 to more than 250 million t in 2020. In contrast to rice production, the increase came almost entirely from crop area expansion, with just a 3 percent increase in yield per hectare—a clear case of extensification.[23] A century earlier, the westward expansion of cropping in the United States was extensification.

Both intensification and extensification approaches have increased food production in a range of settings. These same concepts can be applied to livestock and fisheries. Consider intensive feedlot production systems in contrast to free-grazing herds. In fisheries, existing aquaculture systems can be intensified or expanded.

Both approaches to increased production have environmental implications. Intensification of food systems has been associated with excessive input levels, overexploitation of natural resources, biodiversity loss, and pollution. In the case of intensive livestock production, nutrient runoff, animal welfare, and greenhouse gas (GHG) emissions have also emerged as concerns. Extensification has been more commonly associated with biodiversity loss, desertification, land degradation, and increased GHG emissions.

Trends over the past several decades show that extensification is on the decline as a means of raising production. The average rate of cropland expansion over the period 1961–2010 was 0.24 percent per year; however, for the final fifteen years of that period, it was only 0.04 percent per year.[24] Intensification has largely replaced extensification as a result of new technologies coupled with a slowing of population growth since 1970, effectively saving millions of hectares of forest from conversion to agriculture.

A study in 2014 by Derek Byerlee, James Stephenson, and Nelson Villoria provides a deeper analysis of the role of intensification in minimizing cropland and slowing deforestation. They distinguished between technology-induced and market-induced intensification. They found that land area expansion can result from intensification close to the farmland–forest frontier and where market expansion creates increased profitability. To improve the land-saving benefits of intensification, Byerlee and colleagues argued for complementary targeted policies to improve land and forest governance, market certification to meet minimum social and environmental standards, and payments for environmental services, including support for biodiversity conservation and carbon sequestration.[25]

If we accept the premise that more food is required to meet the food security needs of ten billion people in 2050, the extra food, clearly, will have to come mainly through intensification rather than from opening up new land. And even if we take a highly optimistic view of the countering forces of shifts to plant-based diets and food loss and waste reduction, it is simply implausible and indeed highly risky to believe that we can rely on existing systems of production to sustain our global food demands.

This is where sustainable intensification (SI) comes in.

DEFINING AND APPLYING SUSTAINABLE INTENSIFICATION

The UN estimates that some form of environmental degradation is affecting the well-being of an estimated 3.2 billion people—about 40 percent of the world's population.[26] As was revealed earlier, particularly in chapters 3 and 4, the current environmental impact of agriculture is unquestionably large and growing. Agriculture (crops and pastures) now occupies about one-third of the ice-free land on Earth.[27] Global estimates of land degradation vary from less than one billion hectares to six billion hectares, the latter estimate exceeding the current land area dedicated to agriculture.[28] These large—albeit difficult to verify—estimates of land degradation place a heavy burden on agriculture and the broader food system, including livestock and fisheries sectors, to reduce their environmental footprints.

Conversion of land to agriculture, forestry, and urban development has been an essential component of human civilization for millennia. As population pressures mount, however, the costs of unintended ecosystem damage in the forms of land degradation, biodiversity loss, and GHG emissions must weigh heavily on our decision-making.[29]

In most cases, the impacts of agriculture on the environment are the result of efforts to increase food production, either through intensification or extensification. Despite the progress made in supplying food for a hungry planet, we are clearly not passing the tests of sustainability and intergenerational equity. One cannot feed ten billion people, however, without an environmental footprint. The challenge is how best to achieve universal food security with the lowest possible impacts on the environment.

In broad terms, SI aims to meet our food requirements while minimizing the collateral environmental damage. The term *sustainable intensification* was first coined in the context of agricultural development as a means of achieving food security with both economic and environmental benefits. The concept was characterized by "regenerative, low-input agriculture, founded on full farmer participation in all stages of development and extension" supported by "local processes of innovation and adaptation" to enable reduced dependency on external inputs.[30] At that point of its development, SI appeared to be more of an ideology than a practical alternative for farmers.

Over time, academics and development agencies have more broadly accepted an alternative approach to SI. This sees SI as a means to increase production with more efficient use of resources and reduced environmental damage, rather than focusing on low-input agriculture and reduced dependency on external inputs, as originally described.[31] Johan Rockström and colleagues define SI as "adopting practices along the entire value chain of the global food system that meet rising needs for nutritious and healthy food through practices that build social–ecological resilience and enhance natural capital within the safe operating space of the Earth system."[32]

Concepts consistent with the SI narrative have been articulated in the Sustainable Development Goals. Productivity improvement is a key element of SDG 2, Zero Hunger, with Target 2.3 aiming to "double the agricultural productivity and incomes of small-scale food producers" and Target 2.4 aiming to "ensure sustainable food production systems and implement resilient agricultural practices that increase productivity and production, that help maintain ecosystems, that strengthen capacity for adaptation to climate change, extreme weather, drought, flooding and other disasters and that progressively improve land and soil quality."[33] These two targets, combined with SDG 13, Climate Action; SDG 14, Life Below Water; and SDG 15, Life on Land, collectively provide a compelling global case for a sustainable pathway toward a food-secure world.

Despite the multitude of well-principled definitions of SI, it is a challenge to define what it is in practice. Yet, as a practitioner, I draw on the concept as both an aspiration and an inspiration. An SI strategy should improve food availability while reducing our environmental footprint to the point where the food security of present and future generations is not compromised. That does not mean that every field and farm should intensify. We should instead see SI as a transformation process *in aggregate*, whereby combinations of the following are realized:

- Some farms increase output with more efficient use of inputs (e.g., increased use of fertilizer and improved seeds, as illustrated in the Malawi case).
- Some farms maintain current output but with a reduced environmental footprint (e.g., reduced water use, reduced energy consumption).
- Some farms may reduce output or go out of production altogether, perhaps shifting to enterprises with less environmental impact or returning the land to natural ecosystems.
- Some abandoned and unproductive lands are restored through strategic use of critical external inputs, such as fertilizer, agroforestry, and irrigation.

- Natural ecosystems are protected and valued for their environmental services (e.g., carbon sequestration and biodiversity conservation, including crop pollination by beneficial insect species).
- By reducing postharvest losses in storage, transport, processing, marketing, and consumption, we generate more net food supply from the land.

These six processes can be integral components of a comprehensive and dynamic process of SI, applied contextually in different food system settings.

As noted, SI should not focus exclusively on production. In his Farrer oration, Timothy Reeves, former director general of CIMMYT, observed: "Sustainable intensification cannot stop at the farm level and is essential throughout the whole agri-food value chain. Hard won efficiencies at the farm level cannot be squandered at the landscape, processing, or consumption levels and vice versa."[34]

Right now, we have a global food system that, to a large degree, (1) inefficiently applies inputs, (2) degrades soil and land resources, (3) leaks nutrients and other toxins into our water systems, (4) releases GHG emissions at a rate that is accelerating global warming and climate change, (5) stands vulnerable to stresses and shocks emanating from climate change and other factors, (6) emphasizes food quantity over quality and safety, (7) pays inadequate attention to food losses and waste, and (8) rewards near-term profitability.

In response to these eight challenges to food system sustainability, I pose eight related questions that farmers, practitioners, and policy makers who bear responsibility for increasing food supply should ask:

1. **How can we improve the efficiency of external inputs, especially nonrenewable and high-demand resources?**

Contrary to some interpretations, SI does not mean a shift to low-input agriculture. Low-input agriculture, which is widespread across Africa, is more often the *cause* of unsustainable food systems and food insecurity. Practical examples of improving input efficiency include solar-powered drip irrigation for high-value crops, alternate wetting and drying techniques in rice, integrated pest management, and precision fertilizer applications. For livestock systems, SI can be promoted through improvements in conversion of feed to meat, milk, and fiber, and through breeds better adapted to biotic and abiotic stresses.

2. **How can we improve the quality of soil and land?**

Our soil and land will be passed to the next generation. As noted in chapter 3, good soil health is the foundation of productive, profitable, and sustainable agriculture.[35] In many parts of the world, unsustainable nutrient mining is the norm—it is agriculture, not mining or logging, that is most likely Africa's largest extractive industry. Fertilizer subsidies, illustrated earlier in the case of Malawi, can be used to kick-start nutrient replenishment efforts in degraded soils. Soil and water conservation practices (e.g., reduced tillage, cover crops, and vegetative contour strips) and nutrient replenishment using both inorganic and organic sources, including nitrogen-fixing legumes, are critical requirements for SI of crop farming. No-till farming is emerging as a high-potential intervention for soil and water conservation in many areas. By recycling nutrients, livestock can play an important role in improving soil health.

3. How can we reduce pollution emanating from the production system?

Agricultural systems are leaking nutrients into groundwater and river systems with detrimental effects on health and livelihoods. In many intensively farmed areas of the world, excessive applications of inputs can be rolled back with little or no yield reduction. Xiao-Tang Ju and colleagues, in a widely cited study in 2009 on intensive cropping systems in China, found that nitrogen fertilizer rates could be reduced by 30 to 60 percent without loss of yield, substantially reducing pollution.[36] Legislative and regulatory actions with appropriate penalties and rewards, along with more effective extension advisory services, can be used to support corrective action.

4. How can we reduce net GHG emissions?

Climate-smart agriculture offers opportunities to reduce GHG emissions (see chapter 6). Increases in carbon sequestration through farming will contribute to climate change mitigation. We are continuing to identify practices that reduce GHG emissions and sequester carbon. Carbon farming through practices like agroforestry and conservation farming are important practical examples of SI. With the growth of carbon markets, we can expect to see carbon farming become more profitable and even emerge as an important driver of SI.

5. How can we enhance system resilience to stresses and shocks?

We know that agriculture will be adversely affected by long-term stresses and short-term shocks accompanying climate change as well as from nonclimate factors. For intensification to be sustainable, we must improve agriculture's ability to resist, recover, and reorient, to reduce the vulnerability of the food system to stresses and shocks. Greater enterprise diversification and increased use of stress-tolerant crop varieties and livestock breeds are examples of how to improve resilience.

6. How can we improve nutritional quality and product safety?

From a food systems perspective, sustainability requires attention to the quantity and quality of agricultural outputs. In assessing progress toward SI, it is important to apply the quality lens to agricultural systems. Production of more nutritious and safer products is one dimension of SI. A shift to nutrient-rich foods, like orange-fleshed sweet potato, which is high in beta-carotene, is a good example of quality-driven SI. Greater use of legumes in farming systems can improve nutrient yields, help restore soil health, and improve system resilience.

7. **How can we reduce postharvest losses and waste?**

One-third of the food we produce never gets consumed. Postharvest losses and waste occur across all food value chains, with impacts on physical availability (amount) and market access (cost). There is also a high environmental cost when we consider the land, fertilizers, pesticides, energy, and water that was used in production of food that is not consumed. We will dive deeper into this topic in chapter 11, where I share best practices for reducing losses and waste.

8. **How can we improve the profitability and adoption of SI innovations?**

Historically, the search for higher profits is often associated with unsustainable practices. Most farmers make decisions based on an expected return on their investment, be it their land, their labor, or their financial resources, tempered by risk considerations. If we apply any or all of the preceding seven questions for assessing SI and the result is not profitable, we cannot expect sustained adoption. However, given that some of the benefits are positive externalities, such as reduced off-farm pollution and reduced net GHG emissions, adoption will require public policy intervention to improve profitability, for example, through payment for environmental services.

The answers to those questions will likely change over time and place, as food supply and demand systems interact in a setting of unpredictable climate events; as new technologies and policies emerge, succeed, and fail; and as our true capacity for change is revealed. We have no choice but to advance these ideas in good faith, drawing on the best that science and experience can offer. Reeves adds his perspective: "The clear message for me is that agricultural sustainability is a 'moving target' and that no single system has remained sustainable for one reason or another, and nor will it in the future."[37]

SI can be applied to any setting where we are committed to improving food availability while reducing our environmental footprint. A long history of research provides us with important knowledge, principles, and practices for achieving SI for almost any food system. The eight questions I have posed provide a framework for defining SI principles. In response to each of these questions, I provide a few illustrative examples of practical SI innovations that can enhance the productivity and sustainability of Malawi's food systems (figure 9.2).

SUSTAINABLE INTENSIFICATION PRINCIPLE	ILLUSTRATIVE INTERVENTIONS IN MALAWI
Improve input use efficiency	• Rainwater harvesting • Location-specific fertilizer recommendations • Well-adapted varieties
Improve the quality of soil and land	• Legumes in rotations • No-till farming (conservation agriculture) • Balanced fertility improvements using inorganic and organic sources
Reduce pollution from production systems	• Improved timing of nutrient applications • Slow-release fertilizers • Monitoring of water resources
Reduce net greenhouse gas emissions	• Agroforestry for carbon sequestration • Relay cropping with reduced tillage • Improved water management (rice)
Enhance resilience to stresses and shocks	• Drought-tolerant maize varieties • Income diversification • Crop insurance
Improve nutritional quality and safety of products	• Diversification, including legumes and livestock • Orange-fleshed sweet potato • Improved drying and storage to reduce aflatoxin
Reduce postharvest losses and waste	• Improved drying and storage infrastructure • Warehouse receipting systems • Electrification
Ensure profitability and adoption	• Improved roads • Market information services • Extension advisory services

FIGURE 9.2 Sustainable intensification principles and illustrative interventions for the Malawi food system.

Faced with increased food demands by 2050, SI embodies an essential strategy for universal food security. Ken Cassman, a professor at the University of Nebraska and a former colleague and head of agronomy at the International Rice Research Institute, has long argued that we should stay carefully focused on crop yields and their contribution to global food supply. In a paper published in *Nature Sustainability* in 2020, Cassman and Patricio Grassini maintain there is a strong scientific consensus on three points in relation to SI:[38]

- Meeting the food demand of 10 billion people by 2050 in a sustainable fashion is one of humanity's greatest challenges.

- SI is the path to meeting that challenge by reducing further conversion of natural ecosystems to farmland.
- SI should be assessed by the nutritional value and associated environmental metrics of production, rather than by the farming approaches or types of inputs used.

They argue persuasively that amid the contextual diversity and uncertainties facing the global food system, SI efforts should prioritize the following:

- major crops and cropping systems in the world's major "breadbaskets," in contrast to some calls to shift focus away from these staple commodities like rice, wheat, and maize
- regions with large gaps between existing and potential yield, as in sub-Saharan Africa
- regions with potential to expand irrigated agriculture, including some parts of sub-Saharan Africa and South America

Cassman and Grassini called for an acceleration and prioritization of research and development efforts to develop better crop-management practices in these environments:

> Although achieving SI to ensure global food security by 2050 is a massive scientific challenge, it is not beyond reach if there are well-prioritized national and global R&D agendas with a ruthless focus on the dual objectives of achieving large increases in yields on existing farmland coupled with substantial improvement in environmental performance that adequately protects natural resources, environmental services and minimizes GHG emissions. A 50% yield increase on existing farmland in tandem with a 50% decrease in negative environmental externalities provide useful initial targets for establishing national SI research portfolios.[39]

SI can be applied to any setting where we are committed to improving food availability while reducing our environmental footprint. A long history of research provides us with important principles, knowledge, and practices for achieving SI for almost any food system. Increased investment and prioritization of research and development efforts are needed to develop better crop-management practices in support of SI.

The inclusion of sustainable intensification as one of the Big Five investment areas for universal food security counters a commonly held misconception that food availability is no longer a priority for achieving food security. Productivity improvements will always be needed, in no small part, to offset and reverse the past and ongoing degradation of production environments. In addition, SI provides us with greater opportunity to spare land, conserve biodiversity, and reduce GHG emissions. The value of SI, however, cannot be achieved in isolation from other investments. SI at the farm level will be of limited benefit to our goal of universal food security without investments that connect farmers to markets, reduce postharvest losses and waste, result in healthier, more sustainable diets, and improve ways to reach the unreached through social protection.

10

MARKET INFRASTRUCTURE

My terms of reference were clear: "Your job is to show farmers how they can benefit from the new roads we are building." In May 1977 I arrived in Pagadian City, a sleepy township in Mindanao, in the southern Philippines, and started work as an agronomist with the Zamboanga del Sur Development Project. It was so much the centerpiece of Australia's aid to the Philippines that it went by the acronym of PADAP: in full form, the Philippines–Australia Development Assistance Programme. Two large Australian firms, the Snowy Mountains Engineering Corporation (SMEC) and Leighton Contractors, were hired to build hundreds of kilometers of bitumen-sealed ("black top") roads across the province with the purpose of opening markets and services to one of the country's poorest and most conflict-prone regions.

Mindanao was always the land of great potential. The province of Zamboanga del Sur, occupying the southern half of western Mindanao, was blessed with fertile volcanic and alluvial soils and an extraordinary nine-month rainy season that enabled two successive crops of maize to be grown most years. Situated 7 to 8 degrees north of the equator, the province was generally free of typhoons that annually caused havoc further north. Relatively recent waves of deforestation for logging and slash-and-burn farming had transformed the landscape into denuded hills with low-intensity, maize-dominant farming on the gentler slopes and ever decreasing forest remnants on the steeper more remote parts. As the topsoil eroded and fertility declined, some areas were abandoned to unproductive grassland. But land remained in abundance in the late 1970s.

This was indeed the Promised Land, aside from the lack of roads and a long-standing conflict between the government and local separatists, then known as

FIGURE 10.1 A project road under construction in Zamboanga de Sur, Philippines (1977).

Credit: Glenn Denning.

the Moro National Liberation Front. It therefore seemed entirely sensible to undertake an integrated rural development project that combined roads with agricultural improvements, including an irrigation scheme designed to boost rice production. SMEC subcontracted an Australian consultancy firm (ACIL Pty Ltd) to handle agriculture. And ACIL found me.

The three years I spent in Mindanao had a lifelong impact on my understanding of rural development, in particular the importance of physical infrastructure investment. I was introduced firsthand to the value of roads in connecting farmers to markets—input markets, financial markets, and product markets. What started out as a somewhat ill-conceived demonstration project, designed to introduce farmers to new varieties, fertilizers, and other modern practices, transitioned quickly into a multifaceted effort to enable farmers to access inputs (mainly seeds, fertilizer, and pesticides) and to market the crop surpluses being generated. The missing component was finance—an essential part of the soft infrastructure needed for market development. How do poor subsistence farmers pay for the inputs? Without that essential complement to roads, there would be no surpluses to transport.

SEEKING CREDIT

This period—the late 1970s—coincided with the Green Revolution in full flight, in the Philippines and elsewhere across Asia. President Ferdinand Marcos had a firm hand on the country; martial law had been in place several years and would be lifted only in 1981. Marcos made agriculture and food security a high priority. Masagana 99 was a government farm-credit program that promoted new rice-production methods with a focus on irrigated land and utilizing varieties developed by the International Rice Research Institute (see chapter 2).[1] Rice farmers obtained credit and technical advice (a package of support known as supervised credit) to boost adoption of the new rice technologies. Extension workers trained farmers in the new methods and oversaw delivery of inputs. Early maturing rice varieties enabled farmers to grow two crops in a year, mainly in the large rice-growing areas of Luzon, the country's most populated island.

There were only a few small pockets of Zamboanga del Sur (and indeed Mindanao) where irrigated rice was grown and where the Masagana 99 program had penetrated. One of these was the Sibuguey River valley, where our project developed a series of water-control structures, including weirs and canals. The more expansive upland environment, the rolling hills dominated by maize, were unable to access funds from Masagana 99. In 1981 Marcos launched a supervised credit program to boost maize production, modeled on Masagana 99 and called Maisagana. In 1977, however, there were no government programs supporting crop diversification in western Mindanao. Leguminous crops like mung bean, cowpea, and peanut were not deemed a priority. The government's focus was rice.

A decade earlier, legendary agricultural development specialist Arthur T. Mosher had highlighted the role of farm credit as an "accelerator of agricultural development." Mosher argued: "to produce more, farmers must spend more—on improved seeds, pesticides, fertilizers and implements."[2] As I went about redesigning my role in the project, I was also influenced and encouraged by the writings of Hubert Zandstra, a Canadian soil scientist and agronomist who, during the early 1970s, had worked on the Caqueza Project. This was a research initiative to support rural development in Colombia with the support of Canada's International Development Research Centre.[3]

I first met Zandstra in late 1977 during a visit to IRRI, where he had led the institute's Cropping Systems Program since 1975. Zandstra generously and enthusiastically shared his experience working with smallholder farmers in the Colombian Andes—in particular, seeking to understand the barriers to technology adoption—in a region east of Cundinamarca known as Caqueza. I identified parallels with the challenges we were facing in Zamboanga del Sur, among them the need to identify appropriate technology, while understanding the risks faced by smallholder farmers, the importance of market access, and the critical role of farm credit. Zandstra and colleagues recognized that incomes from smallholder farming barely met consumption requirements, and there was little prospect for savings to serve as a source of capital:

> These results do seem to indicate that scarcity of capital does limit adoption rates, and that access to capital is limited because of the low levels of collateral available associated with low wealth levels. This is particularly true in the case of low-income corn producers. Experience with these producers in the project indicates that if they are given credit and technical assistance, as well as the security that they will not become liable for their losses, most of them will not hesitate to increase their production.[4]

Recognizing the importance of credit to support farmers in diversifying their crops and improving their incomes, I quickly established a small revolving fund, using some of the budget we had been assigned for field demonstrations. The project essentially played the role of personal banker to a select group of upland farmers. Those farmers were identified by several local extension workers who had been assigned to the project by the Bureau of Plant Industry, the agency that served as our local counterpart on agriculture. The revolving fund proved to be a nightmare to administer, not least the risks in Zamboanga del Sur associated with carrying around large sums of cash. It was obvious that a more sustainable solution would be necessary. I visited the local branches of two government banks in Pagadian City: the Philippine National Bank (PNB) and the Land Bank of the Philippines, both of which were directly supporting Masagana 99. The local PNB manager, Dionisio Gran, showed greatest interest and referred me to the PNB vice president, Manny Soliven.

The project director (Ian Aberdeen, a sheep farmer and farm management consultant from Kilmore, Australia) and I paid Soliven a visit at the plush Manila headquarters of PNB. In hindsight, it is a little surprising that he was spending time on such a trivial proposal in the overall operations of PNB, which was tasked with running Masagana 99 and other flagship programs of President Marcos. Perhaps he was curious to see what these Australians were up to in far-off Mindanao, where few Manila residents dared to tread. In any case, after our short, enthusiastic pitch, Soliven flatly rejected the idea, unimpressed by the weak financial analysis and concerned about the riskiness of a focus on upland farming operations—both lessons I have never forgotten.

We went back to the drawing board and assembled more data on crop performance and market potential to prepare for another meeting with Soliven. Critically, when we returned to Manila, we had in hand a proposal for the Australian government to set up a guarantee fund (A$25,000) that would ensure repayment of any loan defaults up to 50 percent of the value of loans given.[5] With little to lose, Soliven approved the proposal. The program had now become institutionalized and ran for at least another ten years beyond the formal end date of our project. More than a thousand farmers participated in the first few years. Repayments were about 90 percent, so little of the guarantee was ever required. Access to supervised credit did make a difference and was proven to be an important complement to the roads, though I sometimes wonder whatever happened to the A$25,000 guarantee deposited with PNB.

Access to credit, offset by a guarantee fund, played an important role in offsetting the risks associated with smallholder farming in our project area. More recently index-based crop insurance has emerged to encourage farmers to adopt technologies in the face of climate risk.[6] Loss-based insurance is not feasible for remote smallholder farmers because of the high cost of loss verification. Instead, with index insurance, payouts are made based on verifiable adverse weather conditions measured by an index. India's National Agricultural Insurance Scheme has reached millions of farmers by bundling the insurance with farm credit and a government subsidy on the insurance premium. Reporting on five case studies, including India's experience, the CGIAR Research Program on Climate Change, Agriculture, and Food Security concluded that index insurance shows high potential for promoting technology adoption by smallholder farmers and pastoralists.[7]

With credit and technical advice, farmers in the uplands of Zamboanga del Sur responded by increasing their production of maize, upland rice, and legumes. Roads made it possible for buyers to reach more farmers and support their transformation from subsistence to commercial producers. Trucks came from hundreds of kilometers away to buy mung beans, a crop that we introduced to the province. This high-protein legume was in strong demand in urban areas throughout Mindanao and beyond. Government extension workers were assigned to previously remote areas. The electricity grid followed roads. Agribusiness outlets decentralized, and farm inputs were more readily available to farmers. Schools and health clinics were built and staffed in towns connected by our new roads. We were seeing firsthand the power of roads-led rural development.

ROADS TO PROSPERITY

Recognizing the need to invest in roads was not new to the Philippines. More than a decade earlier, Ferdinand Marcos himself campaigned on the promise of more roads, more rice, and more schools.[8] Infrastructure and roads, in particular, were a visible means of getting the economy moving, while demonstrating a government commitment. While Mosher saw farm credit as an "accelerator" for the adoption of improved practices, he regarded transportation and related market development as "essentials." Mosher argued that other critical investments could not be provided effectively without transportation. He astutely identified the necessity of a well-integrated system of transportation infrastructure and services, extending to water and rail transport.[9]

In their classic work, *Transforming Agrarian Economies: Opportunities Seized, Opportunities Missed,* Tom Tomich, Peter Kilby, and Bruce Johnston observed that "investments in rural roads provide a rare opportunity for policymakers to please almost everyone: reducing transport costs lowers marketing margins, allowing consumer prices to fall and producer prices to rise without squeezing private traders' incentives."[10] Like Mosher, they argued that efficient transport enables the transition from traditional to science-based agriculture by enabling surpluses to reach consumers. Throughout the 1970s and 1980s, roads were central to the integrated rural development approach that spread widely with the support of the World Bank and bilateral donors.[11]

American diplomat and former World Food Prize chair Kenneth Quinn drew similar conclusions from his experience in Vietnam and Cambodia. As described in chapter 2, during the late 1960s agricultural improvement in the Mekong Delta was seen as a key strategy in winning the support of the Vietnamese people. Quinn observed that new rice technologies spread with road improvements: "Where the road improvements stopped, though, so did any increased agricultural productivity."[12] Though this "roads and rice" strategy fell short of larger U.S. military expectations in Vietnam, Quinn argues that a similar approach played an important part in ending the Khmer Rouge insurgency in Cambodia. He views roads as the entry point for peace and prosperity.

The United Nations Millennium Project Task Force on Hunger recognized the vital importance of connecting farmers to markets.[13] The Hunger Task Force argued that functioning markets are critical in ensuring that people generate income, obtain farm inputs, and sell their produce at fair prices. The task force called for a campaign to increase road building, especially in Africa. Several other interventions were recommended to complement physical infrastructure, including agro-dealer networks, financial services, farmer organizations, market information, and storage and processing facilities.[14] These recommendations were echoed and amplified by then UN secretary-general Kofi Annan in his call to action for a uniquely African Green Revolution, at the High-Level Seminar on Innovative Approaches to Meet the Hunger Millennium Development Goal in Africa, in Addis Ababa on July 5, 2004, detailed in chapter 2. Annan remarked: "It makes little sense to help with soils and water, while leaving impoverished villages without improved roads, energy or seeds."[15]

Recent analyses continue to recognize the value of investing in transport for stimulating rural growth, structural transformation, and economic development. A World Bank review of transport policies in 2015 highlighted the potential for achieving positive impacts in low- and middle-income countries, arguing that transport investments can support the process of structural transformation (see chapter 8).[16] The report noted that transport benefits to agricultural commercialization were strengthened when accompanied by soft-infrastructure investments in agricultural extension services, telecommunication services, and improved access to farm credit. By investing in roads and other transportation infrastructure and policies, the production benefits of sustainable intensification can be transformed into economic benefits. The corollary to that argument is that failure to connect farmers to urban consumers through markets will stall sustainable intensification and successful structural transformation.

In a recent comprehensive review of food system performance and trends in the Asia–Pacific region, the UN Food and Agriculture Organization noted that road density had increased in nearly all countries.[17] Farm-to-market roads were noted as essential components of improved value chains, through their role in encouraging greater involvement of the private sector in supplying inputs and procuring farm surpluses.

Roads also remain a priority for the major international finance institutions operating in Asia. In its Strategy 2030, the Asian Development Bank (ADB) committed to improving market connectivity and agricultural value chain linkages by investing in rural roads, market infrastructure, and agricultural logistics centers.[18] ADB has been a major supporter of regional economic integration through infrastructure investment, especially roads. The same is true for the World Bank and the Asian Infrastructure Investment Bank. Regional trade in agriculture and food have benefited from investments in road, rail, ports, electrification, and information and communication technology.

In recent years China has invested heavily in transport infrastructure through the Belt and Road Initiative (BRI).[19] Launched in 2013 by President Xi Jinping, the BRI has embarked on infrastructure investment initiatives from East Asia to

Europe. An example of a BRI investment seeking improved regional integration is the Lao–China Railway. The 414 km high-speed railway connects the Lao People's Democratic Republic (PDR) capital of Vientiane with the city of Boten at its northern border with China. At Boten, the railway connects with the BRI network at Kunming, China, through another 595 km railway link. This investment has the potential to transform Lao PDR from a landlocked to a *land-linked* economy.

With support from the World Bank, I have worked with a team of consultants and students to identify opportunities for smallholder Lao farmers to benefit from this significant infrastructure investment. Access to the massive China market is not a given. Farmers need to produce and deliver their products competitively and sustainably. But much needs to be done to deliver on the promise of low-cost, high-speed connectivity. Our research has included value chain analysis for several commodities, including rice, maize, cassava, banana, tea, melon, coffee, legumes, and pork. But simply building the transport infrastructure is not enough. Our assessments have included economic, social, and environmental aspects in alignment with the 2030 Agenda for Sustainable Development.[20] Our initial recommendations include:

- improving customs clearance processes
- ensuring that transport infrastructure exists between farms and the railway, known as *last mile* connectivity
- implementing and monitoring environmental regulations
- supporting smallholder farmer cooperation
- strengthening extension services for smallholder farmers
- developing and promoting more-sustainable production systems
- integrating digitization into value chains
- establishing cold chains to support marketing of perishable goods
- increasing awareness of rail services and facilities

It is difficult to contain my enthusiasm for roads and other transport infrastructure investments that improve market connectivity. Three years in Mindanao left a deep impression. But my subsequent field experience in the Mekong River basin and in Africa have only reinforced my impressions of the centrality of transport networks in transforming agriculture and rural economies. In forty years I have never met a farmer who said, "Please don't improve the roads." As Tomich and colleagues noted earlier, "Rural roads provide a rare opportunity for policy makers to please *almost* everyone" (my italics).[21] But I cannot leave this topic without recognizing some of the reasons why roads are not always greeted with thunderous applause.

I will begin with another anecdote from Mindanao. One day, while undertaking a farm survey, my team and I came across a large basalt boulder in the middle of a barely passable buffalo track. Painted on the rock in English was "Keep Out, Bangsa Moro Liberation Army." A new road in this remote part of Zamboanga del Sur was not welcome—by some folks at least. Roads might open up markets. But they can be bad news for a separatist group wanting control over the local population and the freedom to operate without military interference.[22] We initially took heed and retreated, though later consultations with communities, brokered by some remarkable Australian priests from the Missionary Society of St. Columban, led to the said buffalo track becoming a tarmac road.

The second possible reason for reservations about investing in roads is the cost, both the initial capital cost and ongoing maintenance. The latter was commonly underestimated in the case of projects supported by aid programs, as donors almost always viewed maintenance as the responsibility of the recipient government. If not undertaken, the assumed benefits from lower transportation costs are quickly lost. Concrete roads, which have a much longer life span than bitumen roads, and lower-cost gravel roads are alternatives where maintenance budgets are limited or uncertain. A second Australian aid project in Northern Samar—a typhoon-ravaged province in the Visayas region of the Philippines—opted for concrete roads to reduce the risk of negligence in road maintenance. Beyond the fiscal challenges associated with the initial capital cost and maintenance, there is a related concern about corruption because of the opportunities to pad and extract resources from such large infrastructure projects.

A third reason for caution on road investments relates to their environmental impacts. Environmental costs are rarely factored adequately, if at all, into benefit–cost analyses. Roads that penetrate forested and other wilderness areas have often resulted in loss of biodiversity and other ecosystem services.[23] An International Energy Agency study estimated that construction in non-OECD countries will account for 90 percent of road additions between 2010 and 2050.[24] Frontier areas of the Amazon, Papua New Guinea, and the Congo Basin are threatened.[25] Loggers and miners often lead the way. While economic and social benefits are possible through agricultural development, the time has come to better account for the trade-offs associated with road penetration into environmentally vulnerable areas. As discussed in chapter 9, success in sustainable intensification can reduce deforestation, provided that complementary land and forest protection policies are in place and enforced.[26]

Other cautionary considerations when planning roads include:

- understanding the differential benefits of being nearer to the roads and the potential biases in assessing benefits
- ensuring fair compensation to owners and users of land acquired for roads
- accounting for road safety implications[27]

BANKING ON MARKETS

As a member of the Hunger Task Force and an organizer of the event where Kofi Annan made his call to action, I was committed to putting these ideas into practice. To that end, and with the support of Jeffrey Sachs and Pedro Sanchez, the Millennium Development Goals Centre for East and Southern Africa (MDG Centre) was established in Nairobi, Kenya, in 2004. I served as its founding director from 2004 to 2009. Part of our mandate was to support the recently established Millennium Villages Project (MVP), a bold but controversial effort to demonstrate the transformative potential of simultaneous investments across several sectors in ten sites across ten countries in sub-Saharan Africa.[28]

I was personally most involved in the MVP sites in Kenya, Tanzania, and Malawi, where the main crop was maize. The production increases—doubling or tripling of yield in the first few years—were inspiring, giving rise to belief that the African Green Revolution was indeed possible. As described in chapter 2, the drivers of these early results were production increases brought about by a combination of high-yielding varieties (mainly hybrid maize) and inorganic fertilizer (mainly nitrogen and phosphorus).[29] Farmers received their inputs on credit from the project and repaid in kind after the harvest, albeit at a subsidized rate. The repaid grain was used for school meal programs in villages. Some of the production surpluses above the repayments were sold locally to meet cash needs of farming households. Over time, farmer organizations—another component of soft infrastructure—were formed, and surpluses were stored in anticipation of better prices. The work of these farmer organizations continues. Some have connected to large institutional buyers such as the UN's World Food Programme. A recent evaluation of the MVP found the greatest impacts were on agriculture and health.[30]

Despite the excitement of these quantum yield increases, and like all development projects before it, the MVP struggled to connect farmers to markets in a sustainable and scalable manner. Just as I found in Mindanao thirty years earlier, it is not difficult to double or triple production with the right agronomy, generally improved seeds and fertilizer. In Mindanao, I managed to convince the Australian government and the Philippine National Bank to support a mechanism for connecting farmers to the input market through credit. But there were no banks to be found operating in the African villages where I worked, at least none that would lend to small-scale subsistence farmers. Smallholder agriculture was seen as too risky. As we found in Mindanao and Africa, roads, seeds, fertilizers, and farmer organizations are all necessary, but they are not sufficient for a *sustained* transformation of agriculture. Institutionalized farm credit is needed to complement the other investments.

In October 2009, in the wake of sharp food price increases and panic over world food security, Bill Gates addressed the annual World Food Prize Symposium in Des Moines, Iowa.[31] Speaking to a largely converted audience, Gates argued for improved smallholder productivity through the application of modern science. Gates also highlighted the need to make investments across the value chain to ensure that productivity gains translate into increased farmer incomes. I have to say, listening to Gates, then the world's richest man, speak comfortably about *SUB1* rice, nitrogen fixation, and no-till farming was inspirational and a reflection of the times and the man.

At a reception for Gates Foundation partners at this event, my former IRRI colleague Prabhu Pingali (then with the foundation's agriculture program) invited me to share with Gates my experience with fertilizer subsidies in Malawi. Well, why not? After a cocktail or two, I gave my elevator pitch on what I viewed as encouraging signs of the emerging African Green Revolution that Norman Borlaug, Kofi Annan, Gates, and not least African farmers were seeking. Apparently not a big supporter of subsidies, Gates immediately countered with something like, "If the technology is so productive, why don't the farmers simply borrow to pay for the inputs? Why the need to subsidize?" I offered two answers, hoping one would stick: First, there are no banks in Malawi serving these farmers. Second, even if there were, these farmers are so small in scale (half a hectare, often less) and so disconnected from markets that they would be unlikely to obtain, let alone sell, much of a surplus above their subsistence needs. Failure to repay would inevitably lead banks to flee the scene. With others hovering around for a break in the conversation, our meeting ended there. I did not get the impression that I had convinced Gates. In the years since, my views have not changed a lot on this topic. Connecting farmers to markets (both for buying inputs and selling outputs) is easier to say than do.

MORE THAN ROADS

Many development organizations have adopted market-based approaches in their efforts to advance rural development and achieve food security in Africa and around the world. The emphasis on improving market connectivity in Africa has been advanced in recent years by the African Development Bank (AfDB), the Alliance for a Green Revolution in Africa (AGRA), the U.S. Agency for International Development's Feed the Future initiative, and many other organizations. The AfDB, under Akinwumi "Akin" Adesina's presidency, has placed major emphasis on stimulating agricultural productivity through improved market access and the modernization of agricultural value chains. Building on his prior experience with AGRA and as Nigeria's minister of agriculture, three of Adesina's High 5 priorities—Feed Africa, Integrate Africa, and Light Up and Power Africa—were firmly based on need to invest in market connectivity.[32]

Electrification is another essential investment for accelerating both market connectivity and nonfarm rural enterprises.[33] Like roads, access to reliable, affordable electricity can attract businesses to process and package food products, opening up new products and new markets and therefore expanding the horizons of producers and increasing options for consumers. Refrigeration for milk and other high-value perishables like fish, fruit, and vegetables improves product storage life and quality at all stages along the value chain, from the producer to the consumer.[34] Postharvest losses and food waste can be reduced with access to electricity.

Through its New Deal on Energy for Africa, AfDB plans to contribute and catalyze massive investment in on-grid and off-grid power generation, unlocking huge renewable energy sources, including hydropower and solar.[35] Farmers, food processers, wholesalers, and retailers can take advantage of these investments to add value and raise incomes while meeting increasing consumer demand for perishable foods. While electrification will support market connectivity in all parts of the world, the most pressing need is in Africa, where more than 600 million people, the majority of them in rural areas, live without electricity.[36]

In the past two decades, information and communication technologies have opened up a range of exciting possibilities for improving market connectivity. Rapid growth of the internet and associated digital technologies, including mobile phones, are critical to helping farmers obtain information on market demand, in terms of volume, product specification, and prices. The high costs associated with sourcing market information constrain opportunities in developing countries, where markets are often poorly integrated. The World Bank's *World Development Report 2016*, focusing on the digital divide, cited numerous studies indicating that the introduction of digital technologies, including mobile phone services, can improve delivery of advice to farmers as well as facilitate market transparency.[37] Just four years later, through the *World Development Report 2020*, the World Bank again highlighted the importance of digital technologies in transforming global value chains, including agricultural value chains.[38] These innovations include big data, cloud computing, the Internet of Things, platform firms, e-commerce, and recently the use of blockchain technologies. Although still in early days of implementation, blockchain could be a game-changer in terms of market connectivity.

Blockchain is a decentralized, distributed database—a digital ledger—that is used to record transactions chronologically across many computers. Records of transactions are open to all participants, and, once verified, they cannot be altered retroactively. Encryption ensures a high level of privacy among the participants. Blockchain technology reduces the risks associated with central control of the transaction records. The consortia of participants have oversight of all transactions. Blockchain shows potential for reducing transaction costs while providing greater transparency and traceability, thereby improving the efficiency and security of agricultural commodity markets.[39]

For agricultural value chains, blockchain technology can be used to trace the provenance and enable trustworthy certification of food commodities and products, potentially bringing benefits to producers and consumers alike. Increasingly, consumers are demanding and are willing to pay for reliable information about the quality, safety, geographical source, and even production ethics of their food. Deloitte describes blockchain as "a trust machine enabling transparency through collaborated efforts," and based on early indications, blockchain has the potential to be transformative in the agribusiness sector.[40]

But how can we realize that potential?

If I were to start a new job in rural Mindanao today, the core challenge would be the same: connecting farmers to markets. But the tools at my disposal have expanded from what I had in May 1977. The new terms of reference might be: "Your job is to show farmers how they can benefit from the new *digital technology systems* we are building." Instead of relying wholly on the government, as I did then, I would advocate the establishment of digital innovation hubs that support entrepreneurial start-ups in designing and deploying innovative business models.[41] Such an approach would require a full value chain mapping and an analysis of opportunities within value chains for improving competitiveness. The initiation and organization of blockchain networks and other digital technologies requires a trusted public- or private-sector network initiator with interest and incentive to act.

Notwithstanding the current excitement around technology-enabled interventions, they cannot be relied on as the panacea for food systems transformation. Complementary investments are still essential in physical infrastructure, in enabling regulatory policies, and in capacity building, including literacy. The World Bank cautiously notes: "The versatility and near-constant innovation that characterize digital technologies can sometimes be a distraction that can cause interventions to focus more on the technology than on the demands and priorities of the intended beneficiaries and the trade-offs imposed by resource constrained environments."[42]

Since its inception in 2006, AGRA has emphasized market connectivity as a vehicle for unleashing the economic potential of African smallholder farmers. Inspired by Kofi Annan's call to action and by his subsequent leadership role as its founding chair, AGRA has sought the transformation of smallholder farms from subsistence to business entities and has supported efforts to finance farmers and small and medium enterprises to improve competitiveness and market access. USAID's Feed the Future program works through partnerships with governments, communities, and the private sector to tackle hunger and poverty with the view that "lasting change requires progress driven by private sector investment, a healthy business environment, and strong market systems."[43]

TechnoServe is a nonprofit, launched in 1968, with a mission "to build competitive farms, businesses, and industries."[44] It has worked in more than forty countries in Africa, Asia, and Latin America, deploying business approaches and solutions to reduce poverty. TechnoServe uses a three-pronged approach:

- developing the capacity of individuals and businesses
- promoting connections among market participants
- improving the business environment

Among the many organizations out there promoting market connectivity, I know of none that matches TechnoServe's professional and practical approach to transformation. Its work with farmer organizations and banks to identify and mobilize finance for smallholder farmers is especially innovative and important. I often think how exciting and productive it might have been had the Millennium Villages Project and TechnoServe come together to tackle food security. Between these entities, we could have arrived at a better answer—a more actionable response—to the question Bill Gates posed to me in Des Moines.

Mobilizing the business instincts of smallholder farmers and other value chain actors is the key to achieving financial viability and sustainability. Public-sector organizations and public-interest not-for-profit organizations are essential partners, enablers, and, in some cases, the best initiators of these efforts to improve market connectivity. Since 2015 the World Bank has assessed the role of governments in enabling agriculture and agribusiness.[45] Using globally comparable data to measure progress and obstacles to market integration and entrepreneurship, the research has revealed large disparities among countries and scope for improvement in such areas as access to inputs, credit, and markets.[46]

Almost all organizations working to improve market infrastructure recognize that there is no silver bullet for enhancing market efficiency and connectivity. While my own experience has drawn me toward better transport infrastructure and finance as essentials for market development, we need to invest in a system of interlinked and synergistic components that also include electrification; digital technologies like big data, cloud computing, the Internet of Things, and blockchain; and creative institutions such as market exchanges and market-enhancing policies. The local context will determine the exact mix of investments in any setting, which will reflect several factors, including (1) the current state of physical infrastructure, (2) the proximity to local, national, and international markets, (3) the state of the business environment, (4) the potential for increased public investment, and (5) the state of public institutions. Investing in hard and soft market infrastructure is essential for improving the availability and access of food to consumers and for achieving universal food security.

11

POSTHARVEST STEWARDSHIP

The determined efforts to increase maize production in Malawi initiated by President Bingu wa Mutharika and his government have attracted worldwide attention. In 2021 Malawi's farmers produced twice as much maize as they did in any of the three years leading up to the introduction of the input subsidy program in 2005.[1] This raised the next challenge: managing grain storage, particularly protecting such a hard-earned achievement from grain pests.

Through the Millennium Villages Project, I was exposed firsthand to the challenges faced by smallholder farmers in securing their stored grain, especially maize, that was the result of unprecedented harvests. Doubling or even tripling yields at our sites in Kenya, Tanzania, and Malawi placed stresses on the ability of farmers to handle, store, and market their harvests. Desperate to secure these surpluses, farmers stacked their grain in their houses, community centers, and schools.

In one sense, it was a nice problem to have. But it was heartbreaking to witness the disillusionment of farmers whose maize grains were being hollowed out by the larger grain borer (*Prostephanus truncatus*), a storage pest accidentally introduced to Africa in the late 1970s.[2] Bumper harvests were reduced to dust. In 2008 I initiated a study by CABI Africa to document the maize storage losses due to pests and to make recommendations for cutting losses.[3] The study reported losses of 30 to 60 percent in Malawi, Tanzania, and Kenya. Maize losses experienced by farmers varied, but all farmers confirmed that losses can be up to 100 percent if maize is not protected with insecticides before storing.

Recommendations from the report included improved drying methods, fumigation, and training of extension workers and farmers.[4] The need for chemicals, with all its attendant knowledge and training needs, is not unique to farmers in developing countries. It is a factor that even the largest grain growers in the most advanced grain-growing countries find a perpetual challenge. And it does not get easier, with strict regulations governing chemical use, and the pests themselves able to quickly evolve resistance to the few pesticides or fumigants farmers are allowed to use.

The government of Malawi responded to the importance and urgency of this problem by providing subsidized postharvest chemicals and distributing improved storage bins under the 2008–2009 input subsidy program. But, as demonstrated by the inadequacy of traditional drying and storage facilities, postharvest management was a lower priority than production, and much more was needed to help farmers manage their unprecedented surpluses and reduce such demoralizing harvest loss.

SETTING A TARGET

A landmark study in 2011 commissioned by the United Nations Food and Agriculture Organization (FAO) highlighted that food losses and waste amounted to 1.3 billion metric tons annually, representing about one-third of the edible parts of all food produced for human consumption.[5] FAO estimated the value of this food to be about US$1 trillion at retail prices. Bear in mind that this estimate did not include agricultural products that are intentionally produced and directed to animal consumption and bioenergy. The study highlighted the food security and environmental implications of reducing these food losses and waste.

For more than a decade the estimate of one-third of food being wasted has been widely cited as a call to action. More recently, FAO and the United Nations Environment Programme (UNEP) have firmed up their estimates of postharvest losses and waste. In 2019 FAO's flagship annual report, *The State of Food and Agriculture*, found that about 14 percent of global food produced is lost from the postharvest stage up to, but excluding, the retail stage.[6] The UNEP *Food Waste Index Report 2021* found that a further 17 percent of global food production is wasted—11 percent in households, 5 percent in food service, and 2 percent in retail.[7] The resulting total of 31 percent—postharvest loss plus waste—remains sufficiently close to the one-third estimate to justify its continued use as a valid reflection of the world's lack of postharvest stewardship.

So let us pause and consider this sobering statistic: one-third of the food we produce for human consumption never gets consumed.

FIGURE 11.1 Poor storage conditions result in postharvest losses and reduced food security for smallholder farmers.

Credit: Jessica Fanzo.

Beyond the impact on food security through physical availability (less food per capita) and market access (higher price per unit of food), we should reflect on the resources that were consumed in the production of the food that was never consumed: land, fertilizers, pesticides, energy, water, and labor. As discussed in chapter 9, land will need to be cleared (extensification) and/or made more productive (intensification) in order to meet the world's growing demand for food. Inefficient conversion of production to human consumption has consequences on land degradation, biodiversity loss, and greenhouse gas emissions. The impacts of food losses and waste have far-reaching economic, social, and environmental consequences.

The Intergovernmental Panel on Climate Change (IPCC) has estimated that food losses and waste contribute 8 to 10 percent of total anthropogenic greenhouse gas emissions, almost a third of the emissions that come from the global food system (see chapter 6).[8] By cutting losses and waste, we can—in principle—reduce the amount of food that needs to be produced compared to business as usual. By reducing food loss and waste, we can further reduce pressure on deforestation. Sustainable intensification and postharvest stewardship therefore are powerful *complementary* strategies for achieving food security. Neither one alone is a silver bullet for achieving universal food security.

For these reasons, food losses and waste have attracted increasing attention over the past decade. The pinnacle of recognition was the inclusion of this issue in the Sustainable Development Goals in 2015. Under SDG 12, "Ensure sustainable consumption and production patterns," Target 12.3 is "By 2030, *halve* per capita global food waste at the retail and consumer levels and *reduce* food losses along production and supply chains, including post-harvest losses" (my italics).[9]

One may quibble at the unexplained absence of a quantitative target for food losses and, notwithstanding the recent progress by FAO and UNEP, the practical challenges of applying meaningful indicators to measure progress toward the target. The emergence of an explicit SDG target on postharvest stewardship, however, was an important breakthrough in terms of recognizing the problem and challenging all stakeholders to take action.

I would have preferred to see postharvest stewardship as a target under SDG 2, Zero Hunger, more tightly associating the issue with hunger, food security, nutrition, and agriculture. But we should rejoice that a target now exists and has been agreed by all nations. SDG 12.3 now provides a rallying point for action, exemplified by the establishment of an initiative called Champions 12.3. This is a coalition of senior executives from governments, businesses, international organizations, research institutions, farmer groups, and civil society—all dedicated to inspiring ambition, mobilizing action, and accelerating progress toward achieving SDG 12.3.[10] Postharvest stewardship is now firmly established on the global development agenda.

DEFINING FOOD LOSS AND WASTE

Several studies go to agonizing lengths to define and differentiate what they mean by food loss and food waste.[11] In the end, the terms seem to get used inconsistently and interchangeably in the literature and in popular use. If I lose track of a punnet of strawberries at the back of my refrigerator and the fruit becomes soggy and inedible, is this food loss or food waste? If food is produced with the intent of consumption and it exits the food value chain, intentionally or unintentionally, I would argue that it is *both* lost and wasted. It is lost from the value chain before consumption—my strawberries represent a waste of money, not to mention a waste of the natural and other resources that went into producing these tasty, nutritious treats. For this reason, it is tempting to argue, from a practitioner's standpoint, that we simply use the terms *loss* or *waste* as best fits the context. That said, it always helps to use a common language for sharing knowledge and experience. And for that, I draw on FAO's most recent definitions (figure 11.2), with one amendment of my own to each:[12]

- "**Food loss** is the decrease in the quantity or quality of food resulting from decisions and actions [and inactions] by food suppliers in the chain, excluding retail, food service providers and consumers."
- "**Food waste** is the decrease in the quantity or quality of food resulting from decisions and actions [and inactions] by retailers, food services and consumers."

Note that I have added the possibility of unintentional loss and waste consequences. To be clear, no farmer deliberately offers up their maize for consumption by larger grain borers or rats. And the loss of my strawberries was the consequence of neglect rather than a conscious decision to act recklessly from an economic or environmental standpoint.

Broadly, food loss relates to food supply, while food waste relates to food demand. Wholesalers are considered part of the supply chain, so losses incurred there are recorded as food loss.[13] FAO acknowledges that these differences are difficult to apply in practice, since retailers are also suppliers from a consumer perspective. The convention for now is to measure losses up to, but not including, the point of retail and food service provision. FAO estimates food loss using the Food Loss Index, which considers quantitative losses adjusted by economic value.[14]

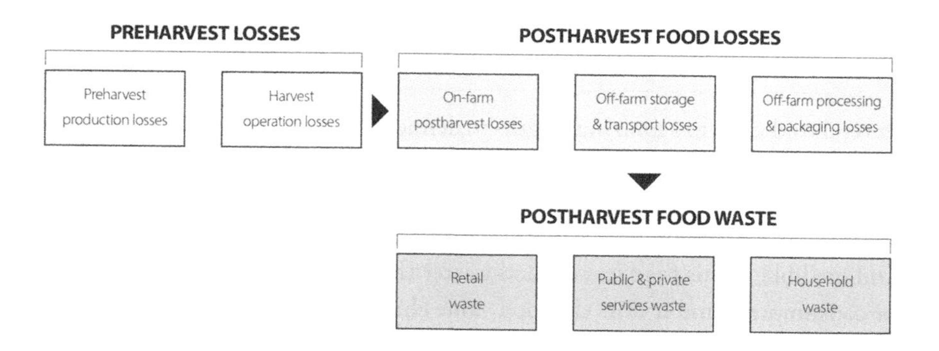

FIGURE 11.2 Food losses and food waste before and after harvest.

Source: Glenn Denning, inspired by FAO, *The State of Food and Agriculture 2019: Moving Forward on Food Loss and Waste Reduction* (Rome: FAO, 2019), 10–13.

In 2021 UNEP released the details of a new Food Waste Index that measures food waste at the retail, food service, and household levels. Unlike the Food Loss Index, the Food Waste Index measures total food waste by sector (retail, food service, and household) without differentiating by food commodity.[15] It is clearly a difficult and unpleasant task to differentiate commodities once they enter the bin or the dumpster. FAO further distinguishes between quantitative and qualitative losses and waste:[16]

- "**Quantitative food loss and waste** (also referred to as physical food loss and waste) is the decrease in the mass of food destined for human consumption as it is removed from the food supply chain."
- "**Qualitative food loss and waste** refers to the decrease in food attributes that reduces its value in terms of intended use. It can result in reduced nutritional value (e.g., smaller amounts of vitamin C in bruised fruits) and/or the economic value of food because of non-compliance with quality standards. A reduction in quality may result in unsafe food, presenting risks to consumer health."

The FAO and UNEP loss and waste methodologies represent important commitments by these agencies to improve our understanding of actual losses and waste, enabling us to take actions and monitor progress.

COUNTING AND CUTTING THE LOSSES

As illustrated in the CABI Africa study cited earlier, estimates of smallholder postharvest losses of maize in sub-Saharan Africa vary greatly. Tadele Tefera summarized research indicating overall cumulative losses of 14 to 36 percent.[17] A meta-study by Hippolyte Affognon and colleagues found maize storage losses in the range of 4 to 21 percent over an average period of seven months depending on the level of postharvest management applied.[18] Using the Food Loss Index, FAO reported that on-farm losses in cereals and legumes during storage vary greatly from one region to another as well as within regions. In sub-Saharan Africa, losses for cereals and legumes ranged from negligible to 18 percent, while for more perishable fruits and vegetables, reported losses were as high as 50 percent.[19]

The wide variation in reported losses during storage implies that the causes and need for intervention must be assessed in context. FAO admits: "To summarize all possible causes would be impossible, as these are highly context-dependent, based on the crop, commodity group and geography."[20] The most likely reasons for loss, however, were harvesting and handling practices, adverse weather, and poor storage and transport infrastructure. Having personally observed close to total loss of grain from the larger grain borer, I know it is necessary to examine the losses critically and take action to reduce losses when needed. Farmer expectations of high losses in storage encourage early sales when prices are low. The corollary is that cutting storage losses can encourage farmers to store grain and reduce low-price sales.

Among the different approaches we tried, including individual bags and small group storage bins, the most promising intervention was a large community warehouse that the Millennium Villages Project constructed in the village of Mwandama in southern Malawi. While some critics initially viewed this structure as an unsustainable "white elephant," the farmers responded with enthusiasm, organizing themselves into cooperative groups to manage the drying, fumigation, storage, and marketing of the surplus maize and ultimately other crops, including soybeans. On a visit to Mwandama in July 2019, I observed the warehouse functioning well as a legal cooperative union of eight constituent farmer groups, several years after the end of the project. The community also established a small shop serving the village. In addition, an international NGO rented an office adjacent to the warehouse and provided support to local farmers in the form of training and credit. The investment had the effect of crowding-in complementary enterprises, a promising step toward sustainability.

In Timor-Leste, maize farmers also struggled to manage surpluses generated by adopting new varieties under the Seeds of Life project, an Australian aid project promoting improvement in staple crop yields that ran from 2000 to 2016. Adopting new varieties had improved maize yield by about 40 percent. With high storage losses from weevils and rats, estimated to be 14 to 17 percent with traditional storage methods, various models of on-farm storage were assessed. The most attractive in terms of effectiveness, cost, and practicality were recycled 200-liter oil drums that were capable of storing 180 kilograms of maize under airtight conditions. This simple technology, first promoted by two NGOs (Drums on Farms and CARE Timor-Leste) and later scaled up by Seeds of Life and the International Fund for Agricultural Development, has improved household food security for tens of thousands of smallholders in Timor-Leste.[21] All the implementing agencies have used subsidies to accelerate uptake in these low-income rural settings.

These two examples are illustrative of the diversity of potential solutions that can help smallholder farmers reduce on-farm storage losses. The ability to store grain safely improves the price that can be obtained by reducing the need to sell produce immediately after harvest. In both cases cited, these innovations were coupled with productivity improvements for the maize crop, creating investment synergies. Governments, NGOs, and aid agencies have developed and introduced scores of methods to improve postharvest management of commodities, not only for staples, but also for horticultural products, livestock, and fisheries. Like other investments across commodity value chains, the key to reducing postharvest losses for smallholders is to understand local context and to assess a range of options for reducing losses for their fit and adaptability.

QUALITY MATTERS

Beyond quantity losses from insects, rodents, and birds, deterioration in grain quality can affect food safety and market price. In such cases, we see a reduction in value, not necessarily in quantity. Mycotoxins are a natural and potentially hazardous contaminant of grains. The toxins are produced by fungi that infect the crops, especially in hot and humid climates. Eating mycotoxin-infested grain can cause serious, even fatal, consequences for people and animals.

Aflatoxins are a class of mycotoxins produced mainly by the fungal species *Aspergillus flavus* and *A. parasiticus*. Among food crops they are found most commonly in maize and peanut. Insect damage (including preharvest infestation), poor storage conditions (high humidity), and drought during ripening can increase aflatoxin occurrence. Long-term exposure to aflatoxin-contaminated food can have adverse health impacts, such as liver damage (including liver cancer), immunosuppression, mutagenic birth defects, and possibly stunting.[22]

The health risks associated with aflatoxin contamination are a growing concern worldwide. Reducing the risk of aflatoxin occurrence requires an integrated, multistage approach spanning the commodity value chain. A good starting point is to select a crop variety that has resistance to the aflatoxin-producing fungi. This approach to reducing aflatoxin damage is showing promise in both maize and peanut.[23] Additional preharvest interventions include irrigation to reduce the risk of drought. But while these are desirable, farmers may not have access to either option. In such cases, timely harvesting and improved storage practices can reduce aflatoxin occurrence.[24]

A key prerequisite for successfully reducing aflatoxin is the ability to rapidly and cost-effectively monitor its occurrence. Aside from the health effects on local communities, contamination of food products can disrupt trade, affecting both producers and consumers. For these reasons, new methods of detection and analysis have been developed and deployed. These range from sophisticated laboratory equipment to rapid test kits for grain silos and shipments. The International Crops Research Institute for the Semi-Arid Tropics (ICRISAT) has developed and distributed a low-cost aflatoxin-testing kit that has helped Malawi's farmers regain their peanut export market. In Timor-Leste, a low-cost Aflatoxin Quicktest has been trialed for maize and peanut.[25]

There are concerns about the decline in nutritional quality during storage. Loss of nutritional quality, however, is difficult and expensive to measure, not least because food of degraded nutritional quality generally looks no different from food at normal nutrition levels.[26] Nutritionally degraded food is still consumed, and the consequences on individuals and populations may be hidden for years, given the chronic effects of undernutrition.

PRACTICAL SOLUTIONS

So where does this leave us in terms of reducing these harmful losses in food quantity and quality that occur along the value chain? Beyond the logical appeal of cutting such losses to improve food security at household and higher levels, we need to establish a practical plan of action to invest resources to improve food security outcomes.

A detailed review, published in 2017 by Megan Sheahan and Christopher Barrett, synthesized and critically examined the literature on postharvest losses in sub-Saharan Africa, which has relevance in most resource-constrained settings.[27] The authors identified a number of abatement interventions being promoted in different situations across the region. They noted that most were farm-level interventions, including improved varieties, extension advice on best practices, chemical spraying of stored products, improved storage of grains through hermetic technologies, and integrated pest management.

Of these, Sheahan and Barrett noted the widespread promotion of hermetic grain storage containers, the best known being the Purdue Improved Cowpea Storage bags, which have been deployed for cowpea and other grains, including maize. The bags are widely thought to suffocate grain pests by restricting the supply of oxygen required for respiration, but Larry Murdock, a professor of insect physiology at Purdue University, found they worked in a different way. When the oxygen level in the bags decreases, the weevils cannot metabolize water from the consumed grain; instead of suffocating, the offending insects die of thirst.[28] The drier the grain to begin with, the better the hermetic system works. Similar experience emerged from the use of airtight drums in Timor-Leste.

While the technical effectiveness of airtight storage is well established, Sheahan and Barrett concluded that the attractiveness of hermetic technologies (and all abatement technologies) depends on the benefits in relation to costs, both of which vary widely across different settings. Beyond calling for more rigorous research and analysis of postharvest loss interventions, Sheahan and Barrett argued for a broader strategic shift to include the role of less direct, off-farm investments, including infrastructure improvements (including roads and electrification, as introduced in chapter 10), warehouse receipt systems, rural finance development, and more efficient food value chains.[29] These investments will likely complement on-farm interventions to improve grain handling and storage.

In earlier chapters the case was made for investment in sustainable intensification and market infrastructure: use resources more efficiently for production and deliver food more efficiently to consumers. It would be illogical to embark on such objectives without a sharp eye on minimizing losses in quantity and quality at all steps along the food value chains. The process must begin with realistic assessments of physical and economic loss. A starting point is to seek estimates from comparable situations, where possible, complemented with locally collected data. Armed with this information, applied researchers should be assigned the task of identifying and assessing "best bet" technologies, just as the Millennium Villages Project and Seeds of Life project did. These interventions should be piloted and assessed with active involvement of the intended beneficiaries, be they farmers, traders, transporters, or processors.

Like all good R&D efforts, the first option piloted may falter for unexpected reasons. It is important, therefore, to get innovations out early in their development to test their effectiveness and acceptability under real-world conditions. The key recommendation here is that efforts to intensify production and improve market infrastructure must be accompanied by an evidence-based action plan to reduce postharvest losses.

Sheahan and Barrett observed that the growing interest in reducing postharvest losses, particularly in sub-Saharan Africa is "largely thought of as the counter-strategy to improving productivity."[30] If correct, that would be a seriously ill-conceived basis for action. Postharvest stewardship *complements* sustainable intensification and improved market infrastructure, and all three strategies are needed to achieve our food security objectives. Though the nature and relative importance of interventions will be context-specific, this approach, drawing on practical, field-tested technologies and policies, should be an integral component of investments aimed at improving food security.

WASTE NOT, WANT NOT

A different set of postharvest challenges occurs at the consumer end of the food value chain, including retail (but not wholesale), food service establishments (food consumed outside the home), and households.[31] Every day in every market and restaurant, and most households, edible food is discarded for a variety of reasons, including its deteriorating quality and appearance, anticipation or passing of "best by" (or "best before") dates, or simply the oversupply of products relative to consumer demand.

"Use by" dates on perishable foods serve an important health safety purpose if applied properly. Decisions, however, should be taken on a product-by-product basis and be supported by relevant evidence of associated health risks.[32] In best-case scenarios, in the absence of health risks, these discarded foods can be made available to needy consumers through food banks and other redistribution schemes. In some settings, the food is fed to animals or used for bioenergy. A distressingly large amount of food finds its way into sewers, streams, and landfills. In the United States, according to the Environmental Protection Agency (EPA), food waste is now the main constituent of landfills, making up 24 percent.[33] All these exit routes reduce the proportion of food that reaches the dinner table for consumption.

The UNEP *Food Waste Index Report 2021* estimated that 931 million metric tons of food waste were generated in 2019.[34] Of that amount, 61 percent emanated from households, 26 percent from food services, and 13 percent from retail. Although not entirely comparable, UNEP's estimate appears to be more than twice that of the earlier (2011) FAO estimate.[35] Surprisingly, per capita household food waste levels were broadly similar across countries ranging from lower-middle income to high income. This finding differs from earlier reports indicating food waste was significantly greater in higher-income countries.[36] It turns out that food waste is a universal challenge that needs to be addressed in every country.

Ultimately, consumer behavior determines the amount of food wasted, with the caveat that consumer behavior is shaped by marketing strategies and investments. Therefore, given that increasing the amount of discarded, unwanted food ultimately raises the price to consumers, why would a consumer behave in such an irrational way? There are a number of reasons:

- **Food is relatively cheap for high- and middle-income consumers.** For many of us, buying 20 percent more carrots than necessary may be a more desirable outcome than running out of carrots. Likewise, ordering an *extra-large* pizza is generally not a budget-crippling decision, especially if the price of a much smaller *large* pizza is often just 10 or 20 percent cheaper. Cheap food leads to poor planning of purchasing decisions and a higher probability of waste.

- **Consumers are concerned about deteriorated food taste and safety.** Better safe than sorry, and it is clearly easier to play safe if food is perceived as cheap. Coupled with this is blind adherence or confusion over food expiry labels. In the UK, a not-for-profit organization—the Waste and Resources Action Programme (WRAP)—estimated that, of the 7.3 million metric tons of household food wasted each year, 2 million metric tons are discarded for "not being used in time."[37] For a third of that food, date labeling is cited as a factor. On that basis, WRAP argues that giving consumers longer to make use of food, when safe, can significantly reduce household food waste.
- **Cultural and social impediments to leaving a clean plate, devoid of any leftovers.** In many cultural and social settings, the host may feel shame at not satiating the needs of the guest. On the other hand, guests may wish to avoid the perception that they are still hungry. In Filipino culture, there is even a derogatory expression *patay gutom* (literally: dead hungry) used for a person who has appeared to be so hungry that they finished everything on their plate. Both dynamics result in wasted food. Cultural and social tendencies to please guests—and even family and friends—often lead to excessive purchase, preparation, and waste of food. Ironically, leaving food on the plate can therefore be viewed as the right thing to do. Compare this with the public disapproval that would be directed to smoking in a no-smoking restaurant, or throwing trash out the window of a car.
- **Ignorance of the sources of food and the journey food has taken to reach the table: out of sight, out of mind.** As the world urbanizes, a smaller proportion of people understand the role of farming, the associated natural resource costs, the food preparation processes, and the destination of food waste. Better-informed citizens will make better decisions on managing their food resources.

Over the past fifteen years there has been a surge of interest and action to reduce the amount of food that is wasted at the retail and consumption stages of the value chain. These efforts have been well supported by governments, business, NGOs, and the public in higher-income countries, most notably Europe.

In the UK, impressive progress has been made through the Courtauld Commitments, a series of voluntary agreements aimed at reducing waste and environmental stewardship across the nation's grocery sector.[38] Courtauld 2030 is a commitment to cut per capita food waste by 50 percent by 2030 using the UK 2007 baseline. With encouragement and financial support from the UK government, WRAP has facilitated sector-wide voluntary agreements that have led to significant reductions in food waste. Around 250 UK businesses have signed up to a commitment "to 'Target, Measure, Act' on food waste, and for public reporting of food waste to become the norm."[39] This public–private partnership is supported by a national consumer awareness campaign launched by WRAP in 2007: Love Food Hate Waste.[40] The approach was to raise awareness on the need to reduce food waste and to promote and support practical action.

For the consumer, the Love Food Hate Waste campaign recommends a practical strategy for individuals who are committed to taking personal action:

- **Plan your buying.** Go with a list based on realistic your meal expectations. Avoid overbuying that may lead to disposal.
- **Make your food last longer.** Use your refrigerator and freezer wisely, and do not take "best by" labels too literally.
- **Make your food go further.** Use leftover food.

These are important actions that can apply at the individual, household, restaurant, and institutional levels. Enlightened individuals and leaders who recognize the reasons for reducing food waste will likely follow them. In a review of household waste practices and policies, drawn mainly from European experience, however, Karin Schanes, Karin Dobernig, and Burcu Gözet argue for a holistic food waste prevention approach that goes beyond putting the responsibility solely on individuals.[41]

Love Food Hate Waste's mission—the end-game—is clear and simple: keep it out of the bin. For some producers and consumers, this is a smart, strategic money-saving approach. For those fortunate enough to have food costs a relatively small part of their weekly budget, other motivations and behavioral nudges may be required—one such would be greater shared awareness of the environmental impacts of food waste. Peer pressure and intergenerational pressure may be important levers to tap as we search for practical ways of reducing the almost one billion metric tons of food we waste every year.

TAKING ACTION

The release of FAO's report in 2011 on food losses and waste reinforced for me the need to place greater emphasis on postharvest stewardship in any food security strategy, be it global, national, or local. I introduced the topic into my Global

Food Systems course at Columbia University in the spring of 2012. Always a little anxious when teaching a new topic—did I have enough material to last a two-hour session?—I was delighted to find how well the topic resonated with my students. I would now rank it at least on a par with GMOs in terms of interest level. In contrast to our annual GMO debate, opinions converged strongly on the need to take action, much of the enthusiasm derived from personal experience. Just as I felt after seeing Malawi's maize harvest being reduced to powder by the larger grain borer, my students felt a sense of outrage that we were not doing more to protect and utilize our precious harvest.

Major synthesis reports published in 2019 by the EAT–*Lancet* Commission, World Resources Institute (WRI), IPCC, and FAO and UNEP in 2021 leave little room for complacency.[42] One of these initiatives, led by WRI with support from the Rockefeller Foundation and in collaboration with food loss and waste experts, synthesizes a useful set of "scaling interventions" that were designed to accelerate the adoption of policies and practices to cut food loss and waste by 50 percent.[43] These interventions, which I have paraphrased, adapted, distilled, and augmented, provide a useful checklist of seven action items for policy makers and development specialists committed to advancing postharvest stewardship:

- **Develop and implement national and local food stewardship strategies.** Strategies should not be limited to government jurisdictions. Private companies, schools and colleges, and other entities should mobilize action around locally owned and executed strategies. Ideally, these strategies should be embedded within broader sustainable development, food security, and climate action initiatives to ensure wider ownership and support.
- **Utilize global, national, and local public–private partnerships.** Postharvest stewardship is a clear case where there are opportunities for both public and private benefits from investment. Where feasible, companies should make a business case, as illustrated in the Courtauld Commitment 2030. But, some benefits are longer term and may require complementary public-sector support in the form of infrastructure and policies.
- **Act across the entire value chain to identify priority points of opportunity to reduce loss and waste.** Identify and address exit points from production to consumption. We know that there are large variations in loss and waste across different commodities in different regions and that virtually all commodity value chains offer opportunities for cutting losses and waste.
- **Mobilize entrepreneurs through education, mentorship, and challenge grants to create sustainable, for-profit enterprises to support postharvest stewardship.** These initiatives should tap digital technologies and

behavioral sciences for maximum impact. Universities and colleges are promising laboratories for generating and testing creative solutions.

- **Shift cultural and social norms through public education and awareness.** Experience from other campaigns on recycling garbage, conserving energy and water, stopping drink driving, family planning, and combating HIV/AIDS tells us it is possible to reshape long-held practices. A behavioral change campaign that values and rewards food waste reduction is needed for each social and cultural setting. Connecting consumers to food production through urban agriculture and farmers' markets are examples. Another important prospect is teaching and encouraging more people to cook.
- **Establish special financing windows and adapt existing mechanisms for supporting postharvest stewardship.** Climate finance is growing in importance and should be adapted and tapped for postharvest initiatives that reduce greenhouse gas emissions. Funds like the Global Agriculture and Food Security Program have focused mainly on production and can be easily adapted to include and encourage postharvest interventions.[44]
- **Advance research and development on postharvest stewardship, including improved quantification of losses and waste.** There is likely a role for digital technologies that can connect consumers to low-cost, about-to-expire food from retail and service establishments, and smart food containers can be fitted with a digital tag that allows them to track food freshness.[45] Genetic engineering may also contribute through products like the Arctic apple, which has been genetically edited to slow deterioration after slicing or bruising.[46]

Throughout this chapter I have offered examples of interventions—from community-managed warehouses to recycled oil drums to voluntary corporate reporting to the Arctic apple—that have worked, or have potential, to reduce postharvest losses and waste in particular settings. Thousands of technical interventions are waiting to be discovered, adapted, tested, and disseminated. Establishing SDG 12.3 was a significant step toward improving postharvest stewardship. The knowledge exists to meet that goal. If we are to move beyond concern and outrage, however, it is important to elevate this agenda to *essential* for action in achieving universal food security and combating climate change.

12

HEALTHY DIETS

On July 12, 2011, at a convention center in Dili, Timor-Leste's national capital, Prime Minister Xanana Gusmão launched the nation's Strategic Development Plan 2011–2030 (SDP).[1] Just nine years earlier, the United Nations had formally declared Timor-Leste (then East Timor) an independent country, following more than four centuries of colonial rule by Portugal, twenty-four years of occupation by Indonesia, and two and a half years of United Nations administration, the latter steering the country to its first national election.

Outlining the strategy's historical context and development priorities, Gusmão acknowledged that the SDP was "an ambitious plan and that the challenges that are inherent to its implementation are enormous," but the former guerrilla leader and national hero had "no doubts that this is one more battle we can win."[2] With substantial oil and gas reserves and strong national pride, Timor-Leste was ready to transition to a new phase of peace and development. The SDP was launched as a plan that "reflects the aspirations of the Timorese people to create a prosperous and strong nation," further declaring that "the true wealth of any nation is in the strength of its people. Maximizing the overall health, education and quality of life of the Timorese people is central to building a fair and progressive nation."[3]

Seated among the assembled government officials, nongovernmental organizations, and development partners, I listened to several high-level presentations outlining and endorsing a strategy that would accelerate national development by investing in social capital (principally education and health), hard infrastructure, and economic diversification across three main economic sectors: petroleum, tourism, and agriculture. Speaking to my own biases and priorities, the strategy contained encouragingly strong commitments to building roads and modernizing agriculture. These explicit investments would surely lead Timor-Leste to a new era of prosperity:

> Our vision for 2030 is that Timor-Leste will have a modern diversified economy, with high quality infrastructure including roads, power, ports and telecommunications. Subsistence agriculture will have been replaced by commercial, smallholder agriculture. Timor-Leste will be self-sufficient in food and be producing a range of agricultural products for world markets including staples, livestock, fruit and vegetables and other cash crops, as well as forestry and fisheries products.[4]

But one aspect of the plan bothered me. Deep into the chapter on health, there was a section headed Nutrition—just 224 words within the 227-page strategy.[5] The word *nutrition* (or *malnutrition*) was mentioned sixteen times in the entire document. I cannot recall nutrition being mentioned at the SDP launch; it was certainly not given any prominence. But within the Nutrition section were some chilling statistics: 58 percent of children under the age of five years were stunted, and 33 percent were severely stunted. I had earlier stumbled across these numbers in the Timor-Leste Demographic and Health Survey 2009–10 (DHS), which also reported that almost 40 percent of children under five and 22 percent of women aged fifteen to forty-nine were anemic.[6]

I shared my findings with Jessica Fanzo, whom I introduced in chapter 7. Fanzo confirmed that these numbers were "off the charts": among the worst, if not *the* worst, stunting levels on a list that included conflict- and drought-affected countries like Afghanistan, Yemen, and Burundi.[7] The rhetorical question I asked was simple: How could Timor-Leste possibly achieve its vision of becoming "a peaceful and prosperous nation by 2030" when 58 percent of its children under five were stunted and, as a consequence, suffering irreversible damage to their brains, impairing their health, education, and employment prospects?[8] Of course, the answer was and remains: *it could not*. No country can sustain peace and achieve prosperity with those levels of undernutrition. From that time, Fanzo and I have committed to supporting the people of Timor-Leste and the agencies working there to find ways to reduce the horrendous malnutrition burden.[9]

In the years following the SDP launch, Fanzo undertook several review and advisory missions to Timor-Leste with Columbia University, Johns Hopkins University, the Seeds of Life project, and the World Food Programme. In a report to the Seeds of Life project in 2013, she and her colleagues drew attention to the lack of dietary diversity.[10] Timorese people consumed a lot of starch—especially rice, maize, cassava, and sweet potatoes. They ate little meat and surprisingly little fish for a half-island nation. Nutrient-dense vegetables were also rarely consumed. As more than 70 percent of the population were subsistence farmers and their dependents, what they produced, by and large, determined what they consumed. Moreover, the meager incomes of these farmers severely curtailed their purchases of more nutritious foods.

In response to Fanzo's initial studies and a series of subsequent research and advocacy efforts to elevate the nutrition issue in Timor-Leste, the attention given to nutrition and diet in the country has increased. This appears to be paying off as the Timor-Leste Food and Nutrition Survey in 2020 showed stunting levels were down to 47 percent. Data from the 2020 survey revealed that stunting levels are correlated with income poverty. Prevalence of stunting in the lowest quintile was 57 percent, compared to 35 percent in the highest wealth quintile. While a reduction in overall stunting prevalence of 11 percentage points since 2010 is encouraging, it is clearly not enough—Timor-Leste continues to have one of the highest stunting rates in the world.[11]

Pointing to a possible contributing factor to the high stunting rate, research from the 2020 TLFNS found that 86 percent of children at the age of six to twenty-three months did not consume a minimum acceptable diet, and 65 percent did not consume the minimum dietary diversity (see chapter 9).[12] Likewise, adults had poor dietary diversity, and Fanzo and others have advocated for a food-based approach to nutrition improvement with increased consumption of legumes, orange- and purple-fleshed sweet potatoes, vegetables, fruits, livestock products, and seafood, as well as micronutrient fortification of rice and cooking oil.[13]

It is impossible to tell how much of the modest stunting reduction was the result of dietary improvements or other investments that less directly improve nutrition, such as clean water and sanitation, improved childcare practices, and more effective health care services. Whatever the contribution of diet, it is indisputable that most Timorese are not consuming the right mix of foods—quantity and quality—for a healthy and productive life. If we are to accept that "the true wealth of any nation is in the strength of its people," as declared by Xanana Gusmão in 2011, increased investment to improve nutrition is needed. My exposure to Timor-Leste since 2010 convinced me that a prosperous nation must be a well-nourished nation.[14] In the most fundamental terms, the question is: How can we deliver the missing nutrients?

PILLS, POWDERS, AND PASTES

Over the years, Fanzo and others have reminded me that nutrition and diets are not the same. As explained in chapter 7, understanding diets provides important insights on nutrition, while understanding nutrition provides the foundation of a healthy diet. The United Nations Children's Fund (UNICEF) framework on the determinants of maternal and child undernutrition highlights the complexity of the interacting factors involved (see figure 7.1). In 2013 the *Lancet* published an acclaimed series on maternal and child nutrition. In what quickly emerged as one of the most widely cited and most influential papers in the field of international nutrition, Professor Zulfiqar Bhutta and colleagues identified ten *direct* nutrition interventions that could save the lives of nearly one million children under the age of five years. These ten interventions were organized in four packages as follows:[15]

1. Optimal maternal nutrition during pregnancy
    - multiple micronutrient supplementation
    - calcium supplementation
    - balanced energy-protein supplementation
    - universal salt iodization
2. Infant and young child feeding
    - early and exclusive breastfeeding to six months, continuing up to twenty-four months
    - complementary feeding, including education and supplements
3. Micronutrient supplementation in children at risk
    - vitamin A supplementation between six and fifty-nine months
    - zinc supplementation between twelve and fifty-nine months
4. Management of acute malnutrition[16]
    - moderate acute malnutrition (MAM) management
    - severe acute malnutrition (SAM) management

By scaling these interventions to 90 percent population coverage, deaths of under-fives could be reduced by 15 percent, stunting could be cut by 20 percent, and severe wasting could be reduced by 61 percent. The *Lancet* study highlighted the importance of community-based delivery strategies whereby community health workers are deployed to mobilize service utilization and support behavioral change at the local level. The estimated additional cost of extending these ten interventions to 90 percent coverage in the study's thirty-four focus countries was $9.6 billion per year (in 2010 dollars), no more than a rounding error in the context of trillions of dollars spent on the COVID-19 economic recovery. The study estimated that this level of investment would save the lives of almost one million children under the age of five each year. Almost half of those lives would be saved as a result of scaling up the therapeutic feeding for severe acute malnutrition.[17]

Unwillingness to invest at this level—less than $10 billion per year—to achieve such important lifesaving and life-changing outcomes is cause for reflection. Why not invest in what appears to be a relatively achievable target with known solutions? I do not have *the* answer, but I suspect there are three possibilities. First, maternal and child undernutrition causing death, stunting, and wasting is most prevalent in the poorest nations, where national capacity is limited and financing health interventions depends heavily on external aid. Second, except in the most extreme cases, such as during droughts and conflict, maternal and child malnutrition is not visible to the public and its leaders. Resources get directed to emergencies where lives and livelihoods are threatened in the short term. But this attention and funding comes at the expense of long-term investments needed to reduce malnutrition. Third, unlike lifesaving food aid in times of disaster, the responses needed to bring about meaningful reductions in stunting are complex and long term, as I shared in chapter 7. One cannot expect short-term impact, thus there is less political imperative to act—a clear failure of true leadership.

A 20 percent reduction in stunting—roughly what we observed in Timor-Leste between 2010 and 2020—is impressive but still not acceptable. These direct nutrition interventions tackle just one subset of the larger set of determinants of undernutrition. Why is it that stunting is near to nonexistent in high-income countries? It is not simply because those most rich nations have achieved complete coverage of these direct interventions. Bhutta and coauthors recognized that the ten interventions must be linked to less direct investments known as *nutrition-sensitive* approaches.[18] These include agriculture, food systems, education, employment, women's empowerment, and social protection—all of which accelerate progress in reducing undernutrition and premature deaths of women and children.

In closing the undernutrition gap, it is critical that we deploy a broader strategy of investments that includes food-based solutions. A major reason why these direct interventions are needed is that populations do not consume healthy diets. In most situations, Fanzo explained to me, if you eat a healthy, balanced diet, you will not need supplements, with the possible exception of vitamin D in sunshine-deprived environments. But the reasons for not consuming a healthy diet are complex and varied and are ultimately tied to the central pillars of food security described earlier in the book: availability, access, utilization, and stability. Fanzo explained that diets have been neglected as a part of nutrition in favor of what is known as the "pills, powders, and pastes" approach. In 2001 Marie Ruel and Carol Levin observed that food-based approaches were perceived to be costly and time-consuming and "have often been overshadowed by national campaigns to reduce micronutrient malnutrition through capsule-distribution and food-fortification programs."[19]

Notably, only two of the ten direct interventions—early and exclusive breast-feeding, and complementary feeding of infants—fall outside the "pills, powders, and pastes" umbrella. Fanzo, however, believes that the importance of a healthy diet, and more specifically dietary diversity, is now emerging as a more potent influence on nutrition policy and investments. This is all the more important with the rise of overnutrition and noncommunicable diseases, for which effective "pills, powders, and pastes" are yet to be developed. For the balance of this chapter, we will examine strategies for achieving good nutrition principally through a healthy diet.

FOOD-BASED APPROACHES

Food-based approaches, also known as dietary modification, have been defined as deploying "a combination of agricultural, educational, and nutritional strategies to increase the production of, availability of, access to, and consumption of micronutrient-rich foods."[20] Recognizing the implementation and sustainability challenges of filling nutrient gaps through supplementation, there has been increasing emphasis on food-based approaches that include dietary diversification and food fortification, including biofortification.[21]

The Global Alliance for Improved Nutrition (GAIN) is a Geneva-based foundation launched at the United Nations in 2002 to tackle malnutrition through partnerships with governments, business, and civil society. GAIN's distinctive contribution has been mobilizing private-sector investment for improving nutrition in low- and middle-income countries. In nutrition circles, GAIN is well known for its support of food fortification: the practice of deliberately increasing the nutritional content of a widely consumed food. Fortification can be done at the point of cultivation (biofortification, e.g., Golden Rice, discussed in chapter 5), at the point of food processing (industrial fortification, e.g., wheat flour enriched with essential vitamins and minerals), or at the point of consumption (home fortification, e.g., micronutrient powders added to home-prepared foods).[22]

The outcome of a successful food-based approach is a healthy diet practiced by all. As introduced in chapter 7, our bodies require appropriate levels of essential nutrients to live a healthy, productive, and long life. Our food systems—extending to the way we produce, distribute, acquire, and consume food—need to be reimagined and redesigned to deliver healthy and sustainable diets for all people at all times.

The best place to begin this reimagining and redesign is the food environment. Drawing on many years of deliberation on concepts and definitions, Christopher Turner and coauthors proposed the following definition: "The food environment is the interface that mediates people's food acquisition and consumption within the wider food system." They elaborate further: "It encompasses external dimensions such as the availability, prices, vendor and product properties, and promotional information; and personal dimensions such as the accessibility, affordability, convenience and desirability of food sources and products."[23]

The food environment is the nexus for transactions between food producers and consumers. Figure 12.1 is my adaptation and simplification of the Innocenti Framework on food systems that was developed by Jessica Fanzo and others.[24] While the Innocenti Framework was developed with a focus on children and adolescents, figure 12.1 aims to include the wider population of food consumers. I have also incorporated elements of the conceptual framework of food systems developed by the High Level Panel of Experts of the UN Committee on World Food Security.[25]

Consumers obtain food in a variety of external food environments and are subject to influences, such as marketing and regulations, that shape the range of possible decisions on what to acquire and consume. Corporate and institutional values also influence the external food environment. In this regard, the level of commitment to healthy and sustainable food varies widely across different commercial and institutional settings.

At the same time, consumers are subjected to a range of internal factors, like knowledge, income level, culture, convenience, and personal values that constitute the internal food environment and further influence and limit the options in making the transaction. For example, gender discrimination can be a powerful influence on household consumption patterns and intrahousehold differences in food security.[26]

The internal and external food environments combine to shape consumer decisions, which in turn determine the quantity and quality of diets. Those dietary decisions by consumers lead to nutrition and health outcomes and, ultimately, to social, economic, and environmental impacts. Peter Timmer called for a better understanding of decision-making in response to policies and market forces: "There is much that individual, informed consumers can do to shape their diets and consequent health outcomes, but it is also true that nutritionally

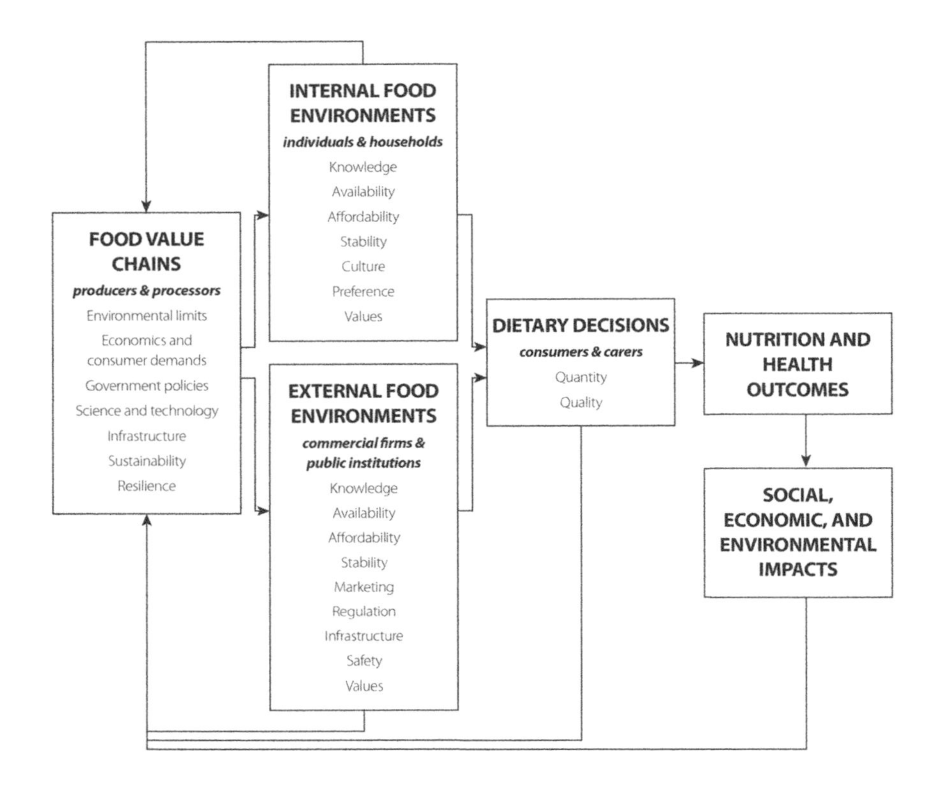

FIGURE 12.1 Food environments at the nexus of food production and consumption.

Source: This illustration combines elements of two earlier depictions of food environments: United Nations Children's Fund Office of Research–Innocenti, *Food Systems for Children and Adolescents: Working Together to Secure Nutritious Diets* (New York: UNICEF, GAIN, 2018), https://www.unicef.org/media/94086/file/Food-systems-brochure.pdf; and High Level Panel of Experts, *Nutrition and Food Systems*, a report by the High Level Panel of Experts on Food Security and Nutrition (Rome: Committee on World Food Security, 2017), http://www.fao.org/3/i7846e/i7846e.pdf.

informed decisions can be very hard to make in today's commercial environment."[27] Understanding the influences at play in both the internal and external food environments, as well as in food value chains, is necessary in identifying policies and programs to achieve healthy diets.

Governments aim to inform consumers through food-based dietary guidelines (FBDGs), described by Anna Herforth and colleagues as "an attempt to translate a vast (and always incomplete) evidence base regarding relations between foods, diet patterns, and health into specific, culturally appropriate, and actionable recommendations." FBDGs have been established in at least ninety countries with the intent to influence consumer behavior and guide diet-related policies and programs. In reviewing these FBDGs, Herforth and colleagues found a high degree of consistency in recommendations across countries in terms of (1) consuming a diversity of foods, (2) consuming abundant fruits and vegetables, (3) including starchy staples, legumes, and animal-sourced foods, and (4) avoiding excessive salt, sugar, and fat. While FBDGs are broadly aligned to the World Health Organization global guidelines, there was less consistency in advice regarding the specific levels of red meat and dairy, and messages related to fats and oils were often unclear. Herforth and colleagues recommended that FGBDs give greater attention to the environmental impacts of diets and to the rapid global dietary transition to ultraprocessed foods.[28]

National dietary guidelines are not immune to the influence of powerful lobby groups and special interests. In a stinging criticism of America's "food pyramid," renowned Harvard nutrition professor Walter Willett and coauthor Patrick Skerrett noted that these guidelines come from the U.S. Department of Agriculture, the agency responsible for promoting American agriculture, not from organizations that monitor and protect public health.[29]

Like many developing countries, Timor-Leste is undergoing the nutrition transition introduced in chapter 7 as a pattern of changing diets associated with economic growth and urbanization.[30] The "Timor-Leste Food and Nutrition Survey" (2020) pointed to high consumption of sugary drinks and foods in both urban and rural areas and across all wealth quintiles.[31] In sub-Saharan Africa, unhealthy dietary patterns are also emerging beyond urban areas with improvements in infrastructure.[32] Road investments and electrification are lowering transaction costs for processors and wholesalers and are contributing to growing consumption of processed foods in rural areas and in small and medium cities.

In the Asia–Pacific region, food consumption patterns have changed in recent decades: consumption of starchy staples has declined; consumption of animal-source foods, fruits, and vegetables has increased; and consumption of ultraprocessed foods, containing high amounts of salt, sugar, and saturated fats, has increased.[33] When coupled with reduced physical activity due to urbanization and related work patterns, these trends have led to high rates of overweight and obesity, leading to increased noncommunicable diseases (NCDs).

So what exactly is a "healthy diet"? Is there such a diet? And how can we ensure that a diet is healthy for people *and* healthy for the planet?

THE PLANETARY HEALTH DIET

"Food is the single strongest lever to optimize human health and environmental sustainability on Earth." So declared the EAT–*Lancet* Commission in launching its report *Food in the Anthropocene: The EAT–Lancet Commission on Healthy Diets from Sustainable Food Systems* in January 2019.[34] The rationale for the commission was that the absence of science-based dietary targets has hindered efforts to transform the global food system. Clearly, if we do not know what a healthy and sustainable diet looks like, how can we take meaningful actions to achieve it? How can we convince consumers to change their behavior? How can we design evidence-based national dietary guidelines? How can we convince governments to invest and to introduce policies that would support the transformation? And how can we influence businesses to develop and market products that lead to healthy outcomes for people and planet?

Cochaired by Walter Willett and Johan Rockström, director of the Potsdam Institute for Climate Impact Research in Germany, the commission convened thirty-seven leading scientists from sixteen countries, in the disciplines of health, agriculture, political science, and environment, with the objective of developing global scientific targets for healthy diets and sustainable food production. Its intent was that these targets would apply "to all people and the planet."[35] The detailed analysis from which the commission drew conclusions was published in the *Lancet*, while the summary report, produced by EAT, contains the key findings and recommendations.[36] In this chapter I focus on the health-related dietary aspects, while the core environmental conclusions and recommended strategies are addressed in chapters 9 and 11.

Based on evidence from controlled-feeding studies, long-term observational studies, and randomized clinical trials, the commission concluded with a high level of certainty that a diet with the following characteristics—the *planetary health diet*—would promote "low risk of major chronic disease and overall well-being." To avoid errors of interpretation, I quote these dietary characteristics directly from the *Lancet* article:

> (1) protein sources primarily from plants, including soy foods, other legumes, and nuts, fish or alternative sources of omega-3 fatty acids several times per week with optional modest consumption of poultry and eggs, and low intakes of red meat, if any, especially processed meat;
>
> (2) fat mostly from unsaturated plant sources, with low intakes of saturated fats, and no partly hydrogenated oils;
>
> (3) carbohydrates primarily from whole grains with low intake of refined grains and less than 5% of energy from sugar;
>
> (4) at least five servings of fruits and vegetables per day, not including potatoes; and
>
> (5) moderate dairy consumption as an option.[37]

The EAT–*Lancet* Commission also referred to this planetary health diet as a *flexitarian diet* because it is compatible with a wide variety of foods, agricultural systems, cultural traditions, and dietary preferences. Ranges of recommended intakes across different food groups were established. The commission argued that such a diet is consistent with many traditional eating patterns, such as the Mediterranean diet. The flexitarian diet can be adapted to a range of intake preferences, including vegan, vegetarian, and omnivore variations. Contrary to the early, sometimes vitriolic critiques of the commission's recommendations, there is no prescribed diet for all the world to follow. Instead, the commission encouraged local interpretation and adaptation of the guidelines to reflect the cultures, geographies, and age distributions of populations and individuals.

As a nonspecialist in the field of nutrition, I found the commission's report to be a valuable synthesis of the science that should drive policies and practices toward the universal adoption of healthy and sustainable diets. I nevertheless highlight the following questions that warrant further consideration and clarification by the commission, EAT, and other organizations seeking impact on food security through dietary changes. To be fair, the report openly acknowledges some of the following limitations:

Diets by demography. The guidelines do not explicitly address the needs of children under two years of age, who have specific requirements for rapid growth and development, including the need for exclusive breastfeeding and complementary feeding regimes. Likewise, adaptation of the guidelines to the different needs of adolescent girls, pregnant and lactating women, and older people would need further consideration.[38] For these vulnerable groups, in particular, we should look to the ten direct nutrition interventions described earlier and the "double duty actions" that are introduced in the next section.

Salt and alcohol. Although added sugar is explicitly addressed, there is no direct mention of added salt (though implied by the restriction on processed food) and no guidance on the impact of alcohol on caloric intake. Both salt and alcohol are prominent ingredients of many traditional and modern diets and cannot be ignored when specifying optimal food intake guidelines. The commission also acknowledges that it did not consider food safety in the guidelines.

Physical activity. There is no direct reference to the role of physical work or elective physical exercise as a consideration in adjusting the overall caloric intake. There is ample evidence that physical activity can in part offset high caloric intake that results in overweight and obesity. Increased urbanization, changes in transport modes, and more sedentary work behavior have been part of the problem that created an energy imbalance between calories consumed and calories expended.[39] Therefore a healthy diet must be considered in the context of physical activity.

Trade-offs and synergies with environment. The commission's dietary guidelines are based entirely on health considerations. These targets were then assessed in the context of sustainable food production based on a set of environmental limits (known as "planetary boundaries").[40] The commission applies these two filters sequentially rather than iteratively. It turns out that excessive consumption of red meat and carbs are bad for human health and can be bad for the environment. For some of the other food groups (e.g., seafood), the synergies and trade-offs between health and environment are less clear and are context-specific. As such, consumers and policy makers will be challenged to make evidence-based decisions that optimize the objectives of human health and environment.

Cost of diets. The commission's dietary guidelines did not consider food affordability. At the outset, the focus was on health considerations, arguing that it is "the job of societies and policy makers to enable people to afford healthy, sustainable diets."[41] Correct, of course, but a brave assumption. Using food price and household income data to estimate affordability of the EAT–*Lancet* target diets, researchers found a global median cost of US$2.84 per day (2011 data), which exceeded the household per capita income of at least 1.58 billion people.[42] The largest share was the cost of fruits and vegetables (31 percent), followed by legumes and nuts (19 percent), meat, eggs, and fish (15 percent), and dairy (13 percent).

A further concern I had—from a practitioner's standpoint—is that the wide ranges of recommended intake levels will be extraordinarily difficult to translate into actionable guidelines for governments, consumers, and other stakeholders.[43] An important next step is to understand the demographic, geographical, and sociocultural patterns of undernutrition and overnutrition. From that information, a methodology could be developed and deployed to enable practitioners, using locally available food sources, to fill the nutrient gaps in intake deficiencies and to remove the intake excesses. This adaptation effort needs to be undertaken at the appropriate national and local levels.

Accepting these points as areas for further work and improvement, the EAT–*Lancet* Commission has provided policy makers, private investors, and individuals with a high-quality evidence base for immediate action toward healthy and sustainable diets. Shifting to the commission's planetary health diet along the lines described would prevent eleven million deaths per year, based on three different methodologies applied by the commission. Tens of millions more would live healthier and more productive lives with the adoption of the commission's dietary advice.

On an aggregate global scale, the planetary health diet, if implemented at scale, will require radical changes in the food system: a doubling in consumption of healthy foods like fruits, vegetables, legumes, and nuts, and a greater than 50 percent reduction in less healthy foods like red meat and added sugars. At the same time, the commission recognizes that local context may justify localized increases in consumption of animal-source foods. As one of its five strategies for transformation, the commission calls for greater national and international commitment to increase the availability and accessibility of healthy foods. It calls for a range of policies and investments in marketing, education, and service provision to promote healthier diets.

ACTION ON OBESITY

According to the *2021 Global Nutrition Report* (GNR), no country has managed to reverse the rising overweight and obesity trend. *No* country. Unlike in the fight to reverse stunting, we have no success stories—at a national scale—from which to draw inspiration. The GNR reported that 2.2 billion adults were overweight and that 190 countries (all of those for which data exist) are considered off course for halting the growth in obesity. About one-third of overweight adults are obese (a body mass index over 30).[44] Overweight, obesity, and diet-related NCDs remain the dominant malnutrition issue in wealthier countries, but as described earlier, low- and middle-income countries are rapidly transitioning toward a double burden of undernutrition and overnutrition.

Data reported by the NCD Risk Factor Collaboration has shown, contrary to popular perceptions, that BMI is increasing at the same rate or faster in *rural* areas than in cities in low- and middle-income regions, resulting in a rural–urban convergence in BMI, especially for women.[45] The authors suggested that this trend reflects the disproportionate improvements in incomes and food expenditures in rural areas, along with the spread of processed food through enhanced access and operations of national and transnational businesses. Overnutrition should no longer be narrowly viewed as an urban issue in the rich world. The authors urged national and international policy makers to enhance access of healthy foods to both poor rural and urban communities and "to avoid replacing the rural undernutrition disadvantage in poor countries with a more general malnutrition disadvantage that entails excessive consumption of low-quality calories."[46]

There are now several recent evidence-based reports offering guidance on how to reverse the alarming trend of increasing overweight and obesity. The most useful among these are the Eat–*Lancet* Commission report (2019), the World Bank report on obesity (2020), and the *2021 Global Nutrition Report*—each of which applied a different lens to the challenge.[47]

As outlined earlier, the EAT–*Lancet* Commission places strong emphasis on the need for dietary change. The World Bank report on obesity focuses attention on the impacts of overweight and obesity on building human capital, ensuring higher economic growth, and sustaining a productive workforce. The *Global Nutrition Report* highlights the need to understand and correct nutrition inequalities within countries and communities to end malnutrition in all its forms. I have reviewed the recommendations of these reports and encourage readers to explore the impressive assemblage of analysis and advice contained in each. Unsurprisingly, there is considerable overlap and convergence of strategies to combat overweight and obesity, as well as some special insights into the causes, consequences, and solutions of what is increasingly viewed with strong justification as a pandemic.

My key policy takeaways from these three reports are as follows:

- **Nothing short of a great transformation is needed to reverse current trends.** We must acknowledge that overweight and obesity—overnutrition and the resulting increase in NCDs—is one of the greatest social and economic challenges of our time.
- **No single agency can create that transformation.** There must be multisector, multi-institutional, and multilevel action, spanning the public and private sectors, sharing a common goal, deploying coordination, cooperation, and partnership, and taking collective responsibility and accountability for outcomes.
- **A wealth of promising policy options and interventions exist**, ranging from "soft options" like consumer advice, information, and education, including food labeling, to "hard options" that include legislation, regulations on marketing and advertising, fiscal measures such as subsidies and taxes, and trade policies.
- **Increased investment and new financing mechanisms are needed for scaling up** these promising policies and interventions, and supporting policy and institutional reforms through multisectoral and institutional engagement.

- **We do not have all the answers. More interdisciplinary research is needed** to provide an evidence base for identifying cost-effective solutions, setting priorities, assessing performance, and supporting innovation and greater impact.
- **Actions are required across entire value chains from producers to consumers.** But effort must be carefully targeted toward communities and people most affected and potentially most influential.
- **Tough, uncompromising actions are needed to harness the potential of business and curb the damage the business sector causes**, especially directed toward multinational food, beverage, and agribusiness firms that influence billions of people through marketing and advertising.

The final item on my list is "the elephant in the room." While each report speaks to the need for regulations and accountability for private actors, we cannot ignore the powerful social, economic, and political influence of the private sector, especially the mega-firms and multinationals. For example, Coca-Cola and Pepsico, the two largest soft drink manufacturers, have each spent about US$4 billion annually in advertising their products worldwide since 2015.[48] To put it bluntly, it is time for governments, civil society, and the public at large to stand up and call businesses *out* for their excesses and simultaneously call them *in* as partners in combating overweight and obesity.

DOUBLE-DUTY ACTIONS

In 2018 the World Food Programme invited me to support the development of its new five-year strategy in Tajikistan. This landlocked, mountainous former Soviet republic had the highest stunting rate in Central Asia. While official surveys pointed to sharp decreases in stunting since 2000, including an impressive decline from 29 percent in 2012 to 17 percent in 2017, the fact remained that one in six children in Tajikistan would suffer cognitive and physical impairment, leading irreversibly to reduced educational performance and economic productivity.[49] Tajikistan's National Development Strategy for 2016–2030 identified a strategic goal to "expand productive employment" and achieve upper middle-income status.[50] However, like Timor-Leste's vision for 2030, described earlier in the chapter, this worthy ambition would likely be undermined by high levels of undernourishment.

The root causes of the high stunting rate in Tajikistan have not been definitively established. The Zero Hunger Country Strategic Review and the Tajikistan Fill the Nutrient Gap (FNG) study, however, both pointed to the critical importance of diets for good health and well-being; the required nutrient intake and the body's utilization of those nutrients; and the likely importance of essential micronutrients, such as iron, vitamin A, and iodine.[51] The FNG report concluded that the current diet of people in Tajikistan included too much "empty calorie foods," with wheat contributing 50 to 70 percent of caloric intake and corresponding low consumption of nutrient-dense foods such as vegetables, fruits, dairy, and other animal-sourced products. While protein and zinc intake were not explicitly addressed in these studies, deficiencies of these two nutrients are candidates as contributors to the high stunting rates.

As told in jest many times during my visits to the country, "We Tajiks eat bread with bread." The high consumption of empty calories has contributed to growing prevalence of overweight, obesity, and NCDs. Nationally, 38 percent of women (aged fifteen to forty-nine) in Tajikistan are overweight. In forty-one- to forty-nine-year-olds, 61 percent of women are overweight. Prevalence is higher in urban than rural areas: 41 percent versus 32 percent. This double burden of undernutrition and overnutrition—introduced in chapter 7—presents a severe challenge to the government and the international development agencies that are supporting the country. Historically, actions to address undernutrition and overnutrition have been undertaken through distinct policies, programs, funding streams, and organizations. To address this challenge, Professor Corinna Hawkes and colleagues have proposed a new *double-duty actions* approach as part of another series from the *Lancet* on the double burden of malnutrition, published in late 2019.[52]

In contrast to previous approaches, double-duty actions simultaneously and holistically tackle the full spectrum of malnutrition problems that are present in dozens of countries around the world. The authors note that these different manifestations of poor nutrition are related to early-life nutrition, diet diversity, food environments, and socioeconomic factors. They further present evidence that actions to address undernutrition, such as the use of energy-dense micronutrient-fortified foods and supplements, have sometimes had unintended negative consequences on overnutrition and NCDs. These outcomes are a major concern in low- and middle-income countries, and in both rural and urban settings, where food environments and dietary patterns are rapidly changing.[53]

Based on the evidence, Hawkes and colleagues identified ten priority opportunities for double-duty actions across four sectors, summarized and paraphrased as follows:

1. Health services
   - Scale up WHO's antenatal care recommendations, including counseling on healthy eating and physical exercise, as well as supplemental feeding where appropriate.
   - Scale up programs to protect, promote, and support exclusive breastfeeding and eliminate the promotion of breastmilk substitutes.
   - Redesign guidance for complementary feeding practices to ensure greater awareness of the risks of foods, snacks, and beverages that are high in sugar, fat, and salt.
   - Redesign existing growth monitoring programs to account for the growing problem of overweight children.
   - Prevent undue harm from energy-dense and micronutrient-fortified foods and ready-to-use supplements.
2. Social protection
   - Redesign cash and food transfers, subsidies, and vouchers to include strong education and communication on healthy diets, physical activity, and preventive use of health services (to be elaborated on in chapter 13).
3. Education
   - Redesign school feeding programs and devise new nutritional guidelines for food in and around educational institutions to encourage and support consumption of healthy nutritious foods.
4. Agriculture and food systems
   - Scale up agricultural development programs that support production and increased consumption of nutritious foods.
   - Design and implement new agricultural and food system policies to support healthy diets, including the promotion and improved affordability of fruits, vegetables, nuts, legumes, and whole grains.
   - Design and implement public policies to improve food environments to tackle all forms of malnutrition.

Of the ten double-duty actions, seven explicitly link to diet. The double-duty actions complement the *Lancet* series' recommendations in 2013 on maternal and child nutrition, only two of which—breastfeeding and supplementary feeding—related to diet. There are circumstances where food is unavailable or inaccessible; as a result, we cannot rely on diets alone, any more than we can rely solely on "pills, powders, and pastes." In combination, however, we now have access to a robust, evidence-based set of generalized actions, focused on low- and middle-income countries facing undernutrition or the double burden of malnutrition. If implemented, we can make important progress toward the elimination of malnutrition.

These recommendations will need to be adapted, contextualized, expanded, and evaluated according to national and local circumstances, drawing on the EAT–*Lancet* Commission recommendations, to arrive at improvements to existing strategies and interventions. To maximize effectiveness in delivery, there will likely need to be changes in program organization and governance, implementation capacity, and financing modalities and capacity at all levels. These issues will be explored in part IV.

CONVERGENCE ON HEALTHY DIETS

In 2019 Timor-Leste presented to the United Nations its first voluntary national review (VNR) on implementation of the Sustainable Development Goals. In his opening statement, then prime minister Taur Matan Ruak declared: "Tackling high rates of child malnutrition and food insecurity and improving access to clean water and sanitation are vital and require sustained investment."[54] The VNR mentioned *nutrition*, *malnutrition*, and related derivatives of the word seventy-six times, compared to sixteen times in the SDP that was launched in 2011; *diet* and its derivatives appeared nine times, compared to a solitary mention in the SDP. As president of the country from 2012 to 2017, Taur Matan Ruak was a tireless advocate of improving nutrition through national campaigns and competitions. Timor-Leste now has a government-led national nutrition strategy, supported by donors, that focuses on reducing maternal and child undernutrition by implementing policies and interventions, including most of those highlighted in this chapter.

Can Timor-Leste reduce stunting and anemia to near zero by 2030 in accordance with SDG 2? As this book goes to press, we have fewer than eight seasons remaining if we expect a significant amount of the diet to be derived from local sources—a daunting challenge in a country with a single, short rainy season and vulnerability to climate fluctuations. We need be realistic. It will take a radical transformation, as foreseen by the EAT–*Lancet* Commission. Crops will need to be more nutritious and drought tolerant. Promotion of and greater accessibility of legumes, vegetables, and fruits, complemented with knowledge on food preparation, will be essential. More livestock products and fish will need to be consumed. Perhaps 2040 is a more realistic time frame. But even aiming for 2040 will need sustained focus, investment, and policies on getting the diets right, complemented by continued development of water, sanitation and hygiene, and enhanced child-care practices. Without these changes, improved diets will have little impact.

As found in most parts of the world, women in Timor-Leste can play an important role in improving nutrition outcomes. Recent research in rural Timor-Leste by Gianna Bonis-Profumo and colleagues has highlighted the importance of maternal dietary quality and educational attainment in determining the diets of children.[55] Based on the Abbreviated Women's Empowerment in Agriculture Index, dietary diversity scores of empowered women and their children were higher than among those disempowered.[56] They concluded that empowering women and promoting agricultural diversification would lead to improved diets of both mothers and children.[57]

At the same time, drawing on the lessons from other countries emerging from poverty and hunger, there is a need for vigilance on the likely rise in overweight and obesity. For women of reproductive age in Timor-Leste, overweight and obesity increased from 5 percent in 2010 to 19 percent in 2020.[58] As noted in chapter 7, stunting from childhood undernutrition can result in overweight and obesity later in life. With such high stunting rates, even today, we can anticipate that the country may experience an explosion in prevalence of overweight and obesity in the years ahead. Cheap, highly processed, calorie-rich, and nutrient-poor foods (high in sugar, salt, and unhealthy fat) are becoming more readily available.[59] Convergence on a healthy diet for all Timorese is perhaps the nation's greatest challenge for the coming decade and emblematic of the challenges faced by emerging economies around the world.

In this chapter we have seen examples of successful policies and interventions that cut undernutrition and show promise in reversing overweight, obesity, and diet-related NCDs. A vision of universal convergence around locally feasible, accessible, and sustainable healthy diets for all is needed to drive coherent policies and investments that will fill the intake gaps and remove the intake excesses in the face of counteracting commercial interests and market forces. Evidence-based policies, education, social mobilization, women's empowerment, and leadership, coupled with creative, sustained, additional financing, will be required to translate that vision into reality.

13

SOCIAL PROTECTION

By late March 2020, as the spread and death toll of COVID-19 infections increased, there were lockdowns and job losses across America. The country's food banks and pantries were besieged by the hungry and fearful, uncertain of what lay ahead. Millions of people were flooding a food assistance system that was never designed for a crisis of this scale. Both food and volunteer workers were in short supply. In early April Nicholas Kulish reported in the *New York Times* that the National Guard had been mobilized in Louisiana to ensure smooth operations in a time of "need, scarcity and anxiety," writing that "uniformed guardsmen help 'take the edge off' at increasingly tense distributions of boxes filled with cans of chicken noodle soup, tuna fish, and pork and beans, said Mike Manning, the chief executive at the Greater Baton Rouge Food Bank. 'Their presence provides safety for us during distributions.' "[1]

Kulish's description triggered for me a flashback to Malawi fifteen years earlier, when an unprecedented dry spell, at a critical time during Malawi's 2004–2005 rainy season, had reduced the national maize yield to just 0.8 metric tons per hectare, 30 percent below the long-term average. As elaborated in chapter 9, this was the country's worst harvest in a decade. And by late 2005, five million Malawians—almost 40 percent of the population—were at risk of acute hunger and possible starvation. BBC correspondent Karen Allen captured the same "need, scarcity and anxiety," reporting from Mulanji in the south of the country:

> It is seven o clock in the morning and many Malawians we stumble across have been queueing for days.
>
> Queueing on empty stomachs and with bare feet. Mothers whose breast milk has dried up due to lack of food, jostle for space, their babies strapped to their backs in the traditional African way.

> Occasionally a scuffle breaks out as some hungry person, accused of pushing in, is plucked from the queue by police officers.
>
> These are Malawi's poorest people—unable to buy maize on the open market where prices have doubled in recent months. Stocks in the main government markets are diminishing fast, so they're starting to impose rations.[2]

The common thread in these two cases—fifteen years apart and in vastly different socioeconomic settings—is that vulnerable people had lost their ability to access food. They had lost their food security. They had lost their "access to sufficient, safe and nutritious food to meet their dietary needs and food preferences for an active and healthy life."[3] In both situations, many of these people had already struggled, sometimes coped, but often failed to meet their basic food requirements, and now they were looking to others for help. When people are unable to provide for themselves, often through no fault of their own, society can and often does step in to protect and support the most vulnerable through systems, policies, and programs that are collectively described as *social protection.*

Many and varied forms of social protection aim to protect the most vulnerable in society: targeting the aged, the poor, the homeless, the unemployed, the disabled, vulnerable children, war veterans, and more. Noncontributory assistance programs are a subset of social protection investments—usually referred to as *safety nets*—that are distinct from social insurance, which includes contributory pensions and unemployment benefits.[4] Virtually all types of social protection affect food security, directly or indirectly. Some programs deal with chronic food insecurity brought about by poverty and inequality, while others are designed to meet short-term emergency needs brought about by crises. Social protection is an acknowledgment by societies of temporary and long-term inequalities in food access. Moreover, the use of social protection and safety nets to improve food security is an acceptance of the human right to food as embodied in article 25 (1) of the Universal Declaration of Human Rights (1948): "Everyone has the right to a standard of living adequate for the health and well-being of himself and of his family, including food, clothing, housing and medical care and necessary social services, and the right to security in the event of unemployment, sickness, disability, widowhood, old age or other lack of livelihood in circumstances beyond his control."[5]

It is more fully elaborated in article 11 of the International Covenant on Economic, Social and Cultural Rights (1966):

> 1. The States Parties to the present Covenant recognize the right of everyone to an adequate standard of living for himself and his family, including adequate food, clothing and housing, and to the continuous improvement of living conditions. The States Parties will take appropriate steps to ensure the realization

of this right, recognizing to this effect the essential importance of international co-operation based on free consent.

2. The States Parties to the present Covenant, recognizing the fundamental right of everyone to be free from hunger, shall take, individually and through international co-operation, the measures, including specific programmes, which are needed:

(a) To improve methods of production, conservation and distribution of food by making full use of technical and scientific knowledge, by disseminating knowledge of the principles of nutrition and by developing or reforming agrarian systems in such a way as to achieve the most efficient development and utilization of natural resources;

(b) Taking into account the problems of both food-importing and food-exporting countries, to ensure an equitable distribution of world food supplies in relation to need.[6]

In this chapter, we will explore a range of options and experiences whereby investments in social protection are intended to meet food security needs and fulfill the unarguable human rights of the most vulnerable people. As I have explained in earlier chapters, we can make major *progress* toward a food-secure world by investing in sustainable intensification, improving market infrastructure, reducing postharvest losses and waste, and supporting dietary shifts. It is unrealistic, however, to believe that investments in any or all of these areas will *guarantee* food security for all by 2050 and beyond. Through repeated crises—global, national, and local—and persistent poverty and inequality, there will always be those in society who cannot provide "sufficient, safe and nutritious food" for themselves through farming or income.[7] For this reason, I have placed social protection as an *essential* investment area for achieving universal food security by 2050.

NUTRITION SAFETY NETS

There are broadly two ways of making food more accessible to vulnerable people: production interventions and consumption interventions. As we saw in chapter 2, the Green Revolution was supported by public investments that led to increased cereal production for farmers and lower real food prices for consumers. Economists like to categorize research, extension, and public infrastructure as *investments*, while viewing more direct seasonal support to farmers such as reducing the cost of inputs and credit, and commodity price support, as *subsidies*. Without entering that debate, it is sufficient to say that these are all designed to stimulate food supply, which, in turn, lowers prices and improves food access by consumers. Production or supply-side interventions (shown in the middle column of figure 13.1) are discussed in earlier chapters. While some public support to farmers is a form of targeted social protection, the scope of this chapter will be limited to consumption or demand-side interventions, which have been estimated to reach 1.5 billion people worldwide.[8]

Governments and donors deploy three broad categories of consumption interventions to directly improve food security for the most vulnerable: direct commodity transfers, food voucher programs, and cash transfers. Each of these may be designed as conditional on actions by the recipient. Of course, not all architects or proponents of these programs are driven by the UN definition of food security, which includes good nutrition. Many programs, especially emergency programs, are broadly concerned with reducing poverty and hunger, and they may have little direct impact on stunting and micronutrient deficiencies. Moreover,

TYPE	SUPPLY INTERVENTIONS	CONSUMPTION INTERVENTIONS
Direct	**LONG TERM** • Research • Extension • Infrastructure (e.g., roads, irrigation) **SHORT TERM** (seasonal subsidies) • Inputs (e.g., seeds, fertilizer) • Utilities (e.g., energy, water) • Credit • Insurance	**PROVISION OF COMMODITIES** • Public food-distribution programs • School meal programs • Food for assets • Emergency rationing **FOOD VOUCHERS** **CASH TRANSFERS** • Conditional cash transfers • Unconditional cash transfers
Indirect	**PRICE SUBSIDIES FOR PRODUCERS AND INTERMEDIARIES** **OPEN-MARKET FOOD SALES** **EXCHANGE RATES, TAX AND TRADE POLICIES**	**PRICE SUBSIDIES FOR CONSUMERS**

FIGURE 13.1 Taxonomy of social protection interventions for improving food availability and access.

Source: Adapted from Harold Alderman, Ugo Gentilini, and Ruslan Yemtsov, *The 1.5 Billion People Question: Food, Vouchers, or Cash Transfers?* (Washington, D.C.: World Bank, 2018), p. 5. License: Creative Commons Attribution CC BY 3.0 IGO. Adaptation of an original work by the World Bank. Views and opinions expressed in the adaptation are the sole responsibility of the author or authors of the adaptation and are not endorsed by the World Bank.

there is also growing concern that food and cash transfer programs may contribute to overnutrition in adults. Research from rural Mexico found that both in-kind and cash transfers led to unhealthy weight gain in already overweight women, with the greatest impact on those who were already obese.[9] In addition to these direct consumption interventions, governments have used price subsidies to make food more accessible to the poor.

The design of each of these consumption interventions varies by country and often shifts over time. The terminologies also vary. However, I will provide a brief description of each intervention type, along with examples that can be explored by readers in greater depth.

Public food distribution programs

These programs unconditionally provide a defined package of free or subsidized food items to eligible recipients. India's Targeted Public Distribution System is the world's largest safety net program, reaching about 800 million individuals who receive a set of subsidized food commodities through a network of more half a million fair-price shops.[10]

In the United States, the Food Distribution Program on Indian Reservations (FDPIR) provides income-eligible households living on Indian reservations with food purchased by the U.S. Department of Agriculture.[11] The administering agencies store and distribute the food packages and provide nutrition education to recipients. Participants in the FDPIR may select from over a hundred products, including fresh produce and traditional foods.

School feeding programs

Around the world, more than 350 million children and adolescents, 310 million of whom live in low- and middle-income countries, receive meals, snacks, and take-home rations at school.[12] According to the World Food Programme (WFP), India feeds more than 100 million children through the Mid-Day Meal Scheme.[13] In Brazil, 48 million children receive meals; in China, 44 million; and 9 million in each of South Africa and Nigeria.[14]

School feeding programs are designed to increase school attendance and performance, improve nutrition and health and gender equality, and in some cases provide a market for local farm production. These programs are conditional on regular school attendance. Programs provide breakfast or lunch, or both. Some provide complete meals, while others distribute nutritious snacks or milk. In 2018 WFP provided school meals or snacks for 16.4 million children, half of whom were girls.[15] WFP provided take-home rations, intended for other household members, to 630,000 school meal recipients. The latter became even more important with school closures during COVID-19 outbreaks.[16] In the United States, about 30 million children receive low-cost or free meals at school.[17]

Food for Assets (FFA) programs

Also known as *Food for Work*, FFA programs provide food or cash in return for the recipients building assets such as roads, irrigation canals, and erosion-control terraces. The rationale for FFA is that recipients improve their livelihoods and their resilience to stresses and natural disasters and ultimately reduce their dependence on emergency assistance.

One of the best-known examples for such programs comes from Ethiopia following the droughts of the mid-1980s and early 2000s. With persistent food insecurity and repeated appeals for emergency food assistance, the government of Ethiopia—with donor support—established the Productive Safety Net Programme (PSNP) in 2005. In contrast to annual emergency appeals, the PSNP was conceived as a multiyear program, mostly comprising payments as food or cash for public works. In their assessment of PSNP, Kalle Hirvonen and John Hoddinott found that most beneficiaries preferred payment as food, especially in times of high food prices.[18]

Emergency rationing

The modality of rationing is used to provide individuals with a specified quantity of food at a subsidized price or free. These quota systems have been long used in times of crisis. Food rationing in Australia, enforced by the use of coupons, was used to manage shortages of tea, sugar, butter, and meat (and occasionally eggs and milk) during and for up to five years after World War II.[19] Introduced to manage shortages and curb inflation, rationing in Australia sought to ensure the equitable distribution of food.

In Sri Lanka (then Ceylon), food rationing was introduced in 1942 as an emergency wartime relief measure during a period of food shortages.[20] The scheme continued until 1979 as a universal ration scheme, reflecting the strong social welfare orientation of the government. With subsequent liberalization of the economy, the ration scheme was replaced by a targeted food stamp program, which, in turn, was replaced by a cash transfer program in 2012.

Rations are also common means of food assistance during crises. WFP varies the size and composition of its food basket according to local preferences, demographic profile, activity levels, climatic conditions, existing levels of malnutrition and disease, and the level of dependency on external food assistance.[21]

Vouchers

Also known as food stamps, vouchers give beneficiaries access to food at predefined private or public outlets, based on a specified value or quantity. Harold Alderman and colleagues view vouchers as lying "midway on the continuum of transfer modalities where cash and in-kind constitute the extremes."[22] Unlike cash transfers, vouchers exert a level of control over the type and source of food assistance.

In the United States, the Supplemental Nutrition Assistance Program (SNAP) serves as a targeted voucher program that aims to supplement the food budget and improve nutrition of needy families. With its origins in providing agricultural surpluses to low-income families in the Great Depression, SNAP emerged as a safety net with wide coverage and impact.[23] During fiscal year 2019 (October 1, 2018—September 30, 2019), SNAP cost US$60.4 billion and accounted for 65 percent of all U.S. Department of Agriculture food and nutrition assistance spending. On average, 35.7 million people (about 11 percent of the U.S. population) participated in the program each month in FY 2019.[24]

Variants of the voucher modality are found around the world. Increasingly, digital technology is used to improve transaction efficiency, transparency, and targeting. For example, Indonesia shifted its food assistance from in-kind provisions to e-vouchers to improve targeting, provide greater choice for recipients, and support local businesses.[25] Food voucher programs generally do not require behavioral compliance beyond specification of eligible foods and points of purchase.

Cash transfer programs

This modality is a form of social protection whereby money is given directly to individuals or households. Most often, cash transfers are provided by governments, donors, and nongovernmental organizations for the purpose of reducing poverty. In the context of food security, those funds enable low-income recipients to increase their expenditure on food. Beneficiaries of cash transfers have increased choice. While there is no direct assurance that the recipient will purchase *nutritious* food, or indeed food of any kind, there is growing interest in using cash transfers to, at least in part, address food insecurity.

There are two broad categories of cash transfers: conditional and unconditional. As the name suggests, conditional cash transfers (CCTs) are provided on the condition that recipients invest in the well-being of their children. These conditions typically include school attendance and/or health care monitoring and compliance. The World Bank and other donors advocated Mexico's CCT program, known successively as Progresa (1997–2002), Oportunidades (2002–2014), and Prospera (2014–2019), and Brazil's ongoing Bolsa Familia for their innovative human capital development. These programs were replicated widely around the world.

Unconditional cash transfers (UCTs) and poorly monitored CCTs give households the choice and flexibility of allocating resources according to their preferences. In a survey of programs in 108 low- and middle-income countries, UCTs covered an average of 7 percent of the population, compared with 3 percent with CCTs. Despite the growing popularity of cash transfer programs, deployment of direct food transfers and voucher programs—with an average 20 percent coverage in the surveyed countries—remains the modality of choice in low- and middle-income countries.[26]

In a review of evidence prepared for the Global Forum on Nutrition-Sensitive Social Protection Programs in 2015, Harold Alderman found little difference between UCTs and CCTs in terms of nutrition outcomes. He concluded that both virtually always augment household food consumption, diet diversity, and participation in preventive health care.[27] As noted earlier, there is evidence from Mexico that CCTs can result in increased overweight and obesity and greater risk of noncommunicable diseases.[28]

General food price subsidies

These consumer-focused subsidy programs have been around for decades. General subsidies aim to lower the cost of food to below free-market prices for entire populations. Their objectives, which vary from country to country and can shift over time, include raising the real purchasing power of consumers, reducing hunger and malnutrition, maintaining low wages, and achieving social and political stability.[29] Critics of generalized food price subsidies have contended that these interventions distort market signals, thereby dampening incentives for local production. In addition, these programs can lead to shortages and black market activity. Moreover, by their general nature, food price subsidies are not targeted to the poor. In any case, there is little evidence to support nutrition benefits from this type of intervention.[30] While there has been a widespread shift from generalized to targeted subsidies, some governments, notably in the Middle East and North Africa, continue to subsidize staple food commodities where "providing food at low (and stable) prices to everyone is regarded as a responsibility of the state and a key ingredient for social contracts."[31]

SOCIAL PROTECTION AND AGRICULTURAL DEVELOPMENT

As described earlier in this chapter, Malawi's maize crop failed during the 2004–2005 rainy season that affected the entire southern Africa region. Kofi Annan, then UN secretary-general, aptly described the situation as "an acute phase of a chronic emergency."[32] In May 2005 the Malawi Vulnerability Assessment Committee estimated that over 4.2 million people required food aid, a number that would increase to 5 million by the end of that year.[33] Working closely with the government of Malawi, the United Nations office in Lilongwe launched what the UN calls a *flash appeal*, a fundraising modality reserved for sudden-onset natural disasters that would require a coordinated response beyond the capacity of the government and any single UN agency. Michael Keating, then UN resident coordinator in Malawi, was adamant that the Malawi 2005 Flash Appeal would deploy an unprecedented two-track approach. The opening paragraph of the appeal read as follows (bold type as in the original): "This is not a typical United Nations (UN) Flash Appeal. It is a 'smart Appeal,' requesting international support for the Government of Malawi's (GoM) leadership in addressing both **immediate humanitarian needs**, and to allow the Government to take action now to **minimise the likelihood of another food-shortage driven humanitarian crisis next year.**"[34]

The first track of the US$88 million appeal was intended to meet the immediate food shortages facing the population. WFP, with a proposed allocation of almost US$50 million, would lead that effort through increased food distribution, voucher schemes, and cash interventions. The second funding track, around US$36 million, would focus on agriculture, primarily in support of subsidizing agricultural inputs for smallholder farmers for the next season. The small balance would support health and education.

Ultimately, according to the UN Office for the Coordination of Humanitarian Affairs (OCHA), US$57 million of the US$88 million was raised.[35] Of that amount, US$49 million went to the first track, mostly as food aid and nutrition support, essentially meeting that component of the Flash Appeal. Only US$7.8 million, however, was raised for the agricultural recovery track, less than a quarter of what was requested. Donors had the appetite to send cash and food to meet the immediate crisis, but they were unwilling to commit to support an agriculture-led recovery the following season. The message was clear: give Malawians food. Keep them alive. But there was reluctance to give them the resources they needed to produce the food themselves. This response by donors led to the bold and ambitious Malawian response that I shared in some detail in chapter 9. As President Mutharika stated, "Let us grow the food ourselves." And indeed they did.

Fifteen years on, I still struggle to understand the rationale for such policy incoherence by the international donor community. Of course, we heard the pragmatic argument that it is much easier to raise funds for food aid than for longer-term development. Political constituents in donor countries are easier to please knowing that food is reaching hungry people. And it is true, those buckets of emergency funds are often available and ready to be deployed at short notice, often left unspent from the previous emergency. In contrast, investing in agriculture sounds more like development, and therefore at least some donors consider it an improper use of emergency funds. It was clear that donors were unwilling to allocate those resources to subsidize agricultural inputs. That would surely be a slippery slope toward dependency, or so the argument was presented. But wait: Those same donors were happy to provide almost US$50 million in food aid without raising concerns of dependency? Consistency and coherence have never been defining features of the international development system.

TOWARD COHERENCE

From the United Nations in May 2020, Under-Secretary-General for Humanitarian Affairs and Emergency Relief Coordinator Mark Lowcock called on wealthy nations to contribute US$90 billion to meet "the cost of protecting the most vulnerable 10 per cent of people in the world's poorest countries from the very worst impacts of the [COVID-19] pandemic."[36] Below an image of a Malawian woman washing her hands as a condition for receiving cash from WFP,

Lowcock's message was that US$90 billion was "an affordable sum of money," being equivalent to just 1 percent of the global stimulus allocations by rich countries. With two-thirds of the money expected to come from the World Bank and International Monetary Fund, donor countries would be asked to contribute the remaining US$30 billion from a one-off 20 percent increase in their aid budgets. The latter strategy was easier said than done at a time when rich nations were sharply focused on their own domestic economic recoveries. Moreover, there were no details on how the US$90 billion might be spent. In any case, Lowcock's bold call for action—a thousand times the amount that Michael Keating had requested for Malawi fifteen years earlier—will likely go down as the largest Flash Appeal in human history. While appeals like this make great headlines, history has shown that only a fraction of funds is ever disbursed, and rarely is there a significant allocation made for building more resilient food systems in the affected areas.

Reflecting on past experience with social protection and agriculture, it is timely to use the COVID-19 crisis and recovery period to rethink the design of food security interventions in times of emergency. How would we redesign food security investments in order to improve resilience and *build back better*, the new mantra of disaster recovery community? The starting point is an analysis of investment coherence in social protection and agriculture. For this, I draw primarily on a peer-reviewed synthesis of evidence on the interaction between social protection and agriculture by Nyasha Tirivayi, Marco Knowles, and Benjamin Davis, published in 2016.[37] Based on this review, country experiences, expert interviews, and policy dialogues, Maja Gavrilovic and colleagues at the UN Food and Agriculture Organization identified the benefits of strengthening coherence between social protection and agriculture, as well as opportunities for achieving those benefits through improved policies and programs.[38] The findings from these two key publications on the potential interactions and mutual benefits are summarized here:

- Social protection investments through cash transfers and food for assets programs can improve agricultural performance through increased input use, asset accumulation, and increased yields.
- Social protection investments can indirectly benefit agriculture by preventing detrimental risk-coping strategies such as distress asset sales and increased indebtedness. Cash transfers can also reduce risks through livelihood diversification, including off-farm investments.
- Social protection through procurement for school meals and food distribution programs can provide markets for farmers, as well as reduce dependency on child labor.

- Taking a longer time frame, social protection can also indirectly improve agriculture through enhanced educational and health outcomes.
- Agricultural investments can improve food security (including nutrition), incomes, asset accumulation, and human capital outcomes like education and health.
- Agricultural investments can reduce vulnerabilities to risks and disasters.

For each of the impacts listed, an important caveat is the word *can*. All these benefits are possible but are not guaranteed. In fact, these two reports cite examples where social protection investments may be detrimental to agriculture and vice versa. For example, food for work programs can divert labor from agricultural work at critical times. It is also well known that some agricultural investments—for example, those focused on cash crops for export—can increase vulnerability to market price fluctuations and lead to greater indebtedness. Thus, in designing for greater coherence, understanding context is critical.

While not highlighted explicitly in the findings, it is clear from the wider literature and my personal observations that women's participation and empowerment are central to both agricultural production and household nutrition. Therefore women have to be both drivers and targets of social protection programs designed to improve household food security. Successful interventions include effectively targeting transfers to pregnant women, households with young children, and women-headed households.[39]

In a review of safety nets for agriculture and nutrition, Daniel Gilligan of the International Food Policy Research Institute observes that few social protection programs successfully integrate agriculture and nutrition. A notable exception is Ethiopia's PSNP, where the government added improved maternal and child nutritional status as a program goal after more than ten years of operation.[40] As a result, women with young children were able to meet their public works requirements by participating in training programs in behavioral change for improved nutrition.

As development practitioners, we need to better understand the interactions, trade-offs, and conflicts between social protection and agriculture, as well as with other sectors like education and health. We must apply a gender lens to these analyses in recognition of the central role of women in food systems transformation. Dr. Jemimah Njuki, a Kenyan national, led the Gender Equality and Women's Empowerment Lever for the UN Food Systems Summit in 2021. With more than twenty years of experience in the agriculture sector in Africa and Asia, working on gender equality and the empowerment of women, Njuki describes women as "the nucleus of our global food system, often taking responsibility for household nutrition and . . . also directly involved in raising livestock and tending to crops."[41]

In an ideal world, our interventions in social protection should generate synergies with other sectors. The UN Hunger Task Force identified homegrown school meal programs as a priority intervention for addressing hunger and stimulating smallholder agriculture.[42] The Hunger Task Force argued for local procurement of school meal ingredients; by guaranteeing a price for smallholders' produce, farmers would be incentivized to make investments in soil improvements, water management, and better seeds. At the same time, school meals can improve school attendance, reduce the gender gap in attendance, improve health and nutrition, and reduce child labor and child soldier recruitment.

While conceptually compelling, homegrown school meal programs are constrained by the financial and operational constraints of going to scale. Responding to the Hunger Task Force recommendations, the Millennium Villages Project included school meals as part of its education strategy. Initially, this component was integrated with the agricultural program in several villages across sub-Saharan Africa. Farmers were provided production inputs (seeds and fertilizer) in return for contributing to a food bank that would be accessed for school meals. In practice, only 11 percent of the input costs were repaid through this mechanism, and

FIGURE 13.2 School meal provision in Ethiopia.

Credit: Jessica Fanzo.

the school meals ultimately proved to be difficult to sustain and take to scale in this way.[43] Moreover, the focus on maize meant that other foods would need to be procured or donated to ensure a healthy nutritious diet for the school meal recipients.

The Committee on World Food Security (CFS), an intergovernmental platform for stakeholders in food security and nutrition that reports to the UN General Assembly, develops and endorses policy recommendations and guidance for UN members and stakeholders on a wide range of food security and nutrition topics. On the issue of connecting smallholder farmers to markets, the CFS argued for expanded institutional procurement for social protection programs, such as school meals and other food assistance, as a mechanism for improving smallholder production, providing a more predictable income, and supporting livelihoods in times of conflict and crisis.[44]

In support of its food assistance operations, WFP pledged to increase its procurement needs from smallholder farmers. This commitment, however, appears to be challenging to implement. In 2018 WFP procured food directly from smallholders in only twenty-nine countries, amounting to just US$31 million of the annual US$1 billion food procurement bill.[45] As I observed in the Millennium Villages Project, efforts to connect smallholders with WFP's Purchase for Progress Project were frustrated by the difficulties smallholders had in aggregating large enough quantities and ensuring consistent quality in accordance with WFP specifications. WFP's cash transfer programs, however, which disbursed US$1.8 billion across sixty-two countries in 2018, are likely to have a much greater impact on local supply, especially if local farmers can be supported in accessing markets.[46]

Brazil, through its National School Feeding Program (Programa Nacional de Alimentação Escolar, PNAE), was the first country to take this local procurement approach to scale. Launched in 1955, PNAE reached forty-three million public school students in 2015.[47] Since 2009, through article 14 of Law 11,947, PNAE required that at least 30 percent of school meal funds be used for procurement from family farms and local rural enterprises. More specifically, the law stated that priority must be given to the resettled farmers (former landless people), traditional indigenous communities, and *quilombolas* (descendants of Afro-Brazilian slaves). Brazil was the first country to legally mandate that school meals be procured, at least in part, from local producers. In their 2016 assessment of PNAE, Corinna Hawkes and colleagues concluded that although the program's goals had not been fully achieved, the practice of procurement from family farms was well established.[48]

Homegrown school meals illustrate both the opportunities and the challenges in generating coherence and creating synergies across investments designed to improve food security. Achieving that potential for coherence requires a more holistic approach to the design and implementation of food security investments and special attention to social inclusion of marginalized people, including women and indigenous communities. In earlier chapters I highlighted four strategic investment areas: sustainable intensification, market infrastructure, postharvest stewardship, and healthy diets. Social protection investments can be designed and implemented in ways that enhance and reinforce each of these other four investment areas and over time reduce the need for social protection.

Food for assets programs can be designed to enhance climate change resilience, for example, by constructing and maintaining irrigation canals, a practice I observed WFP using in Tajikistan. Food for school meals and other public distribution programs can be sourced from farmers who demonstrate practices that are consistent with sustainable intensification, as we piloted through the Millennium Villages Project. Food and cash for assets programs can also be deployed to improve roads and other market-enhancing infrastructure, improving market connectivity, and reducing postharvest losses. At the same time, improvements in market infrastructure can improve social protection delivery. Information and communication technology infrastructure and modern digital systems can be used to enhance efficiency and transparency of cash transfer and food voucher programs. In a study of cash transfer payments to women in Niger following a devastating drought in 2009–2010, researchers found household dietary diversity across ninety-six rural villages was 9 to 16 percent higher among households who received mobile phone transfers than among those who received traditional cash payments. The improved nutrition outcome was attributed to time savings by women and stronger intrahousehold bargaining power for women.[49]

Social protection investments can be designed and implemented in ways that enhance and reinforce investments in sustainable intensification, market infrastructure, postharvest stewardship, and healthier diets. Greater complementarity and coherence between social protection and other investments will require institutional innovation and new approaches to education and training that encourage and reward cross-disciplinary, cross-sector program design and implementation.

Often the constraint is the inability of institutions to act outside their formal mandates, which suggests there is scope for more flexible institutional architecture and operations. In addition, we should reexamine our education and training systems, which encourage and reward disciplinary specialization rather than cross-disciplinary, cross-sector program design. And finally, social protection by definition must be designed with the goal of including those who have been economically or socially marginalized or bypassed by mainstream food systems development efforts. In the pursuit of universal food security, social protection should ensure that no person is left behind.

14

COVID-19 AND FOOD SECURITY

At a media briefing in Geneva on March 11, 2020, Director-General Dr. Tedros Adhanom Ghebreyesus of the World Health Organization declared that COVID-19 should be characterized as a *pandemic*:

> WHO has been assessing this outbreak around the clock and we are deeply concerned both by the alarming levels of spread and severity, and by the alarming levels of inaction.
>
> We have therefore made the assessment that COVID-19 can be characterized as a pandemic.
>
> Pandemic is not a word to use lightly or carelessly. It is a word that, if misused, can cause unreasonable fear, or unjustified acceptance that the fight is over, leading to unnecessary suffering and death.[1]

The following day, Columbia University President Lee Bollinger announced that all classes would be moved to online for the remainder of the spring semester. Fortunately, I had no classes for the rest of the week, after which, spring break would provide the breathing space for some quick lessons in online pedagogy and the mechanics of Zoom.

On what would be my last day in the office for almost eighteen months, I exited the New York subway at 72nd Street and walked toward home. Across from the station, I noticed a long line outside a popular grocery store on the city's Upper West Side. It took a few seconds, but then I made the connection: panic buying had started. In the past I had seen lines of anxious shoppers ahead of approaching snowstorms, but surely there was no need to be alarmed about a disease that, on the day of Tedros's announcement, had just forty-four new cases and no reported deaths in all of New York State. But seeing that long line gave cause to quicken my step, head home, and see where we were in terms of food supplies.

In comparison with climate change, the COVID-19 pandemic presented itself as potentially a more sudden shock to the food system. While there was the immediate priority of stemming the disease's spread, the rush at the grocery store suggested that its impact would likely extend beyond health to other social and economic sectors. I grasped this opportunity to reorient my course syllabus on global food systems to incorporate more prominently the topic of *resilience*. While I had previously referred to resilience as an essential feature of a well-functioning food system, I had climate, not COVID, in mind.

During my first class on Zoom, held the week after spring break, we discussed the impact of the pandemic on food supply chains. I took the case of orange juice, which had jumped more than 20 percent in price from a week earlier.[2] Consumers were looking for OJ's immune-boosting properties. At the same time, anticipated transport disruptions and farm labor constraints dampened the supply outlook.[3] Mexican farmers were rejoicing, in the short term at least, as another health food, avocados, got a boost—surging in price by 60 percent between early March and late April 2020.[4] These early examples, spurred on perhaps by impulsive consumer behavior, provided some early insight into how food supply chains may prove to be vulnerable to a disease pandemic.

DEEPENING AN EXISTING CRISIS

Each year since 2017, an alliance of United Nations agencies and donors, known as the Food Security Information Network (FSIN), has produced a *Global Report on Food Crises*. The report in 2020 announced that 135 million people from fifty-five countries were "acutely" food insecure in a state of "crisis or worse."[5] According to an internationally agreed measure of crisis severity—the Integrated Food Security Phase Classification (IPC)—these 135 million people were at level 3 or higher on the 5-point scale of severity. Level 3 describes a "crisis," while level 4 refers to an "emergency," and level 5, a "catastrophe/famine."[6] The IPC was developed as a tool for improving food security analysis and decision-making by humanitarian agencies and governments. Its objective is to prevent or decrease severe food insecurity that threatens lives and livelihoods.

The FSIN forecasted 135 million people facing acute food insecurity related to the situation *prior* to the arrival of the COVID-19 pandemic. As the impact of COVID-19 spread during 2020, an additional 20 million were recorded as level 3 or above in the 2021 FSIN report.[7] The FSIN determined that another 183 million people, 70 percent of whom were located in Africa, were in IPC level 2 ("stressed") and therefore in imminent danger of slipping into acute food insecurity through shocks and stresses including COVID-19.[8] A year later *The State of Food Security and Nutrition in the World 2021* (known widely as SOFI 2021, reflecting the name of this annual UN report up to and including 2015: the *State of Food Insecurity in the World*), quantified the likely impact of COVID-19 on hunger and food security. The report projected that between 720 and 811 million people in the world faced hunger (energy deprivation) in 2020.[9] Taking the middle of the projected range (766 million), around 116 million more people were facing hunger in 2020 than in 2019.[10]

Based on another indicator—the Food Insecurity Experience Scale (see chapter 7)—2.37 billion people (almost one in three people on Earth) did not have access to adequate food in 2020. This figure had risen by 320 million people (roughly the population of the United States) *in one year*.[11]

While it was impossible to quantify the exact contribution of COVID-19 to the worsening state of food insecurity around the world in 2022, there is no doubt that an existing food security crisis, driven by conflict, weather, and poverty, has been worsened by the direct and indirect effects of the pandemic, as articulated in the FSIN 2021 report: "While conflict continues to displace people, disrupt livelihoods and damage economies, the COVID-19 pandemic and related containment measures have exacerbated pre-existing drivers of fragility, widened inequalities and exposed structural vulnerabilities of local and global food systems, hitting the most economically vulnerable households particularly hard."[12]

The data show that women are more likely than men to be food insecure. The SOFI 2021 reported that the prevalence of moderate or severe food insecurity was 10 percent higher among women than men in 2020. This compares with a 6 percent differential in 2019.[13] COVID-19 was having an impact on global food security, and disproportionately so on women.

In one sense, the virus—severe acute respiratory syndrome coronavirus 2 (SARS-CoV-2)—does not discriminate on who gets infected. One of the earliest celebrity cases was American actor, Tom Hanks, who, along with his spouse, Rita Wilson, came down with the disease in Australia while making Baz Luhrmann's film about Elvis Presley. "Nobody is immune from the disease" was the refrain. Hanks and Wilson spent two weeks at the Gold Coast University Hospital—part of Queensland's public health system—and lived to share their stories.[14] Hanks, at the time a sixty-three-year old with type 2 diabetes and a heart stent, was considered high risk. Since then, we have seen miscellaneous presidents

FIGURE 14.1 Closed markets in La Paz, Bolivia. The COVID-19 pandemic affected livelihoods, damaged economies, and exposed vulnerabilities of food systems worldwide.

Credit: Mateo Romay.

and prime ministers, sports stars, musicians, and other notables laid low with the virus. But almost all survived, many of them prior to the arrival of vaccines. Having a functional and accessible health system clearly helps.

Unlike Hanks and Wilson, most of the billion or so people living with acute or chronic food insecurity live in countries where health systems are already overstretched. Living in New York as the number of new cases in the state rose from 44 on March 11 to more than 7,400 two weeks later, I had growing concerns about the capacity of health facilities to cope, especially in terms of providing access to ventilators, a complex and expensive lifesaving device.[15] As the new cases and death toll rose during March, then Governor Andrew Cuomo declared that the state's twelve thousand ventilators were woefully inadequate to meet the growing demand.[16] On March 25 Cuomo said that an additional thirty thousand ventilators would be needed.[17]

Upon hearing this, I decided to check on the availability of ventilators in Malawi, whose population was about the same as New York State's: nineteen million. A study published in the *Lancet Global Health* reported sixteen working ventilators across the nation's four major hospitals.[18] In April 2020 the *Guardian* reported that "a woefully inadequate health system will leave Malawi unable to cope when Covid-19 arrives."[19] The *Nyasa Times* revealed that Kamuzu Central Hospital in the capital Lilongwe had just five ventilators, one of which was "strictly for Very Important People (VIP)."[20] To conclude the obvious, wealth matters, at least in terms of lifesaving medical devices in the face of a pandemic.

Initially, it seemed that Malawi had been spared from the worst of the pandemic suffering—by late October 2020 the country had just 183 deaths, compared to more than 33,000 in New York State.[21] A reasonable conclusion might be that factors other than ventilators are more critical in resource-poor settings. And perhaps there are some other, as yet unknown, determinants of COVID spread and mortality in Africa. For once, remoteness may have been to the advantage to both communities and countries, at least in terms of buying time to prepare. By July 2021, however, with just 2 percent of the Malawian population vaccinated, the disease had generated a third wave of infections, and cumulative deaths were fast approaching 2,000.[22] The WHO and the U.S. Centers for Disease Control and Prevention designated Malawi a COVID-19 *high-risk country* and warned against traveling there.[23] In all likelihood, despite stepped up efforts of UNICEF and other development agencies, Malawi's worst COVID-19 nightmares were still ahead.[24]

People living in food crises and experiencing undernutrition have compromised immune systems and are at greater risk of all infections, not just COVID-19. Considering the known inflammatory response caused by COVID-19, a weakened immune response and more severe consequences are likely.[25] With a focus on older populations, David Richardson and Julie Lovegrove reported in the *British Journal of Nutrition* that "avoidance of nutrient deficiencies, identification of target groups at high risk of suboptimal nutritional status and the use of practical, safe and effective nutrition policy solutions may help strengthen the resilience of people to the COVID-19 pandemic."[26] Thus the enthusiasm for orange juice and avocados appears—at least as part of a more nutritious and balanced diet—not entirely irrational. Good nutrition and a healthy diet matter in preparing the body for a viral onslaught.

At the other end of the malnutrition scale, Barry Popkin and colleagues have looked at the vulnerability of obese individuals to COVID-19.[27] Through a meta-analysis of seventy-five studies, they found that obese individuals were 46 percent more likely to test COVID positive, 113 percent more likely to be hospitalized, 74 percent more likely to be admitted to an intensive care unit, and 48 percent more likely to die than nonobese individuals. Furthermore, vaccines are likely to be less effective for obese individuals. Government lockdowns have resulted in increased consumption of unhealthy processed foods and decreased physical activity, especially among the poor. Diet-related noncommunicable diseases like hypertension, type 2 diabetes, and heart diseases have also been found to increase the risk of severe complications and death.[28] Healthy diets are an important source of resistance to and recovery from the effects of COVID-19.

We had a nutrition-driven health crisis before COVID-19 appeared on the scene. The failure of the food system to deliver healthy nutritious food for all people and the resulting state of malnutrition, itself a pandemic of sorts, provided a fertile ground for intensifying the severity and extent of COVID-19.

COVID-19 IMPACTS ON FOOD SECURITY

In the preceding chapters I identified five focus areas of investment that would lead to food systems that could enable universal food security. Enter COVID-19. How did the pandemic disrupt our progress across the Big Five? And what strategies and interventions are needed to improve system resilience. The full impacts of COVID-19 will likely take years to understand and address, but what have we learned so far?

Sustainable intensification

In chapter 9 I outlined why more food will be needed by 2050 and showed that we must shift to more sustainable production systems if we are to achieve universal food security. The COVID-19 pandemic had less immediate impact on agricultural production than on the manufacturing and services sectors. At the macro level, early experience suggested that food *availability* has not been seriously affected.[29] The spread of the disease had (to the time of this book's publication) been mainly focused on urban areas, initially sparing rural people from the worst of the infection. Direct impacts on production were caused by border restrictions that disrupted migrant labor and material input supplies.

Consumer demand drove less direct impacts. As lockdowns were implemented, many factories closed and service industries collapsed, leaving millions of people to return to rural areas: a reversal of the structural transformation process described in chapter 8. This trend is reminiscent of the Asian Financial Crisis of the late 1990s, when urban workers, especially in the manufacturing and construction sectors, retreated to rural areas after losing their jobs. These rural returnees have little to offer the economies of their former homes and, in some cases, have only contributed to the spread of COVID-19.

I thought about Bali, the tourist island where I lived in the mid-1970s. Even then the rural exodus to the urban and coastal tourism sector was beginning. Structural transformation was underway. Agriculture was beginning to diversify based on the demands of tourists and the increasingly urbanized and more prosperous population. The arrival of COVID-19 in Bali decimated tourism and created high unemployment. Inevitably, unemployed tourism workers have drawn down their savings, disposed of their assets, and experienced a decline in their standard of living. Anecdotal evidence revealed a collapse of the higher-end markets for fruits, vegetables, and livestock products. Diversification has long been advocated as a measure of sustainability and resilience. But the spread of COVID-19 in Bali demonstrated how diversification can be quickly undermined.

We have seen in Bali and in other parts of the world hard-hit by the contraction of manufacturing and services that rural economies are reverting to the conditions that existed prior to structural transformation. A high level of dependency on tourism can be quickly exposed in a pandemic. As Julia Winterflood reported in the *Diplomat*, Balinese people are starting to reconsider their development options, including a return to agriculture. One specialty coffee farmer and processor told Winterflood, "Coffee is a great option as no matter what happens in the world, people will still drink coffee."[30] Another diversification option being explored in Bali is seaweed aquaculture. Producers' limited technical capacity and access to finance, however, constrains uptake.[31]

Looking back at the Asian Financial Crisis, this reverse migration was relatively short-lived. As the Asian economies recovered, the long-term trend of structural transformation with urban migration returned. It seems reasonable to expect that this will occur again once vaccinations for COVID-19 are widely accessible and used. Meanwhile, in the short and medium terms, we should expect a temporary reversal of development gains in both urban and rural populations.

The short- and medium-term impacts of the pandemic on sustainable intensification come mainly from the disruptions in market demand. Massive urban unemployment and the associated loss of consumer income, along with mandated closures or restricted operations of food markets and restaurants, created market uncertainty for producers as they tried to anticipate market demand. Uncertainty typically results in risk aversion strategies and reduced investment. Until urban employment recovers and food-related businesses reopen, private investment in agriculture will likely be curtailed.

Longer-term impacts may be felt as some producers dispose of their assets, such as land or livestock, in order to survive the loss of their markets. This may lead to fundamental restructuring of the agricultural sector, possibly toward larger-scale, vertically integrated operations that have the assets and access to finance needed to weather uncertainties of the short and medium terms. An alternative scenario is a return to more diversified, subsistence-oriented farming that meets the immediate consumption needs of households and local communities. There are some signs that import-dependent nations such as the Pacific island states are grasping opportunities to reassess and reprioritize domestic production opportunities.[32] At this point, we simply do not know the long-term impacts of COVID-19 and what this means in terms of farm operations, economic viability, and environmental sustainability. Local context and capabilities will be important.

There are also concerns that the need to focus on COVID-19's direct effects on human health and the intense demands on health services can lead to benign neglect of the agriculture sector, as was observed during the HIV/AIDS crisis. Governments and aid agencies are prone to focusing on immediate crises. With lost tax revenues and reduced fiscal capability, it seems inevitable that public investment in long-term sustainability changes will be, at best, postponed. Therefore, while it is likely that food production per se will not be threatened by COVID-19 in the short and medium terms, there are risks that historical trends toward the adoption of more sustainable production systems, including those addressing the adaptation to and mitigation of climate change, may be interrupted as the strategic gives way to the urgent. The fact is that poor farmers make even poorer environmentalists. Faced with hunger and extreme poverty, there is little time or compulsion to think too far into the future. The environmental costs of COVID-19 may ultimately be greater than the direct health costs. COVID-19-induced poverty, in both urban and rural settings, may yet become the enemy of sustainable intensification.

Market infrastructure

Ensuring market connectivity is essential for achieving a food-secure world. Food moves from producers to consumers through commodity value chains that connect producers, input suppliers, processors, transporters, wholesalers, retailers, and consumers. COVID-19 can create disruptions at any point in the value chain because humans are involved. The more labor intensive the process, the more vulnerable the value chain becomes. Where indoor congregation of labor is needed, there is increased risk of virus transmission, as we saw with the early impacts on meatpacking plants. These effects are intensified when migrants, whose movements are constrained by border restrictions, provide labor. For example, in Australia, the closure of international and some state borders created havoc for fruit picking and packing operations that are typically undertaken by visiting backpackers and seasonal migrant laborers from neighboring Pacific countries.

By and large, countries have viewed food industries as essential and have freed them from market lockdowns. Global production of the three most important cereal commodities—rice, wheat and maize—are at all-time highs. In July 2020 the UN Food and Agricultural Organization forecast that global cereal production in 2021 would mark a new record high. FAO's July 2021 forecast for world trade in cereals in 2021–2022 was also the highest ever: 472 million metric tons.[33] Nevertheless, the price of cereals was more than 27 percent higher than a year earlier, and the highest annual average since 2012. Similar trends were observed for other commodities to the extent that FAO's food price index, a weighted composite of major traded commodities, was 31 percent higher across 2021 than in prepandemic 2019, placing pressure on food-importing countries.[34]

It appears that market functionality for the major cereals will be sustained in part because of current high production levels and relative immunity from labor shortages. Exports of wheat and maize are mainly coming from highly mechanized farms. While rice exports are generated primarily from labor-intensive smallholder systems, there has been no indication of labor shortages in major areas of supply. In fact, the reverse structural transformation described earlier may more than compensate for any direct effects of COVID-19 infections on the labor force.

While overall supply for essential food commodities may not be an immediate concern, there are still the issues of demand and the ability of nations to pay for imports. Food-importing nations that rely on manufacturing and services (such as tourism) for hard currency may face difficulty meeting payments, resulting in increasing indebtedness or instability of supply. Also, as was observed during 2007–2008, some export nations may temporarily ban or restrict exports and stockpile imports to ensure national food security. The World Bank argues that such restrictions are unwarranted.[35] At the Extraordinary Virtual Meeting of G20 Agriculture Ministers on April 21, 2020, the World Bank, along with FAO, the International Fund for Agricultural Development, and the UN World Food Programme, issued a joint statement arguing that these "panic-driven policy responses" only exacerbated market disruptions. Instead, these agencies called for collective action "to ensure that markets are well-functioning, and that timely and reliable information on market fundamentals is available to all."[36]

The risks caused by COVID-19 for food security can be reduced by investments in market infrastructure and supportive policies, as detailed in chapter 10. Functioning transport systems are essential for the smooth flow of food from producers to consumers. Stable access to credit and insurance across the commodity value chains can also improve resilience to shocks caused by COVID-19 and any future pandemics. In addition, the COVID-19 crisis has highlighted the value of digital technologies in enhancing online and contactless market access throughout the food system. Digital payment systems and e-commerce platforms have flourished during the pandemic, including in emerging economies.[37]

Postharvest stewardship

An alarming headline in the *New York Times* on April 11, 2020, read, "Dumped Milk, Smashed Eggs, Plowed Vegetables: Food Waste of the Pandemic." The article, shared with its mainly urban readers, describes some stark consequences of disrupted food supply chains: "In Wisconsin and Ohio, farmers are dumping thousands of gallons of fresh milk into lagoons and manure pits. An Idaho farmer has dug huge ditches to bury 1 million pounds of onions. And in South Florida, a region that supplies much of the Eastern half of the United States with produce, tractors are crisscrossing bean and cabbage fields, plowing perfectly ripe vegetables back into the soil."[38]

Closures of restaurants, schools, and hotels slashed demand. Massive quantities of waste occurred before produce even left the farm. The *Times* article cited Shay Myers, a third-generation onion farmer from the border of Oregon and Idaho: "People don't make onion rings at home." Eating habits radically changed at short notice. Changes in work and schooling patterns, shopping patterns, and cooking and consumption patterns all have potential impacts on food waste.

In late June 2020 my spouse and I began fourteen days of mandatory hotel quarantine at a Gold Coast (Australia) hotel. Three meals a day were delivered at the room door with a firm but friendly knock. Given the crisis at hand and the fact that the government was meeting the costs (a policy they would soon withdraw), it did not seem appropriate to complain about the food. We did appeal to room service to reduce the quantities, especially the junk food (crisps, chocolates, and the like). But to no avail. The guidelines were set and should be followed. The waste generated—both food and packaging—was embarrassing, even by rich-world standards.

Standardized food box distributions—whether for COVID-19 refugees like myself or for food relief and institutional service situations—can be wasteful. Drawing on experience from the U.S. Department of Agriculture's Farmers to Families Food Box program, which distributed more than thirty-two million food boxes to food-insecure individuals in May and June 2020, Brian Roe, Kathryn Bender, and Danyi Qi concluded that foods selected by the supplier *rather than by the consumer* are likely to increase waste compared to emergency food outlets that permit patrons to select their meal combinations.[39] Customization of meals and greater agency for the patrons of course sound reasonable. But they take a level of operations skill that is not always at hand in emergency settings that are often staffed by volunteers.

The overall impact of COVID-19 in rich-country settings indicates a complex picture of increased food losses and waste, coupled with opportunities for positive changes post-COVID-19. Just as I observed outside the grocery store on Broadway and 72nd street, the onset of COVID-19 can lead to panic buying, poor alignment with changing consumption needs, and increased household food waste.[40] Research from Spain showed a 12 percent increase in household food waste during the early weeks of the COVID-19 lockdown.[41] This amount, however, was similar to waste usually generated from nonhousehold venues (such as school canteens); as a result, overall food losses were about the same as pre-COVID.

Beyond the initial panic stage, it remains possible that the stresses associated with reduced income, reduced opportunities for food procurement, and improved knowledge of in-home food preparation *could* lead to greater efficiencies of food purchase and food use and reduction in food waste. The net impacts of these forces remain unknown.

Healthy diets

We saw earlier how unhealthy diets, both undernutrition and overnutrition, can exacerbate the extent and severity of COVID-19. In addition, there is mounting evidence that the pandemic's disruptions to food supply chains and changing food environments increase all forms of malnutrition. It is a vicious cycle: poorly nourished people are more vulnerable to COVID-19, and the pandemic's disruptions to the food system are causing more people to be malnourished.

As Engel's law tells us, low-income consumers tend to spend relatively more of their disposable income on food. With rapid loss of employment associated with COVID, budgets become tighter and low-income consumers cut back on more expensive nutrient-dense foods like meat, vegetables, and fruits. This will likely reduce nutrient intake and increase deficiencies of protein and micronutrients. Derek Headey and colleagues, on behalf of the Standing Together for Nutrition consortium, anticipated an increase in child malnutrition, including wasting, resulting from income declines, reduced affordability of nutritious foods, and disruptions in health, nutrition, and social protection programs.[42] Initial analyses of the consortium suggest that increased poverty caused by COVID-19—through lost employment, remittances, and increased food prices—is the driving force toward undernutrition and additional deaths as follows:[43]

- Even short lockdowns, coupled with mobility and food systems disruptions, could reduce per capita income, in the order of 8 percent.
- This level of income decline could cause a 14 percent increase in moderate or severe wasting for children under five years of age. This would translate into an additional 6.7 million children with wasting, 58 percent in South Asia and 22 percent in sub-Saharan Africa.
- With this level of increased wasting, coupled with a 25 percent reduction in nutrition and health services, Headey and colleagues projected about 130,000 additional deaths for 2020, about half of them in sub-Saharan Africa.

These projections were made in the early months of the pandemic, and the actual impact on incomes, malnutrition, and lives lost was not expected to be known for years. The consortium estimates were conservative and did not account for the possible extended duration, repeated waves of new variants, and delayed arrival of the pandemic in various parts of the world. As we saw earlier, sustained food deprivation can also lead to child stunting, with long-term and intergenerational impacts on health, education, and incomes. Morbidity and mortality on this scale call for a rapid commensurate response by national governments and donors at a time when budgets are under stress from reduced tax revenues and competing demands for social protection. In anticipation of the pandemic's impact, the heads of four UN agencies—UNICEF, FAO, WFP, and WHO—issued "a call for action to protect children's right to nutrition," emphasizing access to affordable diets, improved maternal and child nutrition, early detection and treatment of wasting, school meals for vulnerable children, and expanded social protection.[44]

In higher-income settings, the job losses and lockdowns also affected diets and nutrition. There are potentially both positive and negative impacts on diet. On the downside, the disruption of daily routines and extended confinement can lead to unhealthy weight gain through increased consumption of processed high-calorie comfort food, reduced access to fresh fruits and vegetables, and reduced exercise.[45] Stress and anxiety related to the pandemic's health and economic impacts can exacerbate this trend. That may be countered in some households by increased home cooking and attention to healthier eating. A survey conducted by Eat Well Tasmania found that 53 percent of respondents spent more time cooking after the COVID-19 restrictions; 33 percent became more aware of food waste, and 29 percent were eating more fruit and vegetables.[46]

Farah Naja and Rena Hamadeh outlined a framework for action, arguing that we must view the pandemic's impact on nutrition and dietary intake at multiple levels: "A particular feature of this pandemic is highlighting the interdependence of these various levels, whereby the health of the individual became a direct function of his own awareness and choices, the unity of the community, the preparedness of the government, and ultimately the global engagement vis a vis this threat."[47] It is clear that the specific contexts of individuals, households, communities, and government support will determine the extent to which the pandemic adversely affects diets and nutrition.

While market disruptions and income losses provide cause for increased vigilance on potential impacts of COVID-19 on the poor, it is important to examine each situation in full context. In Ethiopia, for example, a study undertaken by the International Food Policy Research Institute (IFPRI) found no evidence of a decline in food security by August 2020 among the 577 households surveyed in Addis Ababa.[48] While Addis Ababa did not go into lockdown during the survey period, residents reported lost jobs and lower incomes. The IFPRI researchers found, however, that food security, as measured through calorie intake, was largely unchanged five months into the pandemic, though they did observe a shift in consumption toward staples and away from vegetables. At least in the case of Ethiopia, and specifically Addis Ababa, the research suggests that food value chains were largely resilient to the shock associated with the pandemic. This finding is seemingly at odds with global projections regarding income declines and appears in contrast with much of the international narrative related to COVID-19 and food insecurity. These findings in Ethiopia should not be a cause for complacency, but rather a call to better understand the complexities and contextual nature of behavior of people and markets during crises.

Social protection

When people are unable to secure access to safe and nutritious food, there is an important role for social protection. In chapter 13 I described a variety of social protection programs designed to address both chronic and acute food security.

COVID-19 disrupted existing mechanisms of social protection that provide essential food access for the most vulnerable. As noted earlier in this chapter, COVID-19 sharply increased the number of food-insecure people and exacerbated the need for social protection.

In the Asia–Pacific region, most governments have ramped up their social protection measures in response to the pandemic. For existing programs, governments have increased how much beneficiaries receive and the number of people receiving benefits and have accelerated disbursements.[49] Government social protection programs, however, are threatened by sharply reduced budgets as well as the logistical challenges. School closures and supply chain disruptions hampered the delivery of school meals, an important instrument for supplemental nutrition for children around the world. In April 2020 WFP and UNICEF warned of "devastating nutrition and health consequences for the 370 million children missing out on school meals amid school closures."[50] By late 2020 more than 250 million children in ninety countries were missing out on school meals due to COVID-19.[51] Cash transfers and take-home rations were being implemented as an alternative in many of these countries.

In 2020 the Social Protection Interagency Cooperation Board (SPIAC-B)—a consortium of UN system agencies, other multilateral and bilateral development agencies, donor governments, and civil society observers—issued *A Joint Statement on the Role of Social Protection in Responding to the COVID-19 Pandemic*, calling for these urgent actions:

- prioritizing the most vulnerable to COVID-19 and its related social and economic impacts
- ensuring access to health services and supporting people in adopting COVID-19 prevention measures
- providing income security and access to essential goods and services and protecting human capabilities and livelihoods
- mobilizing necessary domestic and international financing to meet needs while not crowding out other essential services
- continuing and scaling up coordinated delivery capacities for social protection and humanitarian crisis response
- designing immediate crisis response measures and strengthening existing social protection systems in the medium and long terms[52]

The joint statement highlighted the need to protect and support all people throughout the COVID-19 pandemic. SPIAC-B called for social protection responses that are sensitive to the gendered burden of care-work, noting that women carry disproportionate responsibilities in unpaid and informal care-work.

National governments and the international community appear to be responding to this call, but only time will tell whether their actions will be sufficient to limit the damage to food security caused by COVID-19 and the additional supply chain disruptions caused by the Ukraine–Russia conflict. An even greater ongoing challenge is the redesign of social protection to be more effective in anticipating future shocks, including but not limited to pandemics.

LEARNING FROM THE PANDEMIC

The COVID-19 pandemic exposed and amplified existing vulnerabilities of food systems around the world. The rapid spread of the disease also tested the resilience of individuals, communities, businesses, governments, and humanitarian and development agencies to adapt and respond. Lessons are being learned and should be applied now and revised as needed as our experience grows.

David Dawe has been working on food security in the Asia–Pacific region for more than three decades. I have long admired Dawe, a colleague of mine at the International Rice Research Institute during the 1990s, for his clear thinking and writing on agriculture and food systems development. Dawe identified several key takeaways from his experience in the Asia–Pacific region that I will paraphrase and augment.[53] I believe that these recommendations are sufficiently robust to be relevant in most settings and provide direction in the event of future pandemics:

1. **Stop the virus spreading.** With the slow rollout of vaccines, especially in some low- and middle-income countries, we have no choice but to rely heavily on nonpharmaceutical interventions: social distancing, face masks, handwashing, and yes, business and school closures. We must suppress the spread of COVID-19. A prerequisite for a functional food system is confidence that the disease is under control.
2. **Take policy actions to ensure the smooth operation of food supply chains.** These actions require cooperation between governments and business and extend from production to consumption. They include special arrangements to facilitate continued functions of migratory farmworkers, domestic and international traders, transport operators, and food wholesalers and retailers. Aggregate

food supply is not threatened in the short term. However, disruptions and inefficiencies in the flow from producers to consumers cause unnecessary localized shortages and price spikes.

3. **Strengthen productivity and resilience of local food production and distribution systems.** The COVID-19 pandemic provided a wake-up call on vulnerability and dependence on global supply chains. This should not be taken as a cue to curtail international trade through export restrictions and a focus on national food self-sufficiency. The pandemic, however, provides cause for revitalizing local production to improve food systems resilience. To this end, we must increase cooperation among governments and between levels of governments, and we must foster partnerships with the business sector and communities.

4. **Strengthen social protection and economic stimulus policies** to be more responsive to short-term emergency needs, with special focus on the most vulnerable communities and demographic groups, while creating systems that build resilience and self-reliance of recipients in the short and long terms.

5. **Educate and support food consumers on healthy eating and food stewardship.** For many households, movement restrictions have led to greater awareness of our diets, our physical activity routines, and the food waste we generate. Governments and public-interest organizations, in partnership with businesses, should use this opportunity to build knowledge and support actions through policies (e.g., on advertising of unhealthy food).

As this book goes to press, the full impacts of the pandemic on food systems and food security remain unknown. The crisis in Ukraine has further disrupted an already fragile system of food supply and distribution. Food prices are touching all-time highs. As with the food crises of 1972–1975 and 2007–2012, there are several interacting factors involved, including the climate-related production impacts, high energy and related fertilizer and transport costs, freight disruptions, and geopolitical tensions.[54] COVID-19 has exposed the fragilities and inequalities of our food systems. The impacts of the pandemic vary across countries, sectors, and socioeconomic groups within countries. Low-income food-importing countries face the greatest challenges—poor urban and landless rural consumers depend heavily on wages for their food security. The macroeconomic shock that aggravates the problems these countries face makes it difficult to finance adequate social protection measures to offset high food prices. In contrast, high-income countries and net food exporters have greater scope for shielding poor consumers through cash transfers and other assistance. We must all continue to learn from this experience to be better prepared to withstand the next pandemic and other shocks that can disrupt our strategies and plans for achieving universal food security.

PART IV

Implementation

15

MORE THAN A MIRACLE

In synthesizing knowledge and best practice, I have shared some personal experiences across different scales and diverse contexts—each a challenge, each an opportunity. In the Philippines (Mindanao), in Cambodia, and in Malawi, the need to act was obvious. In these three cases, and in many other instances over the years, I experienced mixed emotions of anxiety and exhilaration as I confronted a common question: How do we improve lives and livelihoods? This was not an abstract question. The stakes were high: in some cases, life and death.

The smallholder farmers I met in Mindanao, Cambodia, and Malawi were struggling to survive in the face of isolation, conflict, and drought, respectively. Unsurprisingly, perhaps, in all three cases my solution—the *agronomist's* solution—involved seed. In Mindanao, it was the introduction of mung beans, a legume crop that would enrich soils, improve diets, and attract buyers. Our mission in Cambodia was to rebuild the country's rice production, starting with the return of thought-to-be-lost traditional seeds and the introduction of high-yielding varieties. In Malawi, we would couple more-productive hybrid maize seeds with subsidized fertilizer. But as I embarked on these three journeys, the challenges of implementation unfolded. How would we deliver results that mattered to farmers and that would be sustained over time? Would it take *a miracle*, recalling the term the *Manila Bulletin* used in June 1966 to announce the release of the first Green Revolution rice variety, IR8: "Marcos Gets Miracle Rice"?[1] Miracle *mung*, perhaps.

Mung beans, however, were not only prized by farmers—this nutritious, nitrogen-fixing crop also proved popular with insect pests that quickly pounced on and devoured the leaves, flowers, and pods, and even the stored grains. As a result, an aggressive extension effort was required to reduce pest losses in the field and in storage. Furthermore, farmers needed cash to pay for seeds, fertilizers, and chemicals, so a credit scheme was developed in partnership with the Philippine National Bank (see chapter 10). There was more to mung than met the eye.

Similarly, reintroducing Cambodia's traditional rice varieties, in the aftermath of the Khmer Rouge regime, required several seasons of research to match the seeds—repatriated from the International Rice Genebank in the Philippines—with the country's diverse rice-growing environments and to multiply and distribute these seeds to thousands of traumatized and displaced farmers (see chapter 2). Investments in improving soil fertility, land leveling, and pest management were also needed for farmers to gain full benefit from the rescued seeds. Again, seeds alone were not the solution.

In Malawi, I quickly discovered that farmers' traditional storage systems were inadequate and postharvest losses were diminishing the anticipated benefits. Moreover, the wide coverage of the Malawi input subsidy program revealed deficiencies and inefficiencies in infrastructure and institutional support. There was no obvious exit strategy for the seed and fertilizer subsidies. In these circumstances, must we insist on an exit strategy? And, if so, what should be the time frame? We were forced to rethink our concept of sustainability in the context of extreme poverty and food insecurity (see chapter 9).

In summary, delivering results that made a difference was not straightforward, especially when facing a deadline and budget limitations. I came to Mindanao on a two-year contract and ultimately stayed three years; project management (all of whom were civil engineers) viewed this as an unreasonably long time to get a job done. After all, how could handing out seed be more difficult than designing and building a road or a bridge? The sense of urgency was even greater in Cambodia and Malawi as they each emerged from nationwide crises affecting the lives and livelihoods of millions of people. As an agronomist, I confess that I placed a lot of confidence in seeds and their increased supply as the solution. Throwing seeds at the problem, however, or even more broadly raising food production, was no panacea.

In my experience, no single intervention—be it a new technology or a new policy—would ever prove to be *the* answer to achieving food security. It did not take long to realize that smallholder credit, previously unheard of for upland farmers, was needed to reach and sustain technology adoption by the farmers in Mindanao. And Cambodia would need its own capacity to develop and disseminate technologies and policies, not only for rice but for other agricultural enterprises, to improve rural productivity and national economic development. Eventually this realization led to the establishment of the Cambodian Agricultural Research and Development Institute.[2]

Malawi can be regarded in many respects as an implementation success. A steely determination to reduce dependency on food aid, and bold policies that were not well received by donors, resulted in a quantum leap in the nation's maize production. Unlike many African nations, Malawi was spared the destructive riots that accompanied the food price spikes during 2007–2012. But this came at a cost to other much-needed public investments in education, health, and roads. And even with improved access to inputs, Malawi's maize farmers obtained yields that were less than a third of the global average, despite having the potential to further double national production through additional investments in technologies, policies, and infrastructure.

We can draw encouragement from these three cases and the many thousands of other examples of partial and often short-lived successes in implementation. But whatever progress we may claim, the harsh reality is that half of the world's population is consuming a less-than-optimal diet for good health—either too little or too much, as highlighted in chapter 12—and our systems of food production are degrading the resource base on which sustained production depends. To reiterate the simple but stark warning of the EAT–*Lancet* Commission: "Food systems have the *potential* to nurture human health and support environmental sustainability, however our current trajectories threaten both," concluding that a "Great Food Transformation" to healthy diets from sustainable food systems is "both necessary and achievable" (my italics).[3]

There is near universal agreement that our global food system must transform, and there have been repeated public commitments to achieve this. But in the words of Pope Francis to the United Nations General Assembly in 2015: "Solemn commitments . . . are not enough, even though they are a necessary step toward solutions." While not one to dismiss the power of miracles, Pope Francis called for "effective, practical and constant, concrete steps and immediate measures" in the pursuit of sustainable development.[4] And there can be no truly *sustainable* development without universal food security.

As the characters in Sidney Harris's most famous cartoon remind us (figure 15.1), we could step back and wait for "a miracle" to occur or we could be a little "more explicit." For better or worse, in Mindanao, Cambodia, Malawi, and parts in-between, I have consistently opted for the latter. In this chapter and the next, I will present my thoughts on how to be *more explicit* about implementation

"I THINK YOU SHOULD BE MORE
EXPLICIT HERE IN STEP TWO."

FIGURE 15.1 More than a miracle.

in our pursuit of a food-secure world, beginning with three key determinants of success:

- a vision of success: universal food security
- an agenda for action: the Big Five investment areas
- a whole-of-society approach: agents of transformation

A VISION OF SUCCESS

Donella Meadows, lead author of *The Limits to Growth* report, wrote in 1987: "Visions alone don't produce results, but we'll never produce results that we can't envision or that we don't like when we do envision them. The absence of vision is one of the main reasons there is still hunger on this earth."[5] In that article, Meadows, like John Lennon,[6] called on her students to "imagine" a world without hunger: "Imagine, really imagine, the whole world with no one living in dire need. Think how it would feel to be part of a society that had taken on and solved, permanently, the problem of hunger."

Throughout this book, I have argued that our ultimate destination must be universal food security by 2050: there should be *no one left behind*, to borrow from a core principle of the 2030 Agenda. To visualize what that looks like, I revisited and modestly adapted (in chapter 8) the powerful description of food security that was first articulated at the 1996 World Food Summit held in Rome: "Food security exists when all people at all times have physical and economic access to sufficient, safe and nutritious [and sustainably produced] food to meet their dietary needs and food preferences for an active and healthy life."[7]

I added the phrase *and sustainably produced* to capture more explicitly the vital importance of sustaining the environment that we depend on for food production. At the same time, I recognize and accept that some readers will argue that the phrase *at all times* implies the objective of sustainability. But I wanted to be absolutely clear and unambiguous that there is no food security if the supply cannot be sustained over the long term.

While I applaud the clarity, detail, and longevity of the World Food Summit definition, I find it a little long and unwieldy to articulate to most audiences, and even more challenging having added my own phrase. So I distilled the definition to its essence—a more succinct and memorable vision statement that describes the food-secure world I believe we all want: **a world where every person enjoys a healthy diet derived from sustainable food systems.**

Pause for a moment and reflect on this proposed outcome of our interventions and investments in transforming the global food system. We now have a vision of *universal* food security from which everyone—I hope—can draw inspiration. This vision sets the goal of an integrated and coherent agenda for action.

AN AGENDA FOR ACTION

In part III, I proposed and elaborated on a transformation strategy that would lead us to universal food security. The strategy describes the importance of an integrated and coherent agenda for action across the Big Five food system investment areas:

- **Sustainable intensification**. We need to produce sufficient, safe, and nutritious food for all, while halting and reversing environmental damage.
- **Market infrastructure**. We need to invest in infrastructure so that we can move food more efficiently from where it is produced *sustainably* to where it is consumed *sustainably*.
- **Postharvest stewardship**. We need to reduce postharvest losses and waste across food value chains, thereby increasing food consumption as a proportion of food production.
- **Healthy diets**. We need to achieve dietary convergence toward context-determined healthy diets that include all essential nutrients in the right amount and balance.
- **Social protection**. We need to protect and support the most vulnerable and needy people, for whom the markets do not work, and for whom healthy diets are out of reach, physically or economically.

Across the Big Five are important *cross-cutting* investment areas, including education, gender and women's empowerment, public health services, clean water and sanitation, climate change mitigation, good governance, and ending civil conflict. Action in each of these complementary areas will enhance outcomes across our Big Five investment areas.

I acknowledge that there will always be a diversity of views on which area of intervention and investment is most important. No doubt, others would organize these five areas of action in different ways. As I have emphasized throughout the book, each context demands a unique blend of actions in each of these areas. But beyond coherence, or simply recognizing complementarities, we are best served by seeking synergies and spillovers, whereby investments in one area yield positive impacts in another. For example, school meal programs can deliver healthier diets while also increasing farmers' incomes by sourcing locally. Investments in transport infrastructure and mobile phone communications can enable sustainable crop intensification, reduce postharvest losses, and improve delivery of social protection programs. Rainfed intensification programs can promote legumes and other nutrient-dense crops, thus lowering the cost of a nutritious diet.

As I have emphasized throughout this book, context and location matter. Achieving food security in New York City requires a different mix of actions from those needed in Malawi or Timor-Leste. In all cases, however, once we agree on *which* interventions and investments are needed, a diverse range of actors and stakeholders must act to ensure coherence, recognize and minimize trade-offs, and exploit synergies toward food system transformation. Transformation requires both individual and collective actions. In delivering results—in being more explicit—it is important to identify and mobilize these agents of transformation.

AGENTS OF TRANSFORMATION

Transforming food systems requires a whole-of-society approach. We cannot leave the challenge of universal food security to any one sector or level of engagement. For the purposes of implementing the agenda for action, I have identified five broad groupings of society as the essential agents of transformation:

1. **Public sector.** These institutions are mandated (albeit sometimes self-mandated) and funded to act in the public interest. These actors are normally governments and their constituent institutions, groups of governments such as the United Nations, or other intergovernmental initiatives such as multilateral development banks.
2. **Private sector.** These firms seek profits for their owners and shareholders through market forces, enabling competition and innovation. These actors may range from large multinational corporations to smallholder commercial farms. Farmers' organizations play an important role in advancing farming household livelihoods through collective action in accessing information, inputs, finance, and markets. They also advance the interests of producers through public and political advocacy.
3. **Third sector.** Not-for-profit organizations, often described as the *third sector*, are entities that seek social and/or environmental outcomes but are not part

of the formal public sector and are not explicitly seeking financial profit for their owners and supporters. Within the third sector, three broad types of actors can shape the transformation, alongside the public and private sectors:

- **Social entrepreneurs.** These entities use business approaches to address social issues.
- **Community-led organizations and social movements.** These entities seek to act in the public interest or for specific subgroups. These organizations can mobilize local awareness and empower communities to hold governments and businesses accountable for their promises and results.
- **Wealthy individuals and families.** These philanthropic entities claim a mandate to act in the public interest or for specific subgroups.

Not all organizations fit neatly into these categories. Understanding the nuances and motivations of their contributions, however, can help design implementation strategies. In addition, I highlight two additional agents of transformation—the higher education sector and individuals—that do not perfectly align under the previous three sectors outlined:

4. **Higher education sector.** Universities and other institutions of higher learning are often underutilized actors in supporting food system transformation. They operate through a wide range of models but increasingly constitute a hybrid model that serves the public interest while seeking profits. Universities traditionally contribute to society through their education, research, and public service functions. With a major focus on agriculture, America's land-grant universities and colleges aim to advance the relevance of education and scholarly research through engagement with society.[8] In September 2019 Columbia University president Lee Bollinger proposed an even greater role for Columbia in meeting the needs of humanity: the *Fourth Purpose*. Bollinger describes this Fourth Purpose as "to extend Columbia's abilities to bring the extraordinary knowledge and capacities of the University in tandem with the wider academic community and actors beyond the campus to more effectively address pressing human problems."[9] Toward this goal of global engagement and impact, Bollinger championed the establishment of the Columbia Climate School.[10]

5. **Individuals.** Ultimately, nothing happens without individual people, acting independently or in concert with others, including through the internet

and other digital communication technologies. Individuals now have enormous power to transform food systems in the following ways:

- **Giving electoral and financial support to other individuals and to political parties** who commit to positive changes in food systems. The food riots across the world, in response to rapidly escalating prices during 2007–2012, spotlighted the position of food as a political issue. Individuals must hold their leaders accountable for the decisions and commitments made at the World Food Summit in 2021.
- **Making buying and consumption decisions** that shape private-sector policies and investment decisions. Individual consumers decide on what to eat and drink within the context of their food environments, financial capabilities, and cultural norms.[11] The growing interest of plant-based alternative foods reflects consumer support for climate action and healthy diets.
- **Volunteering with and giving financial support to third-sector organizations** that act in ways to advance food system transformation. There is growing support for social activist organizations like Global Citizen that mobilize individuals—particularly youth—to end extreme poverty and tackle climate change, with a focus on taking concrete actions.[12]
- **Educating themselves on how to become more informed and effective agents of transformation** through their own behavior and their encouragement and support of others. People who read this book and others on food systems have chosen to expand their understanding in ways that improve their capabilities as agents of transformation.
- **Choosing careers and jobs** that lead to work that directly advances the agenda for action. Based on my experience with graduate students during the past thirteen years, I see there is growing interest among young people to pursue careers in the agriculture and food sectors, across the public, private, and third sectors.

Notwithstanding the power of the individual, transforming food systems to achieve a food-secure world requires *collective action* to advance interventions and investments across the Big Five investment areas. Those actions have to come from coherent and, where possible, synergistic policies and actions from the public sector, the business sector, and the third sector, at three levels of engagement:

- **International.** These initiatives are implemented by organizations that operate beyond national boundaries, often undertaking multicountry cooperation to achieve common goals and address transnational problems and opportunities.
- **National.** These programs and policies are implemented by organizations at the national level with the purpose of achieving national goals.
- **Local.** These programs and policies are implemented at various subnational levels, such as states, provinces, districts, cities, and towns, right down to the village and community levels.

Those actions can take the form of individual policies or programs executed by a single entity or, increasingly, through models of cooperation that range from a simple bilateral public–private partnership to complex coalitions and consortia spanning sectors and levels of engagement. For the remainder of this chapter, I will elaborate on opportunities for delivering results across each of these three broad levels of engagement.

INTERNATIONAL ACTION: NECESSARY BUT NOT SUFFICIENT

Many global forums assemble organizations and convene deliberations on the larger goal of a food-secure world. I will not attempt to provide an encyclopedic listing of such assemblies. Instead, I will focus on a few I have observed to have the potential for greatest impact.

Rome is the headquarters of the three most important UN agencies that deal with food security: The Food and Agriculture Organization, the World Food Programme, and the International Fund for Agricultural Development (IFAD). Known colloquially in development circles as the Rome-based agencies (or RBAs), these organizations each contribute knowledge, technical expertise, and finance, along with forums to discuss technical and policy issues related to food security. Their governing councils meet annually and bring together high-level representatives of the member countries. The Committee on World Food Security (CFS)—with 133 member countries in 2022—was established in 1974 as a UN forum for review and follow-up of food security policies.[13] The CFS receives scientific and technical support from the High Level Panel of Experts on Food Security and Nutrition (HLPE), a science–policy interface established to facilitate policy debates and inform policy making.[14]

Multilateral development banks (MDBs) provide finance—as loans and grants—and give technical and policy advice to member countries across a range of sectors in support of development. The major MDBs are the World Bank and regional development banks, including the African Development Bank, Asian Development Bank, European Bank for Reconstruction and Development, and Inter-American Development Bank. They contribute to food security through investments in agriculture and rural development as well as more broadly through poverty reduction and economic growth.

The World Economic Forum (WEF) aims to advance global food security through public–private collaboration and facilitation of networks at global, regional, and national levels. At its most publicized annual meeting at Davos, and in other more focused gatherings, WEF convenes the world's foremost political, business, and cultural leaders to shape industry agendas. Established in 1971 as a not-for-profit foundation and headquartered in Geneva, WEF organizes various multistakeholder platforms on food security, including regional partnership initiatives known as Grow Africa and Grow Asia.

The major international player in food systems innovation is the Consultative Group on International Agricultural Research (CGIAR). The CGIAR is a network of fifteen international agricultural research centers that aims to provide the science, technologies, and tools that are needed to improve agricultural productivity and sustainability. The work of many of these centers has been featured in this book. Having worked eighteen years with the International Rice Research Institute and six years with the World Agroforestry Centre, I have seen firsthand the power of innovation as a driver of development and a catalyst for improving food security. But much more is needed as we address the failure of food systems to achieve our vision of universal food security, a challenge even more daunting as we increasingly face the effects of climate change, which has brought new threats to the productivity, stability, and sustainability of our food supply. During 2021 and 2022 the CGIAR was undergoing a reorganization that promised "a unified and more impactful One CGIAR."[15] One CGIAR promises "a dynamic reformulation of CGIAR's partnerships, knowledge, assets and global presence, aiming for greater integration and impact in the face of the interdependent challenges facing today's world."[16] In the next section I will share my thoughts on the risks and rewards of reorganizing the CGIAR.

At a broader international level, the UN General Assembly convenes nations annually to discuss progress in all areas of economic and social development. It serves as the main deliberative, policy-making, and representative organ of the United Nations. Comprising all 193 Member States, the General Assembly is a unique forum for multilateral discussion of international issues, one of which is food security. Since 2015 there has been a strong focus on the Sustainable Development Goals. The High-Level Political Forum on Sustainable Development provides the central platform for follow-up and review of the 2030 Agenda for Sustainable Development and the SDGs.

In September 2021 Secretary-General António Guterres convened the UN Food Systems Summit to launch new actions to deliver progress on all seventeen SDGs, each of which relies to some degree on healthier, more sustainable, and equitable food systems. SDG 2 could not have been clearer when it called on the world to "end hunger, achieve food security and improved nutrition and promote sustainable agriculture" by 2030. The summit sought to "awaken the world to the fact that we all must work together to transform the way the world produces, consumes and thinks about food."[17] It was yet another call to action, a reminder of past commitments and unfilled promises.

The skeptic in me asks: With all these commitments and calls to action, why does the global food system remain in such bad shape? Why are four billion people consuming *less than* healthy and nutritious diets? Why are our food systems contributing about a third of the world's greenhouse gas emissions? Why are our land, water, and biodiversity resources all deteriorating? Why, despite all these global and regional actions, do we continue to fall short in delivering universal food security?

In these situations, it is surely necessary to explore improvements in the global institutional architecture. For example, should the three RBAs and others be aligned more formally to create a *World Food Commission*? Recognizing the multidimensional nature of food security, we could add the UN Children's Fund, World Health Organization, UN Environment Programme, World Trade Organization, UN Refugee Agency, and UN Women to this group. The new coalition could be given its *commission* by the UN General Assembly. To ensure an appropriate profile, the World Food Commission could be guided by a board comprising selected heads of state, perhaps representing major regions of the world on a rotational basis, as well as including the UN secretary-general or deputy secretary-general and the heads of the RBAs. All those UN agencies would continue as separate recognizable entities with a high degree of independence and accountability to their respective national constituents. The proposed commission would form a strong coalition rather than a merger. The key difference is that the World Food Commission would provide additional *collective accountability* to the UN through the General Assembly.

Drawing on lessons from other UN-led and -centered initiatives, and especially my own experience with the UN's High-Level Task Force on Global Food and

Nutrition Security (described in chapter 1), I would argue for an even bigger tent for the commission that would include the following:

- **Multilateral development banks.** At a minimum, I would include the World Bank and the regional development banks.
- **International business sector.** The WEF is well organized, experienced, and clearly committed to mobilizing the expertise of the private sector in the area of food systems.
- **Philanthropic foundations.** The Bill and Melinda Gates Foundation and Rockefeller Foundation are obvious candidates for membership of the commission, given their historic leadership in the areas of agricultural development and food security.
- **Civil society.** We should consider including Global Citizen, an active, results-oriented youth group mentioned earlier in the chapter. In addition, the proposed World Food Commission could draw on apex organizations representing food producers and consumers.
- **Research institutions and universities.** The CGIAR is an obvious candidate for bringing to the table the knowledge, skills, and experience in food systems science and technology spanning more than five decades. The Sustainable Development Solutions Network, established in 2012 by then UN secretary-general Ban Ki-moon, would provide access to additional global scientific and technological expertise to promote practical solutions for sustainable food systems transformation.[18]

How would it work? I envisage the commission being supported by a small secretariat and convened annually in association with the UN General Assembly, as well as periodically in each of the world's major regions and subregions. The business of these meetings would be straightforward:

- Organize the work.
- Mobilize the finance.
- Review the progress.

As one of its first tasks, the World Food Commission could establish a new food security fund that builds on the model already established by the Global Agriculture and Food Security Program (GAFSP), which I described in chapter 2. A refreshed and replenished GAFSP would support national and regional food security initiatives. The most obvious institutional home for GAFSP would be IFAD—an international finance institution that was established in 1977 following the 1972–1975 food crisis. IFAD has been effective in connecting smallholder farmers to markets and services through concessional loans, grants, and technical assistance. While working at the World Agroforestry Centre (1998–2004), I found IFAD to be highly effective and well respected by governments, especially in sub-Saharan Africa.

The focus of the World Food Commission would be unambiguous—universal food security—and the deliberations and decisions of the commission would be guided by the best available scientific and policy advice. To that end, there are growing calls for an independent intergovernmental scientific advisory body to serve such a purpose.

Inspired by the Intergovernmental Panel on Climate Change, whose assessments have increasingly attracted the attention of policy makers, an independent science panel on food systems would provide impartial evidence-driven assessments of the state and direction of world food security and would synthesize the available science and technology in addressing food insecurity. Reflecting on the IPCC experience, a *Nature* editorial in 2021 noted the challenge a food systems scientific panel would have in maintaining integrity in the face of vigorously contested views and the high degree of trust required among participants and stakeholders, including industry, governments, nongovernmental organizations, farmers, and consumers.[19] Nevertheless, given IPCC's remarkable success in alerting the world to climate change and the opportunities for climate action, there is no downside in organizing a similar body to advise all agents of transformation on the science and policy options, *based on evidence*.

Considering the relationship between conflict and food insecurity, there is also a strong case for the commission to report periodically to the UN Security Council.[20] Regrettably, some members have opposed past efforts to elevate the profile of food security on the Security Council agenda.[21] But as we look to secure peace in conflict-affected regions of the world, including most recently in Syria, Yemen, Afghanistan, and Ukraine, and to prevent conflict elsewhere, the Security Council would be well advised to include the role of food security in its mandate.

Is there an appetite for creating such an all-embracing, inclusive World Food Commission? Is there the political will to enlarge GAFSP beyond its currently modest—one might argue, token—scale of operation and support to governments? Do we have the energy and patience to engineer a credible and effective intergovernmental science panel on food systems?

I can already hear howls of protest from those who have convinced themselves that the international system is doing just fine as it is: that UN agencies, MDBs, the private sector, and other actors are working seamlessly in the pursuit of a food-secure world, and there are joint statements and MOUs to prove it. But there are many like me who beg to disagree. Four billion malnourished people, a deteriorating natural resource base, and relentless climate change would suggest there is considerable room for improvement in the existing institutional architecture. Others will argue that we do not need more layers of bureaucracy and more consultation processes. The same arguments were made in opposition to the 2030 Agenda for Sustainable Development and the Paris Agreement on Climate Change. But, against the odds, these two global frameworks for action have emerged and show promise. There is always inertia to protect existing institutions, with their positions of power and influence. It would require uncommon leadership among world leaders to agree to step up and become collectively accountable for actually *delivering* on the vision of universal food security by 2050.

FOOD SYSTEMS INNOVATION: THE ROLE OF THE CGIAR

As noted earlier, the CGIAR has been an international powerhouse of agricultural and food systems innovation for more than fifty years. The search for a more integrated, global architecture, known as the "One CGIAR," drives the latest of many reorganizations of the CGIAR undertaken over its history. I have experienced most of them at close range. Past efforts to reshape the system have sought to improve coordination and collaboration among the constituent research centers. Cross-center, system-wide initiatives have been encouraged and supported through a multitude of time-bound programs and projects. One CGIAR goes a step further in explicitly seeking "fewer institutional boundaries" and a "unified governance structure."[22] One CGIAR is seeking to reconfigure the institutional structures that have provided the home for innovation and capacity building, along with well-respected nodes for complex, long-term scientific and institutional collaboration.

My plea to those leading the charge toward radical institutional reform of the CGIAR is to consider an important and undeniable historical fact: almost all successful innovations emerging from the CGIAR system have been the product of individuals or groups of researchers within individual centers, usually working in close collaboration with national ministries of agriculture, food, fisheries, and the like. Almost all the innovations listed in the CGIAR's shortlist of "best scientific breakthroughs of the past fifty years and beyond" come from scientists working at and identified with centers in partnership with national researchers.[23]

Based on almost two decades working at the International Rice Research Institute, I observed that IRRI's scientists and support staff—whether international or national—invariably demonstrated a deep sense of purpose and pride and a knowledge of history as they contributed to the institute's mission. The globally recognized scientific depth and apolitical stature of IRRI positioned it to convene countries in politically neutral forums and provide policy recommendations free of political considerations. We cannot afford to lose these intangible assets at IRRI or at any of the other centers. A strong system requires strong centers.

The *system* that is the CGIAR has clearly played a valuable role in mobilizing and sustaining donor support around the work of the centers falling under the system umbrella—just ask those centers not admitted to the CGIAR. Without doubt, collaboration and coordination can be improved and a nimble and responsive One CGIAR entity can help facilitate those outcomes. World-class scientific research is required to address the extraordinarily complex biological and social challenges facing agriculture as it undergoes profound transformations in the twenty-first century. This kind of research can only be conducted in well-funded centers of excellence with scientists and support staff committed to their work, which may well extend over decades in partnership with national research and development organizations. Facilitation of funding and partnerships by One CGIAR could be transformational. But, in any scenario, the quality of the science and supporting infrastructure in the research centers will determine success. And these centers are ultimately recognized, valued, and supported by their national partners based on the quality and impact of their research. A significant loss of the identity of the CGIAR centers is an inevitable consequence of their loss of independence and risks undermining the core strength of the system.

My advice to those "reformulating" the CGIAR is to proceed with caution and sensitivity, and to give weight to the perspectives of those who are the direct clients of the centers' research—the national agricultural research partners and national government ministries that are ultimately accountable to their citizens for delivering food security. Yes, a strong system requires strong centers. But a strong system also requires national research and development partners that value and contribute to the work of centers. Consultation with national partners is needed if the CGIAR is to remain relevant to their needs in advancing universal food security.

In 2020 the CGIAR's total revenues were reported as US$736 million, down from a high of US$1.08 billion in 2014.[24] Science and technology development to underpin the food requirements of ten billion people by 2050 and beyond requires much more than what is currently allocated. The CGIAR has announced its intention to raise US$2 billion per year by 2030.[25] That seems a modest ask. The right amount is a matter for deeper analysis, but a figure of at least ten times the 2020 funding level—to support the CGIAR and its collaboration with national research and development partners—would surely not be excessive in a world where the two largest producers of sugary drinks spend $8 billion a year on marketing alone.[26]

If we learned anything from the COVID-19 experience, it was (1) the fragility of biological systems, in that case, human susceptibility to a novel virus; (2) the importance of nimble and well-resourced research capability spanning public and private sectors, which led to the development of vaccines; and (3) the inequality of access to the benefits of science and technology, notably the case of slow vaccine rollout in low-income nations. For a moment, imagine a pathogen spreading through the world's rice, wheat, or maize fields with a ferocity similar to that with which COVID-19 infected the human population. The near total demise of the widely grown Gros Michel banana variety due to Panama disease (caused by a fungus) provides insights into the catastrophic effects of a virulent plant disease on food production, trade, and livelihoods.[27] Are we comfortable dismissing such a possibility as science fiction? If even a quarter of the world's food harvest were lost to a pest or disease, it would require a rapid and coordinated response to avoid a crisis without precedent in human history. We must be better prepared at the global level to tackle the food systems' equivalent of the COVID-19 pandemic. The CGIAR is readymade to serve that purpose, and its greatest assets—the research centers—should not be imperiled.

WELL-NOURISHED NATIONS

Translating global commitments like the 2030 Agenda and the Paris Agreement into actions and impact has never been a straightforward task. What happens in New York often stays in New York. Most countries already have national development plans that are shaped by national priorities and politics and are constrained by fiscal limits. At a Columbia University event immediately following the UN Sustainable Development Summit in September 2015, I asked the president of an Eastern European country how he would respond when asked by his people: So, what happens next, now we have signed on to the SDGs? With a wry smile, he responded that he was not sure anyone would ask! That brought me back to earth after all the back-slapping and high-fiving that immediately followed the summit declaration. We need to be realistic. Global commitments, compacts, and resolutions are merely words on a page unless there is willingness to translate those noble words into national plans, policies, and investments.

Notwithstanding the overarching global commitments on SDG 2, to which all nations have agreed (in principle), the 2030 Agenda is emphatic about the responsibility of *individual* nations to translate these ideals into normative policies and operational programs. For me, the most important and relevant statements in the agenda are found in paragraphs 55, 59 and 63:

> The Sustainable Development Goals and targets are integrated and indivisible, global in nature and universally applicable, taking into account different national realities, capacities and levels of development and respecting national policies and priorities. Targets are defined as aspirational and global, with each Government setting its own national targets guided by the global level of ambition but taking into account national circumstances. Each Government will also decide how these aspirational and global targets should be incorporated into national planning processes, policies and strategies.
>
> . . .
>
> We recognize that there are different approaches, visions, models and tools available to each country, in accordance with its national circumstances and priorities, to achieve sustainable development
>
> . . .
>
> Cohesive nationally owned sustainable development strategies, supported by integrated national financing frameworks, will be at the heart of our efforts.[28]

Therefore, in response to *global* intent and commitments, there is more than enough encouragement in the 2030 Agenda to plan and invest with purpose within a national context—to design and execute comprehensive and coherent *national* food security strategies and plans.

The foundation of a successful national food security strategy is a vision of success. We can reframe our definition of universal food security to define that vision: **a *nation* where every person enjoys a healthy diet derived from sustainable food systems**—in short, **a well-nourished nation.**

We have seen that poor nutrition leads to poor outcomes in terms of costs to the health system, educational performance, and ultimately prospects for economic advancement. Therefore a *poorly* nourished nation is unlikely to become a prosperous nation. In contrast, a *well-nourished nation* provides the foundation for a prosperous nation. And to return to Donella Meadows and her colleagues and draw on the inspirational words in *The Limits to Growth*, a well-nourished nation would be one where "each person has an equal opportunity to realize [their] individual human potential."[29] To me, that would resonate in any national political agenda.

Achieving the "human potential" of a society should be the aspiration of every national government leader and every business leader seeking opportunity and growth of a nation. A well-nourished nation requires national-level actions across the Big Five investment areas that are highlighted throughout this book. The complementary and synergistic nature of food systems policies across the Big Five requires an integrated *whole-of-government* approach that uses and adapts existing structures and systems.[30] Led and coordinated at the highest national level, ministries responsible for agriculture, health, education, infrastructure, environment, women, social welfare, and finance will need to collaborate and be held jointly accountable for national food security, which should be in the interest of all citizens. Informed, responsive, and courageous national leadership is essential.

As noted earlier, businesses bring important knowledge, skills, and resources to the challenge of national food security. But they are fundamentally driven by the needs of their owners and shareholders. Governments must actively engage with business to ensure the continued benefits that emerge from innovation and competition in the private sector. Governments are needed to set enabling policies, build infrastructure, and generate public goods to create and sustain opportunities that will serve the public interest for the long run.

An educated, demanding, and organized public can also be a powerful driver of transformation by holding governments and businesses accountable for results. Large-scale societal change requires social movements, public activism, and awareness campaigns that can serve as a trigger for government policy changes.[31] No single ministry or business or individual can create a well-nourished nation. No nation can afford to wait for a miracle to occur. And no responsible leader can afford to ignore the consequences of food insecurity. As we have seen throughout this book, the consequences of delayed action are often irreversible.

PILOT PROJECTS: LEARNING TO FLY

There are myriad actors and approaches as we move to local levels of societal organization. We can take the *well-nourished nation* concept and apply it to a province, a state, a city, a town, or even a village. In practice, as we localize the elements of food system transformation in pursuit of food security, we need to specialize, differentiate, and prioritize actions and investments. For example, a small town may have little need to address sustainable intensification but would have good reason to execute investments in reducing food waste, promoting healthy diets, and improving social protection. But our overarching vision of food security remains the same: a place where **every person enjoys a healthy diet derived from sustainable food systems.**

Successful transformation of our food systems requires horizontal and vertical collaboration across all levels: from global to national to local. Local decisions become aggregated to create tangible impact at scale when local experience is understood, adapted, and replicated through policies, programs, and investment decisions. Once again, we should not leave this to chance and wait for a miracle to occur. Opportunistically or by design, we can take local experience as *pilot projects* that can be taken to scale.

While local initiatives can provide local benefits and demonstrate the *potential* for delivering results at national, regional, and even global levels, too often these investments fail to deliver on their promise for wider impact. The road to agricultural development and food security is littered with promising pilot projects that failed to fly. Reflecting on my own experience in Mindanao in integrated rural development, in agroforestry with the World Agroforestry Centre, and later with the agricultural component of the Millennium Villages Project, I have identified five key strategies that lead to success in scaling up results from pilot projects:[32]

1. **Communicate the purpose, design, results, and full implications** of a pilot project from the outset.
2. **Anticipate and plan for the institutional and technical support** that contributed to local impact of the pilot project.
3. **Undertake research during the pilot phase** of projects to maximize the learning that would lead to improved design.
4. **Allow sufficient time for learning and adapting** to the realities of implementation, including the crucial need to develop and sustain partnerships.
5. **Involve higher-level political and policy leaders** in the planning and assessment of the pilot projects, to foster ownership and lasting commitment.

Of course, there is no inherent requirement for local initiatives to be pilots for anything larger. Their impacts can be, and often are, limited to the local communities they were intended to benefit. That said, our failure to achieve anything approaching universal food security—or even identify a *single* well-nourished nation—would suggest that more should be done to learn from actions and impacts at local levels to inform efforts to generate impact at scale.

Successful transformation of food systems inevitably involves decisions by very large numbers of individual people. These individuals make decisions on behalf of their households, their communities, their governments, their firms, and all other organizations that play a role in transforming food systems. This brings us to the final question: Who will make those critical decisions and take actions that will lead to positive outcomes at the international, national, and local levels?

The answer is *people* who are willing and able to lead—we need bold and informed *leadership* across public, private, and third sectors, at all levels. But what are the characteristics of leadership needed to transform complex and dynamic food systems? And how can we prepare and nurture transformational practitioner-leaders? We will tackle those questions in the final chapter.

16

LEARNING TO LEAD

In this final chapter I return to our common vision of universal food security: *a world where every person enjoys a healthy diet derived from sustainable food systems.*

Depending on where we sit, that "world" may be the planet—or it could be a nation, a province, a city, a community, or even a single household. In the spirit of the 2030 Agenda for Sustainable Development, attaining a food-secure world requires that we renew our pledge that "no one will be left behind."[1]

The challenge of universal food security is daunting. In terms of consuming a healthy and nutritious diet, four billion people—around half of the world's population—are "left behind." As I noted in chapter 7, some 2 billion are undernourished by one measure or another, and 2.5 billion or more are eating in excess of what is healthy. Each subunit of our world, from nation to household, can make its own calculation.

At the same time, in pursuit of our consumption needs and wants, we are degrading the environment on which the productivity of our food systems depends. We have seen the disastrous effects of our food systems: soil and land degradation (chapter 3), water resource depletion (chapter 4), loss of genetic resources (chapter 5), and climate change (chapter 6). I noted that our food systems generate a third of the greenhouse gases that contribute to climate change and the further degradation of our natural resources. And if that were not enough, almost a third of the food we produce is lost or wasted and never consumed, and that amount excludes food that is produced specifically for animal production and energy.

Universal food security is a fundamental requirement for a peaceful, productive, and sustainable planet. Of course, we could take a "glass half full" perspective and seek comfort in how much worse it could have been without the various Green Revolution initiatives described in chapter 2. Throughout part III and in chapter 15, I have maintained that we have the knowledge and the practical solutions with potential to "*fill* the glass." I have argued for food systems investments in five key areas: sustainable intensification, market infrastructure, postharvest stewardship, healthy diets, and social protection. I provided examples where these investments have improved food security at various locations and over a range of timescales. I have identified the need for different agents of transformation—public sector, private sector, third-sector institutions, universities, and the public at large—working concurrently, collaboratively, and coherently from global to local scales. My firm conclusion is that we have the knowledge—the *know-how*—to achieve our common vision.

As Norman Borlaug would say, "The potential is there, but you can't *eat* potential!" (my italics).[2] This analysis inevitably brings us to ask: How do we realize that potential? How can we ignite and sustain the necessary processes and motivate the key actors who will transform our food systems and lead us to a food-secure world?

WHY ACHIEVING FOOD SECURITY IS SO HARD

In 2015 Professor Peter Timmer, whose research I have shared throughout this book, published a masterpiece on food systems economics entitled *Food Security and Scarcity: Why Ending Hunger Is So Hard*. Drawing on more than five decades of research and practice, he argued for the centrality of understanding markets and finding the most appropriate public policies for ending hunger. Why does Timmer feel that ending hunger (and achieving food security) is so hard? Two of his statements sum it up: "Ending hunger is hard because it is complicated and deeply entwined with the organization of economic activities and their regulation through public policies," and "Because governments and markets need to work together."[3]

Timmer concluded, as I have, that the private and public sectors, along with the third sector, must work together collaboratively and coherently. The public and third sectors, through their policies and investments, must act to unleash market power in the public interest. Furthermore, Timmer argues that efforts to improve food security must take place mainly at the national level. International, regional, and local efforts must be mobilized to strengthen and augment the capacity of national institutions to undertake needed food systems transformations—consistent with the well-nourished nation paradigm introduced in chapter 15. Referring to the role of donors, though relevant to all food system investors,

Timmer says, "They will be successful only in environments in which political determination is coupled with good governance, high-quality data, and analysis to illuminate and guide effective policy. Only in those societies able to put all three pieces together—political determination, good governance, and effective policy—will it be possible to end hunger."[4] I maintain that these principles are applicable at *all* levels, from global right down to the individual household.

Drawing on the multitude of examples and experiences described in this book, it is clear that we have made important though imperfect progress in increasing agricultural productivity, reducing malnutrition, and avoiding the most severe famines. When Malthus first raised the alarm, the world population was less than a billion. With varying degrees of success, we now somehow manage to feed around eight billion people, even if we do not meet our definition of universal food security. But as we proceed on track to reach a world population of almost ten billion by 2050, we are decidedly *off track* in our trajectory toward a genuinely food-secure world.

All evidence presented in these pages indicates that the business-as-usual approach—tinkering at the margins—will leave us in an even worse state in 2050 than now. The most compelling argument for a radical transformation of our food systems is the irreversibility of—or, at best, the extraordinary cost of reversing—the most severe consequences of our inaction, including our failure to deal with stunting, land degradation, water resource depletion, biodiversity loss, and climate change.

How, then, do we change course, increase our level of ambition, and deliver on the promise of universal food security? How do we ensure that by 2050, no woman, man, or child will be left behind?

At this point, I have observed a common tendency to engage in what I regard as *rhetorical retreat*, coupled with the ubiquitous *call to action*. These often—but not always—well-meaning exhortations entail a foggy declaration that "all we need to do is [*reader to fill in the space*]." For example:

> *We already produce enough food to feed the world, right?*
> *All we need to do is* distribute the food more equitably.
> *All we need to do is* cut our food losses and waste.
> *All we need to do is* change our diets.

This notion that there is an obvious universal solution is appealing but predictably leads individuals and organizations to adopt evangelistic, myopic approaches that fall short in delivering our goal. To address the question of *how* to transform our food systems, I propose that we pause and step back from a solutions-driven strategy, while recognizing that context-specific interventions do exist and need to be deployed. In place of a technocratic solutions-driven approach, I offer a set of eight interrelated essentials in a constellation of *effort and impact* that apply to all areas of food system transformation (figure 16.1):

- **leadership:** providing direction and inspiration
- **governance:** ensuring systems of accountability
- **policies:** evidence-based decisions and pathways for impact
- **finance:** mobilizing funding to enable effort and accelerate impact
- **capacity:** individual and collective ability for effort and impact

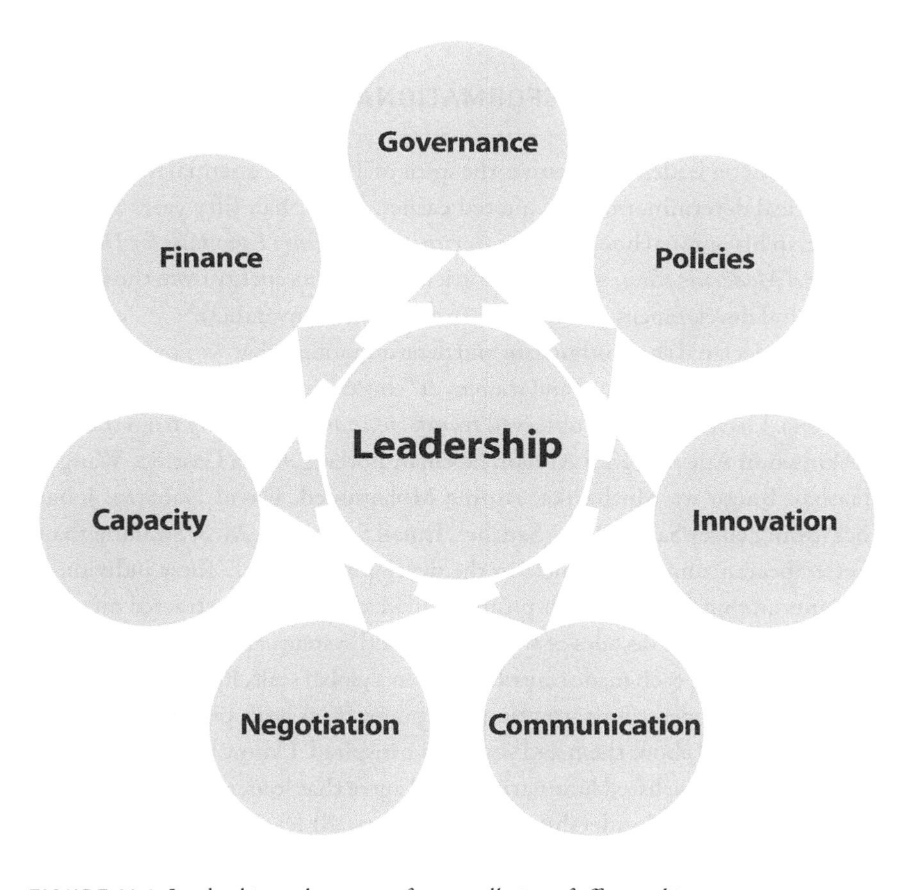

FIGURE 16.1 Leadership at the center of a constellation of effort and impact.

- **innovation:** knowledge to inform action and impact, now and in the future
- **communication:** providing an understanding of issues and opportunities
- **negotiation:** building alliances and commitments

I have placed leadership at the center of the constellation because, without leadership, it is unlikely the other seven areas of effort and impact will occur.

We are not starting from scratch. We already have models of success to inspire new levels of ambition and achievement. For each of these essentials, irrespective of the setting, there is important ongoing work, but major deficits of effort and impact must be filled. I have deliberately coupled effort with impact in recognition that effort will ultimately be judged by its impact on food security.

THE TRANSFORMATIONAL LEADER

The emphasis on leadership captures the spirit of Timmer's essential requirement of "political determination," as quoted earlier. More than fifty years ago, Art Mosher, in his seminal book *Getting Agriculture Moving: Essentials for Development and Modernization*, observed: "Agriculture gets its energy from the sun but agricultural development gets its drive from *persons*" (my italics).[5]

Mosher declared that enthusiasm and determination, what we now often refer to as *passion*, are "the engine" and sources of "constructive contagion."[6] I call this *leadership*. I have personally observed *transformational leadership* from the likes of Akinwumi Adesina, Kofi Annan, Norman Borlaug, Gelia Castillo, Wangari Maathai, Bingu wa Mutharika, Amina Mohammed, David Nabarro, Johan Rockström, Jeffrey Sachs, Pedro Sanchez, Ismail Serageldin, M. S. Swaminathan, Josette Sheeran, and many others in the development world. These individuals are giants among dozens of high-profile individuals who have attracted my great admiration over four decades of working on food systems and sustainable development. They have each made contributions on a global scale, inspiring and mobilizing action and deserving attention and praise from both present and future generations. Read about them and you *will* be inspired. I know for sure, however, that every one of the listed luminaries would agree that leadership cannot be left to those at the top. Leadership must extend to all levels in the public, private, and third sectors, and throughout society at large.

One of those transformation leaders, M.S. Swaminathan, would often remind us that there are two fundamental requirements for transformational change:

know-how and *do-how*. In a commentary for *Science* magazine in 2012, Swaminathan remarked: "If synergy can be created among scientific know-how, political do-how, and farmers' participation, it should be possible to achieve the goal of overcoming chronic and hidden hunger in large sections of the population of developing countries during this decade."[7]

Swaminathan's call to end hunger "during this decade" was not realized, nor will it be realized by 2030. But it is entirely possible to end hunger and achieve our wider vision of universal food security by 2050 *if* we invest heavily and cleverly in advancing both the know-how and do-how of food systems transformation.

In pursuit of a food-secure world, a transformation leader must draw on practical policies to implement the know-how. Most of those policies, though not all, can be drawn from the Big Five investment areas elaborated in part III. To be implemented at scale and speed, food systems policies need to be financed and supported by human capacity. Individuals and institutions must be mobilized and deployed to execute those policies, using the art and skill of negotiation and communication.

A longtime Columbia Earth Institute colleague of mine, Lisa Dreier, now a senior program fellow at the Harvard Kennedy School, has been a leading thinker and practitioner in the emerging field of systems leadership for more than a decade. Dreier founded and led the World Economic Forum's (WEF's) global program on food security and agriculture; developed global leadership initiatives involving more than 650 participating organizations; and catalyzed regional and national initiatives across Asia, Africa, and Latin America. She played a leading role in organizing the Transformational Leaders Network of 150 practitioners and experts. As a network member, I experienced firsthand the power of multistakeholder action programs and the valuable and unique facilitating role of the WEF in bringing together the diversity of complementary knowledge, skills, and authority needed to transform food systems.

With coauthors David Nabarro and Jane Nelson, Dreier has documented research and practice across multiple sectors to identify the core ingredients of effective *systems leadership*. This departs from the traditional top-down, linear model and instead requires "innovative and adaptive approaches that engage broad networks of diverse stakeholders to advance progress toward a shared vision for systemic change." The authors proposed the systems leadership approach for advancing the complex challenges embodied in the 2030 Agenda for Sustainable Development. Along what Dreier and colleagues describe as the "journey of systems leadership," they highlight ten "Aha! moments," five of which I found especially relevant to my journey as a food systems practitioner:[8]

- "That's our North Star." This speaks to the value of having a common goal of universal food security, as I articulated in the vision statement at the beginning of this chapter.
- "Everything is connected." We must appreciate the interconnectedness of policies and the need for coherence and coordination within and across the public, private, and third sectors.
- "No one is in control." It is important and challenging to understand that there is never an individual or single organization with full control over the food system.
- "We'll find a way." In dealing with complexity and interconnectedness, we must emphasize the role of innovation and collaboration in solving problems.
- "I can make a difference." Ultimately, it is up to individuals to accept responsibility for change through their direct contributions, partnerships, and influence on others.

Transformational food systems leaders must be agile and adaptive. As I have noted, the challenge of achieving universal food security and the consequences of failure are great. In fact, we have every reason to adopt a crisis management approach. Responding to the leadership needs of the COVID-19 pandemic, Ben Ramalingam, David Nabarro, and colleagues have identified several key principles to guide adaptive leadership that can be applied directly to the attainment of a food-secure world. They noted that the COVID-19 crisis is constantly evolving, with leaders required to face "unpredictability, imperfect information, multiple unknowns, and the need to identify responses quickly—all while recognizing the multidimensional (health-related, economic, social, political, cultural) nature of the crisis."[9] There could be no better description of the global food crisis. Effective food systems leadership requires the mindset and agility of a first responder while maintaining the focus, commitment, and stamina of a marathon runner.

Building on the preceding contributions to leadership research and practice, and drawing on my studies, experience, and observations as a food systems practitioner, I conclude that a transformational food systems leader will need to do the following:

1. Articulate a compelling vision—universal food security.
2. Connect the dots—take time to appreciate complexity, trade-offs, and synergies.
3. Show courage and diplomacy to raise uncomfortable truths.

4. Craft bold strategies and plans but revise them often.
5. Reflect and pursue the morally correct path.
6. Stress test and pilot promising ideas, not waiting for perfection.
7. Deploy evidence-based learning and adaptation systems.
8. Streamline decision-making with a focus on results and outcomes.
9. Build, nurture, and deploy coalitions of diverse partners for collective action.
10. Demonstrate accountability and have the humility to make corrections as needed.
11. Share and be prepared to sacrifice credit for success.
12. Learn to lead and contribute to the education of others.

Do I regularly score 12 out of 12 on these leadership attributes? Of course, I do not. They are both aspirational and motivational. But I do wish I had this checklist as a guide when setting out on my journey as an agricultural development practitioner.

A transformational food systems leader is most effective when operating within good governance systems that display transparency, inclusion, and accountability. Not everyone has this luxury, so one of the tasks of leadership is to promote and sustain good governance while exercising creativity and agility. If you find yourself in a less-than-enabling or even an oppressive work environment, you need perseverance and courage to exercise all those attributes. The attributes, however, cannot be graded as pass or fail; there are degrees of attainment. Any effort directed toward achieving them is worthwhile. And effective transformational leaders need to live to tell their tales.

LEADERSHIP EDUCATION: ADVANCING THE DO-HOW

One afternoon in January 2009, while at my desk in Nairobi, I received a call from Jeffrey Sachs, then director of Columbia University's Earth Institute, informing me of a new master's program to be established jointly by the Earth Institute and Columbia's School of International and Public Affairs. One thing led to another, and in late August 2009 I was addressing, as founding director, the incoming first cohort of twenty-four students in the Master of Public Administration in Development Practice (MPA-DP) program.

The establishment of the MPA-DP and, in subsequent years, more than thirty similar master's programs across six continents was the outcome of the International Commission on Education for Sustainable Development Practice, an independent group of development and education leaders commissioned by the

Earth Institute and sponsored by the John D. and Catherine T. MacArthur Foundation.[10] Their brief was to analyze existing training and education programs for development practitioners and to make recommendations for the future. The commission, cochaired by Sachs and John W. McArthur (a Canadian economist and director of the Center for Sustainable Development at the Brookings Institution), was inspired by the Flexner Report of 1910, which transformed medical training in the United States and Canada.[11] The commission observed: "Just as the field of medicine suffered from inconsistent and often ineffective medical training prior to the release of that report, the practice of sustainable development suffers from the lack of comprehensive and systematic training to foster the core competencies required of an effective practitioner."[12]

Drawing on the commission's recommendations, the MPA-DP program emphasized an interdisciplinary and multisectoral approach to problem-solving, along with a strong practice orientation. We conceptualized sustainable development as an ecosystem of institutions working toward common goals: initially the Millennium Development Goals and the end of extreme poverty, and progressively transitioning to the 2030 Agenda and the Sustainable Development Goals. By 2022 more than five hundred early-career professionals from eighty countries had graduated from the program, 70 percent of them women.

Since the first year of the program, all my students have been required to take a three-credit course called Global Food Systems (renamed Universal Food Security in 2022). I designed this course to address the challenges and opportunities for achieving a productive, profitable, inclusive, healthy, sustainable, resilient, and ethical global food system. Through lectures, readings, case studies, and classroom debates, we have examined the roles of science, technology, policies, politics, institutions, business, finance, aid, trade, and human behavior in advancing sustainable agriculture and achieving food and nutritional security. We have probed the interactions of food systems with global issues, including poverty and inequality, the persistence of chronic hunger and malnutrition, climate change, environmental degradation, international food business and value chains, genetic engineering, postharvest losses, and food waste. We have also confronted controversies, reflected on historical trends, identified common myths, and surfaced little-known but important truths about agriculture and food systems. In our final sessions, we have addressed the ultimate questions: Can we feed and nourish the world without wrecking it for future generations? Can we achieve universal food security?

Every year, especially during the first few weeks of the course, when I cover the basics of soil science and plant genetics—a totally new language to many of my otherwise well-read students—I have received more than a few queries as to why Global Food Systems would be a *required* course for an MPA in Development Practice. Yes, it makes sense to study economics, statistics, and management. But why study agriculture and food systems, when I am planning to work in public health or early childhood education or human rights?

By the end of the semester, the answer becomes clear to almost everyone. Food systems and food security provide a cornerstone of human development. Food is a fundamental daily need and the right of every person on Earth. Moreover, as I hope this book has shown, an understanding of food systems and food security requires holistic multidisciplinary, multisectoral, and multi-institutional approaches, which make up the core ethos of our MPA-DP program. Indeed, many students have gone on to apply their knowledge and skills through internships and jobs with organizations working on food systems, food security, and rural development.

All this suggests that an understanding of food systems should be a requirement for *all* education programs seeking to prepare practitioners and the wider public to address sustainable development. If we are seriously looking to achieve

FIGURE 16.2 Ban Ki-moon, UN secretary general, accompanied by Jeffrey Sachs, meeting students of Columbia University's MPA in Development Practice program in Mwandama village, Zomba District, Malawi, May 30, 2010.

Credit: Millennium Promise, https://www.flickr.com/photos/millenniumpromise/4721498430/in/photostream/.

a transformation in the way food is produced, distributed, and consumed, surely agriculture, food systems, and an understanding of food security—as articulated in these pages—should be part of all curricula, from kindergarten to graduate school.

Every graduate of the MPA-DP program has taken the Global Food Systems course. I think of each of these graduates as future development leaders finding their niche in the sustainable development ecosystem. They currently work for governments, UN agencies, development banks, investment management corporations, development consultancy firms, and nongovernmental organizations. Others have established social enterprises. A relatively small but ever-growing number of graduates work directly in the food systems sector in organizations such as the UN World Food Programme, the International Fund for Agricultural Development, the U.S. Agency for International Development, and third-sector organizations like the International Rescue Committee and the One Acre Fund. Others have gone on to pursue doctorates in the field of food and nutrition. I know of at least two who have become medical practitioners. All these graduates are leaders or have the potential to exercise leadership in ways that contribute to universal food security.

Reflecting on the successes and failures of the past, drawing inspiration from the many leaders who have influenced my thinking and practice, I have concluded that education is *the* single most important driver of positive change in food systems transformation. A well-informed, reflective *practitioner-leader* has the means to ignite the food systems transformation needed in the public sector, the private sector, the third sector, and the public at large. In the United Kingdom, John Ingram and colleagues from five universities have devised and implemented a similar multidisciplinary, cross-sector approach (the Interdisciplinary Food Systems Teaching and Learning program, or IFSTAL) for developing graduate professionals with the skills, tools, and capabilities to manage food systems' complexity.[13]

Effective implementation of past programs to improve food security has unquestionably relied on the good work of talented and motivated people. By introducing informed and motivated practitioner-leaders into enough organizations, we can transition our food systems inexorably toward our goal of universal food security. A global cadre of informed leaders, characterized by the twelve attributes articulated earlier, is now urgently needed to inspire, direct, and mobilize human endeavor toward our ultimate goal. That leadership will be required at all levels: from the highest echelons of the United Nations to the frontline workers who directly engage with food producers and consumers.

Inspired by what I have seen and experienced at the MPA-DP program at Columbia over more than a decade, I envisage universities playing a critical role in igniting the food systems transformations that we are seeking. For the most part, these institutions have been ignored and untapped by governments and the international community in advancing sustainable development, including food security. Some of the blame lies with the universities themselves as a result of their failure to incentivize work across disciplinary boundaries and to connect their research and teaching with real-world problems. Yet these institutions are the source of the future practitioner-leaders who will be responsible for policies and investment decisions over the coming decades. Columbia University has recognized this explicitly in adopting a Fourth Purpose "to help bring deep knowledge to the world we serve, and, in so doing, enhance the vectors of university research, teaching, service and impact."[14] If we are to achieve universal food security by 2050, targeting today's millennials and postmillennials is highly strategic.

Complementing this university-led approach are emerging institutional initiatives to support in-service development of food systems leadership. The Uongozi Institute (also known as the Institute of African Leadership for Sustainable Development) was established by the governments of Tanzania and Finland in 2010.[15] *Uongozi* means leadership in Kiswahili. Through a variety of learning programs and events, the institute works with both senior and aspiring leaders to advance sustainable development in Africa. Having served on its founding board for six years, I am convinced that the Uongozi Institute can make a major contribution by preparing practitioner-leaders to transform Africa's food systems, lead the continent to food security, and contribute to a food-secure world.

In April 2021 the Alliance for a Green Revolution in Africa (AGRA) launched the Centre for African Leaders in Agriculture (CALA) "to provide hands-on implementation support for African leaders in the agriculture sector." In launching CALA, AGRA president Agnes Kalibata highlighted the central role of leadership in advancing Africa's Green Revolution: "Our experience at AGRA has taught us that achieving Africa's food security targets and achieving inclusive economic growth will require leaders who are responsive, adaptable and collaborative, while also being ready to integrate new strategies for environmentally sustainable agriculture together with a diverse set of stakeholders."[16]

Through its sixteen-month Advanced Leadership Program, CALA recognizes the critical importance of state capability, multisector partnerships, local ownership, and professional leadership in designing and executing food security policies and programs in Africa. The first cohort of eighty current and emerging leaders was drawn from Ethiopia, Ghana, Kenya, Malawi, Nigeria, Rwanda, Tanzania,

UNIVERSAL FOOD SECURITY

A world where every person enjoys
a healthy diet derived from sustainable food systems

FOOD SYSTEMS TRANSFORMATION

LEVELS
- International
- National
- Local

ATTRIBUTES
- Productive
- Profitable
- Inclusive
- Healthy
- Sustainable
- Resilient
- Ethical

BIG FIVE INVESTMENT AREAS
- Sustainable intensification
- Market infrastructure
- Postharvest stewardship
- Healthy diets
- Social protection

AGENTS OF TRANSFORMATION
- Public sector
- Private sector
- Third sector
- Higher education sector
- Individuals

EDUCATION AND DEVELOPMENT OF PRACTITIONER–LEADERS
- Interdisciplinary
- Cross-sectoral
- Practice-oriented

FIGURE 16.3 Achieving universal food security: the theory of change.

and Uganda. Each country team comprised individuals currently working in government, the private sector, and civil society toward improving national food security. With ongoing training, skills development, event participation, and coaching from CALA, the program participants will apply their learnings to spearhead the implementation of high-priority agriculture and food security strategies in their respective countries.[17]

Columbia's MPA-DP program, the UK's IFSTAL program, the Uongozi Institute, and CALA exemplify the growing realization that investing in leadership—at all levels—will provide the impetus to ignite, scale, and sustain food security. These institutional approaches provide models for adaptation and application in all regions of the world and beyond agriculture to incorporate food systems transformation.

In figure 16.3 I summarize the theory of change that underpins the strong emphasis I have placed on developing leadership as the foundation of food systems transformation.

Investing in the education and development of practitioner-leaders will lead to better decisions by individuals, acting on their own and with others through institutions. Those individuals and institutions will directly and indirectly influence the actions and outcomes from the Big Five investment areas. Those actions, in turn, will be manifested in food systems transformation and will lead to our ultimate goal: universal food security.

Thomas Robert Malthus, a man not known for his excessive optimism, observed: "The exertions that men find it necessary to make, in order to support themselves or families, frequently awaken faculties that might otherwise have lain forever dormant, and it has been commonly remarked that new and extraordinary situations generally create minds adequate to grapple with the difficulties in which they are involved."[18] It is now *our* time to "awaken faculties" and "create minds" in our quest for a food-secure world. Like climate change and COVID-19, hunger and food insecurity are not going to just disappear. We have our vision of success. We have the knowledge, the tools, and the resources to be transformational. It is possible to end hunger while protecting the planet. Who is prepared to step up and take the lead?

ACKNOWLEDGMENTS

To write a book on ending world hunger while protecting the planet. Now how hard could that be? The answer is "very hard," dare I say "impossible," without the inspiration, advice, and support of many hundreds, perhaps thousands, of people. There are those who have encouraged and shaped my work over significant and pivotal segments of my forty-plus-year career—they provided me with the opportunities, the knowledge, and the confidence to want to say something. Others stepped in and offered critical advice on topics that stretched my personal expertise and experience. As in any successful book project, there are those who know how to turn a manuscript into a book—I am grateful for their contributions. And then there are the many friends and family members who were just there for me—cheering me on, reassuring me that the finish line was not too far away.

I will start with those whom I regard as my principal mentors and advisers who fundamentally and positively determined the direction of my career and life, more or less in chronological order: Ross Humphreys, Ian Aberdeen, John Flinn, Dennis Greenland, Sisira Jayasuriya, Walter Falcon, M. S. Swaminathan, Peter Timmer, Klaus Lampe, George Rothschild, Pedro Sanchez, and Jeffrey Sachs. Many others were extraordinary colleagues who worked alongside me for years at a time. Without them, the stories in this book could not have been told. Again, roughly in order of appearance, those indispensable colleagues are John Dalton, Colin Piggin, Greg Wells, Fabian Sweeney, Hubert Zandstra, Dan Minnick, Jojo Lapitan, Lydia Garcia, Thelma Paris, Gurdev Khush, S. K. De Datta, Kinlay Dorjee, Harry Nesbitt, John Schiller, Vo-Tong Xuan, Mimi Gaudreau, Vethaiya Balasubramanian, Mayette Nadal, V. Pal Singh, Jeff Zlonis, Catherine Kenyatta, Tiff Harris, Bob Huggan, Amadou Niang, C. J. Jones, Stephan Bognar, Jessica

Fanzo, and Shiv Someshwar. For every one of these amazing individuals I have named, there were many more whose unwavering support and collegiality are also highly appreciated.

I have called on many colleagues, including several just listed, to review one or more chapters. They have applied the rigors of their respective disciplines and unquestionably improved the quality of the final product, though I take full responsibility for remaining errors and omissions. The following people provided valuable chapter reviews in their areas of personal expertise and experience: Jessica Fanzo, Sarah Garland, Sisira Jayasuriya, Heather Kelahan, Pedro Sanchez, Philip Sayeg, Shiv Someshwar, Reiner Wassmann, and Bob Zeigler. I am also grateful for the inputs from three anonymous reviewers appointed by Columbia University Press. In addition to written reviews, I had valuable conversations on key topics with many others, including Frank Costello, David Dawe, Lisa Dreier, Patrick Kabambe, Joel Negin, and Joseph Rickman.

The writing of *Universal Food Security* began in earnest in 2019 while I was on a six-month study leave granted by the Columbia University School of International and Public Affairs (SIPA). I am grateful to Dean Merit Janow and Vice Dean Scott Barrett for their support and encouragement. I would also like to recognize my SIPA colleagues Shiv Someshwar, André Corrêa d'Almeida, and Kendal Stewart, who more than capably "held the fort" while I pursued this project and have been indispensable colleagues throughout my time at SIPA.

I am grateful for the research assistance provided by many students in the program I direct at SIPA—the MPA in Development Practice. Ryo Ogura and Titilayo Ola deserve special mention. My students have provided a valuable sounding board for many of the ideas contained in this book. They have caused me to broaden my thinking, question my assumptions, and sharpen my delivery. They are prominent among the practitioner-leaders to whom this book is directed.

In the journey from writing to production, I worked closely with a highly professional team from Coretext (Melbourne, Australia). Their content knowledge and editorial expertise were invaluable, and I am especially grateful to editor Brad Collis, production manager Christine Fotis, graphics designer Aidan Gifkins, and scholastic copy editor Margie Beilharz. Max McMaster of Master Indexing enhanced the accessibility of the book's content by crafting an outstanding index.

At Columbia University Press, my editor Caelyn Cobb has been the lead on this book throughout. I appreciate the work of editorial assistant Monique Briones, production editor Marisa Lastres, and copy editor Anita O'Brien, who have helped shepherd the book through production to its final form.

Without the support of my family, this book and indeed the journey leading up to it would not have been possible. My parents—Leslie and Joan—made great sacrifices and set me off on the right path. My dear wife Pam has been at my side, providing constant encouragement, insisting on a work-life balance, yet never complaining of my travels and time spent "working on the book." Pam and our kids—Paolo, Kinlay, and Camille—gave valuable feedback on my ideas and, most important, provided much-needed stability, inspiration, and love throughout the writing of this book.

In their own ways, each of these individuals and groups of supporters was vital to the successful completion of *Universal Food Security*. I am deeply appreciative of their contributions. And I share all credit with them.

NOTES

PREFACE

1. Donella H. Meadows et al., *The Limits to Growth* (New York: Universe Books, 1972), 24. The direct quote referred to "*his* individual human potential."
2. Food systems comprise all the actors, activities, and forces influencing the production, aggregation, processing, distribution, consumption, and disposal of food products. See chapter 8.
3. "Universal Declaration of Human Rights," United Nations, accessed August 11, 2021, https://www.un.org/en/about-us/universal-declaration-of-human-rights; "International Covenant on Economic, Social and Cultural Rights," United Nations Human Rights Office of the High Commissioner, accessed August 11, 2021, https://www.ohchr.org/en/professionalinterest/pages/cescr.aspx.
4. "Rome Declaration on World Food Security," World Food Summit, November 13–17, 1996, http://www.fao.org/3/w3613e/w3613e00.htm.
5. World Commission on Environment and Development, *Our Common Future* (New York: United Nations, 1987), chap. 2, https://sustainabledevelopment.un.org/content/documents/5987our-common-future.pdf.
6. United Nations, *Transforming Our World: The 2030 Agenda for Sustainable Development* (New York: UN Division for Sustainable Development Goals, 2015), 3, https://sustainabledevelopment.un.org/post2015/transformingourworld/publication.
7. United Nations, "Secretary-General's Chair Summary and Statement of Action on the UN Food Systems Summit," statement, September 23, 2021, https://www.un.org/en/food-systems-summit/news/making-food-systems-work-people-planet-and-prosperity.
8. The estimate of approximately 2.5 billion people being overweight (including obese) is based on the sum of the following: 2.2 billion overweight adults (Renata Micha et al., *Global Nutrition Report: The State of Global Nutrition*, [Bristol, UK: Development Initiatives, 2021]); 340 million overweight adolescents, five–nineteen years of age (World Health Organization Fact Sheet, https://www.who.int/news-room/fact-sheets/detail/obesity-and-overweight); and 39 million overweight children under five years of age (Micha et al., *Global Nutrition Report*).
9. The UN agencies are the Food and Agriculture Organization, International Fund for Agricultural Development, United Nations Children's Fund, World Food Programme, and World Health Organization. See FAO et al., *The State of Food Security and Nutrition in the World 2020. Transforming Food Systems for Affordable Healthy Diets* (Rome: FAO, 2020), http://www.fao.org/3/ca9692en/online/ca9692en.html#chapter-executive_summary.

10. Isaac Newton to Robert Hooke, February 5, 1675, Simon Gratz collection, Digital Library, https://digitallibrary.hsp.org/index.php/Detail/objects/9792.
11. The thirty-member Open Working Group of the United Nations General Assembly was established on January 22, 2013, and tasked with preparing a proposal on the Sustainable Development Goals.

1. PROPHETS OF DOOM

1. Subsistence farmers are small-scale farmers who usually produce just enough food for their own household needs with little or no surplus for sale.
2. Summarized from John Crowley, William J. Smyth, and Michael Murphy, eds., *Atlas of the Great Irish Famine, 1845–52* (Cork, Ireland: Cork University Press, 2012), principally chapters by William J. Smythe, "The Story of the Great Irish Famine 1845–52: A Geographical Perspective," 4–12, and John Feehan, "The Potato: Root of the Famine," 28–37.
3. Ireland was then part of the United Kingdom of Great Britain and Ireland, a sovereign state that existed between 1801 and 1922 under the Acts of Union 1800.
4. As reported by Sarah Lyall, "Past as Prologue: Blair Faults Britain in Irish Potato Blight," *New York Times*, June 3, 1997, https://www.nytimes.com/1997/06/03/world/past-as-prologue-blair-faults-britain-in-irish-potato-blight.html.
5. Thomas Robert Malthus, *An Essay on the Principle of Population as It Affects the Future Improvement of Society, with Remarks on the Speculations of Mr Godwin, M. Condorcet, and Other Writers* (London: J. Johnson, 1798). Referenced page numbers are from a softcover edition published by the Echo Library in 2006, http://www.echo-library.com/. His reference to "natural inequality" first appears on p. 9.
6. Malthus, 42.
7. Malthus, 55.
8. Malthus, 42.
9. Max Roser, Hannah Ritchie, and Esteban Ortiz-Ospina, "World Population Growth," 2013, most recently revised May 2019, https://ourworldindata.org/world-population-growth.
10. William Crookes, "Address of the President Before the British Association for the Advancement of Science, Bristol, 1898," *Science* 8, no. 200 (1898): 562.
11. Crookes, 573.
12. Carolyn Dimitri, Anne Effland, and Neilson C. Conklin, *The 20th Century Transformation of US Agriculture and Farm Policy*, Economic Information Bulletin no. 3 (Washington, D.C.: U.S. Department of Agriculture, June 2005).
13. Donald N. Duvick, "The Contribution of Breeding to Yield Advances in Maize (*Zea mays* L.)," *Advances in Agronomy* 86 (2005): 83–145.
14. FAO, *The State of Food and Agriculture 1960* (Rome: FAO, 1960), 1–2, https://www.un-ilibrary.org/content/books/9789210472760/read.
15. William Paddock and Paul Paddock, *Famine—1975! America's Decision: Who Will Survive?* (Boston: Little, Brown, 1967), 206.
16. Fairfield Osborn, *Our Plundered Planet* (Boston: Little, Brown, 1948); William Vogt, *Road to Survival* (New York: Sloane Associates, 1948).
17. Paul R. Ehrlich, *The Population Bomb* (New York: Ballantine Books, 1968), 72–80.
18. See Mann's critique of *The Population Bomb*: Charles C. Mann, "The Book That Incited a Worldwide Fear of Overpopulation," *Smithsonian Magazine*, January/February 2018, https://www.smithsonianmag.com/innovation/book-incited-worldwide-fear-overpopulation-180967499/.
19. Rachel Carson, *Silent Spring* (Boston: Houghton Mifflin, 1962).
20. Donella H. Meadows et al., *The Limits to Growth* (New York: Universe Books, 1972), 23–24. The original quote referred to "*his* individual human potential." The Club of Rome was created in 1968 to "address the multiple crises facing humanity and the planet." With one hundred full-time

members—notable scientists, economists, business leaders, and former politicians—it is best known for this, its first report. See https://www.clubofrome.org/.

21. Norman Borlaug, "The Green Revolution, Peace, and Humanity," Nobel Lecture, December 11, 1970, https://www.nobelprize.org/prizes/peace/1970/borlaug/lecture/.
22. FAO, "World Food Situation," accessed October 9, 2021, http://www.fao.org/worldfoodsituation /foodpricesindex/en/.
23. The initial effects of COVID-19 on food prices are briefly discussed in chapter 14.
24. Derek Headey and Shenggen Fan, *Reflections on the Global Food Crisis: How Did It Happen? How Has It Hurt? And How Can We Prevent the Next One?*, Research report 165 (Washington, D.C.: IFPRI, 2010).
25. Headey and Fan, xiii.
26. C. Peter Timmer, "Reflections on Food Crises Past," *Food Policy* 35, no. 1 (2010): 1–11.
27. "Global Fund Overview," Global Fund to Fight AIDS, Tuberculosis, and Malaria, accessed August 10, 2021, https://www.theglobalfund.org/en/overview/.
28. "L'Aquila Joint Statement on Global Food Security," L'Aquila Food Security Initiative, L'Aquila Summit, July 9, 2009, http://www.g8.utoronto.ca/summit/2009laquila/2009-food.html.
29. "G20 Leaders Statement: The Pittsburgh Summit September 24–25, 2009, Pittsburgh," G20 Research Group, University of Toronto, last updated November 29, 2011, http://www.g20.utoronto.ca/2009 /2009communique0925.html, para. 39.
30. "Our Impact," Global Agriculture and Food Security Program, accessed December 30, 2021, https://www.gafspfund.org/impact.
31. International financial institutions (IFIs) include multilateral, regional, and national development banks with international operations. IFIs include the World Bank, the International Monetary Fund, and regionally focused development banks such as the Asian Development Bank. IFIs are major sources of financial and technical support for developing countries.
32. United Nations Integrated Implementation Framework, L'Aquila Food Security Initiative. The data was located at https://iif.un.org/content/laquila-food-security-initiative# in mid-2020 but can no longer be reached. Launched on June 6, 2012, the UN Integrated Implementation Framework was intended to serve as "a concrete global follow-up to the 2010 Millennium Development Goals Summit and the General Assembly mandate to report on progress in the implementation of the Millennium Development Goals," https://www.un.org/press/en/2012/sgsm14334.doc .htm.
33. G7 France Biarritz, *Biarritz Progress Report G7 Development and Development-Related Commitments*, 2019, https://www.diplomatie.gouv.fr/IMG/pdf/rapport-g7-bat-web_cle85fc7e.pdf, 65.
34. United Nations, *Transforming Our World: The 2030 Agenda for Sustainable Development* (New York: UN Division for Sustainable Development Goals, 2015), https://sustainabledevelopment.un .org/post2015/transformingourworld/publication.
35. United Nations, *The Millennium Development Goals Report 2015* (New York: United Nations, 2015), https://www.un.org/millenniumgoals/2015_MDG_Report/pdf/MDG%202015%20rev %20(July%201).pdf.
36. Food and Agriculture Organization, International Fund for Agricultural Development and World Food Programme, *The State of Food Security in the World 2015. Meeting the 2015 International Hunger Targets: Taking Stock of Uneven Progress* (Rome: FAO, 2015), 8, table 1, http://www .fao.org/3/i4646e/i4646e.pdf. The number of malnourished people in the world was subsequently adjusted downward in 2020 based on new data on population and food supply, along with new household survey data that enabled the revision of the inequality of food consumption for thirteen countries, including China. See "Understanding Hunger" in chapter 7.
37. World Commission on Environment and Development, *Our Common Future* (New York: United Nations, 1987), https://sustainabledevelopment.un.org/content/documents/5987our-common -future.pdf.
38. John Beddington, *Food, Energy, Water and the Climate: A Perfect Storm of Global Events* (London: Government Office for Science, 2009).

39. United Nations, *Transforming Our World*, para. 14.
40. "About the Summit," United Nations Food Systems Summit 2021, accessed August 10, 2021, https://www.un.org/en/food-systems-summit/about.
41. Walter Willett et al., "Food in the Anthropocene: The EAT–Lancet Commission on Healthy Diets from Sustainable Food Systems," *Lancet* 393, no. 10170 (2019): 447–92; Tim Searchinger et al., *Creating a Sustainable Food Future: A Menu of Solutions to Feed Nearly 10 Billion People by 2050*, Final Report July 2019, World Resources Report (Washington, D.C.: World Resources Institute, 2019), https://research.wri.org/sites/default/files/2019-07/WRR_Food_Full_Report_0.pdf; International Panel on Climate Change, *Climate Change and Land: An IPCC Special Report on Climate Change, Desertification, Land Degradation, Sustainable Land Management, Food Security, and Greenhouse Gas Fluxes in Terrestrial Ecosystems*, ed. P. R. Shukla et al. (Geneva: IPCC, 2019).
42. "HLPE Reports," High Level Panel of Experts on Food Security and Nutrition, accessed August 10, 2021, http://www.fao.org/cfs/cfs-hlpe/reports/en/.
43. EAT–*Lancet* Commission, *Healthy Diets from Sustainable Food Systems*, Summary Report of the EAT–*Lancet* Commission (Oslo: EAT, 2019), https://eatforum.org/content/uploads/2019/07/EAT-Lancet_Commission_Summary_Report.pdf, 20, 5.
44. Searchinger et al., *Creating*, iv.
45. HLPE, *Impacts of COVID-19 on Food Security and Nutrition: Developing Effective Policy Responses to Address the Hunger and Malnutrition Pandemic* (Rome: Committee on World Food Security, 2020), http://www.fao.org/3/cb1000en/cb1000en.pdf. In this report, the HLPE introduced two new dimensions of food security: agency and sustainability. Previously, food security was most commonly described in terms of four "pillars": availability, access, utilization, and stability (Committee on World Food Security, "Global Strategic Framework for Food Security & Nutrition," 2014, third version, 7).

2. GREEN REVOLUTIONS

1. William S. Gaud, "The Green Revolution: Accomplishments and Apprehensions," address before the Society for International Development, Shoreham Hotel, Washington, D.C., March 8, 1968, http://www.agbioworld.org/biotech-info/topics/borlaug/borlaug-green.html.
2. Johnson's visit to the Philippines, including his visit to IRRI, is captured in a four-minute video clip: "United States President Lyndon B Johnson Visiting International Rice Institute in Los Baños and Cam Ranh Bay in Vietnam, October 26, 1966," Critical Past, https://www.criticalpast.com/video/65675039499_Lyndon-B-Johnson_paddy-field_President-decorating-troops_President-Marcos.
3. Johnson's speech is included in the video clip (note 2); also available at "Redux: U.S. President Lyndon Johnson Visits IRRI, 26 October 1966," IRRI, YouTube, February 25, 2011, https://www.youtube.com/watch?v=WEwdkz7XK2s.
4. Tom Hargrove, "I Remember Honda Rice," *Rice Today* 5, no. 4 (2006): 39–44.
5. Thomas R. Hargrove, *A Dragon Lives Forever: War and Rice in Vietnam's Mekong Delta, 1969–1991, and Beyond* (New York: Ivy Books, 1994).
6. The Viet Cong was the military arm of the National Liberation Front of South Vietnam.
7. Hargrove, "Honda Rice."
8. Robert W. Chandler, *War of Ideas: The US Propaganda Campaign in Vietnam* (New York: Routledge, 2019), 109.
9. Hargrove, "Honda Rice."
10. Robert Flint Chandler, *An Adventure in Applied Science: A History of the International Rice Research Institute* (Los Baños, Philippines: IRRI, 1982).
11. Gerardo P. Sicat, *The Economic Legacy of Marcos*, Discussion Paper no. 2011–11 (Manila: University of the Philippines School of Economics, November 2011), https://econ.upd.edu.ph/dp/index.php/dp/article/viewFile/679/144.

12. C. Peter Timmer, *Food Security and Scarcity: Why Ending Hunger Is So Hard* (Philadelphia: University of Pennsylvania Press, 2015), 41.
13. For additional details on Norman Borlaug's life and work, see Ambassador Kenneth M. Quinn, "Extended Biography," World Food Prize Foundation, 2009, https://www.worldfoodprize.org/en/dr_norman_e_borlaug/about_norman_borlaug/.
14. For an exquisitely researched account of the history of wheat research in Mexico and the role played by Norman Borlaug, see D. Byerlee, *The Birth of CIMMYT: Pioneering the Idea and Ideals of International Agricultural Research* (Mexico City: CIMMYT, 2016).
15. M. S. Swaminathan, "Obituary: Norman E. Borlaug (1914–2009)," *Nature* 461, no. 7266 (2009): 894.
16. "A Digital History: Mexico," Rockefeller Foundation, accessed August 10, 2021, https://rockfound.rockarch.org/mexico.
17. "The Beginning of the Green Revolution," University of Minnesota, last modified June 12, 2004, https://web.archive.org/web/20041227090100/http://www.coafes.umn.edu/The_Beginning_of_the_Green_Revolution.html.
18. Byerlee, *Birth of CIMMYT.*
19. Leon Hesser, *The Man Who Fed The World* (Dallas, Tex.: Durban House, 2006), 80–82.
20. "Sir Charles Pereira," *Times* (London), January 13, 2005, https://www.thetimes.co.uk/article/sir-charles-pereira-vkth68j796r.
21. J. S. G. McCulloch, "Sir Herbert Charles Pereira. 12 May 1913–19 December 2004," *Biographical Memoirs of Fellows of the Royal Society* 54 (2008): 245–56, doi.org/10.1098/rsbm.2008.0001.
22. Warren Weaver and J. George Harrar, "Research on Rice," October 21, 1954, appendix 1 in the Rockefeller Foundation Board Meeting, November 30—December 1, 1954, 49–60, https://iiif.rockarch.org/pdfs/7CGpU9AhiMfZGuQsr5L64K.
23. Chandler, *An Adventure*, 104–8.
24. S. K. De Datta, A. C. Tauro, and S. N. Balaoing, "Effect of Plant Type and Nitrogen Level on the Growth Characteristics and Grain Yield of Indica Rice in the Tropics," *Agronomy Journal* 60, no. 6 (1968): 643–47.
25. Surajit K. De Datta, *Principles and Practices of Rice Production* (New York: Wiley, 1981).
26. Robert E. Evenson and Douglas Gollin, "Assessing the Impact of the Green Revolution, 1960 to 2000," *Science* 300, no. 5620 (2003): 758–62.
27. Peter B. R. Hazell, "Transforming Agriculture: The Green Revolution in Asia," in *Millions Fed: Proven Successes in Agricultural Development*, ed. David J. Spielman and Rajul Pandya-Lorch (Washington, D.C.: International Food Policy Research Institute, 2009): 25–32.
28. Hazell.
29. Prabhu L. Pingali, "Green Revolution: Impacts, Limits, and the Path Ahead," *Proceedings of the National Academy of Sciences* 109, no. 31 (2012): 12302–8.
30. James R. Stevenson et al., "Green Revolution Research Saved an Estimated 18 to 27 Million Hectares from Being Brought Into Agricultural Production," *Proceedings of the National Academy of Sciences* 110, no. 21 (2013): 8363–68.
31. Pingali, "Green Revolution," 12303.
32. FAO, *The State of Food and Agriculture, 2010–11. Women in Agriculture. Closing the Gender Gap for Development* (Rome: FAO, 2011).
33. IRRI, *Annual Report for 1975* (Los Baños, Philippines: IRRI, 1976), 2.
34. This section is an adaptation and update of Glenn Denning's "Fostering International Collaboration for Food Security and Sustainable Development: A Personal Perspective of MS Swaminathan's Vision, Impact and Legacy for Humanity," *Current Science* 109, no. 3 (2015): 447–55. Permission to reproduce part of the article with modest edits and updates was granted by G. Madavan, executive secretary, Current Science Association, in an email to the author dated November 11, 2021.
35. William Shawcross, *Sideshow: Kissinger, Nixon and the Destruction of Cambodia* (New York: Simon and Schuster, 1979), 396.
36. "Cambodia Genocide Program," Yale University, accessed August 10, 2021, https://gsp.yale.edu/case-studies/cambodian-genocide-program.

37. R. Ross Russell, ed., *Cambodia: A Country Study* (Washington, D.C.: GPO for the Library of Congress, 1987), http://countrystudies.us/cambodia/40.htm, Population.
38. David P. Chandler, *Brother Number One: A Political Biography of Pol Pot* (Boulder, Colo.: Westview Press, 1999), 118.
39. FAOSTAT, http://www.fao.org/faostat/en/#data/QCL.
40. Jeffrey Himmel, *Khmer Rouge Irrigation Development in Cambodia* (Phnom Penh: Documentation Center of Cambodia [DC-Cam], 2007).
41. Ben Kiernan, *The Pol Pot Regime: Race, Power, and Genocide in Cambodia Under the Khmer Rouge 1975–79* (New Haven, Conn.: Yale University Press, 2014).
42. Nitya Rao, *M. S. Swaminathan in Conversation with Nitya Rao* (Chennai, India: M. S. Swaminathan Research Foundation, 2013), 31.
43. Don Puckridge, *The Burning of the Rice: A Cambodian Success Story* (Hartwell, Victoria, Australia: Temple House Pty, 2004). Telexes were then the most common two-way form of instant communication between subscribers using public telecommunication systems.
44. Ben Kiernan, *Cambodia's Twisted Path to Justice*, The History Place: Points of View, 1999, https://www.historyplace.com/pointsofview/kiernan.htm.
45. Rao, *M. S. Swaminathan in Conversation.*
46. H. Nesbitt, "Harnessing People Power for Technology Uptake," in *Food for the Future: Opportunities for a Crowded Planet*, ed. A. G. Brown (Fyshwick, Australia: Crawford Fund, 2002), 54–62, https://ideas.repec.org/p/ags/cfcp02/123934.html.
47. H. J. Nesbitt, ed., *Rice Production in Cambodia* (Los Baños, Philippines: IRRI, 1997).
48. Nesbitt, "Harnessing People Power"; Nesbitt, *Rice Production.*
49. G. S. Khush, D. W. Puckridge, and G. L. Denning, *Trip Report to People's Republic of Kampuchea, 23–30 January 1986* (Los Baños, Philippines: IRRI, 1986).
50. Nesbitt, "Harnessing People Power."
51. Khush, Puckridge, and Denning, *Trip Report.*
52. Puckridge, *Burning of the Rice.*
53. "Inauguration," Cambodian Agricultural Research and Development Institute, December 2, 2011, http://www.cardi.org.kh/?page=detail&menu1=6&menu2=69&menu3=76&ctype=article&id=76&lg=en.
54. Rao, *M. S. Swaminathan in Conversation.*
55. IRRI, *Cambodia and IRRI* (Los Baños, Philippines: IRRI, October 2020), http://books.irri.org/Cambodia_IRRI_brochure.pdf.
56. IRRI, *Cambodia and IRRI.*
57. FAOSTAT, http://www.fao.org/faostat/en/#data/QCL.
58. According to Cambodian minister of agriculture, forestry and fisheries Veng Sakhon, reported by Thou Vireak in "2020 Rice Exports Nearly 3.6M Tonnes," *Phnom Penh Post*, January 4, 2021, https://www.phnompenhpost.com/business/2020-rice-exports-nearly-36m-tonnes.
59. Joyce Gorsuch, *Bhutan–IRRI Project: Local Tradition Meets Modern Know-How* (Los Baños, Philippines: IRRI, 2001), http://books.irri.org/9712201589_content.pdf.
60. Liana Williams and Rob Cramb, "Adapting the Green Revolution for Laos," in *White Gold: The Commercialisation of Rice Farming in the Lower Mekong Basin*, ed. Rob Cramb (Singapore: Palgrave Macmillan, 2020), 121–49.
61. G. L. Denning and Vo-Tong Xuan, eds., *Vietnam and IRRI: A Partnership in Rice Research. Proceedings of a Conference Held in Hanoi, Vietnam, 4–7 May, 1994* (Los Baños, Philippines: IRRI, 1995).
62. United Nations, "Secretary-General Calls for 'Uniquely African Green Revolution' in 21st Century, to End Continent's Plague of Hunger, in Addis Ababa Remarks," news release SG/SM/9405-AFR/988, July 6, 2004, https://www.un.org/press/en/2004/sgsm9405.doc.htm; Millennium Development Goals Centre, *Africa's Green Revolution: A Call to Action. Proceedings of the July 5, 2004 High-Level Seminar, Addis Ababa, Ethiopia* (Nairobi: MDG Centre, 2004).
63. United Nations, "Secretary-General Calls for 'Uniquely African Green Revolution.' "

64. UN Millennium Project Task Force on Hunger, *Halving Hunger: It Can Be Done* (London: Earthscan, 2005).
65. Throughout the book, in reporting data on sub-Saharan Africa, I consistently use the World Bank definition comprising forty-eight countries that are largely south of the Sahara Desert, including Sudan. The amount of fertilizer applied is based on FAO statistics for 2004 and is expressed in terms of kilograms of nutrient (combined amounts of nitrogen, phosphate, and potash). FAOSTAT, https://www.fao.org/faostat/en/#data/RFN.
66. Keith Fuglie and Nicholas Rada, *Resources, Policies, and Agricultural Productivity in Sub-Saharan Africa*, U.S. Department of Agriculture, Economic Research Service Economic Research Report no. 145 (Washington, D.C.: USDA-ERS, February 2013), dx.doi.org/10.2139/ssrn.2266459.
67. InterAcademy Council, *Realizing the Promise and Potential of African Agriculture: Science and Technology Strategies for Improving Agricultural Productivity and Food Security in Africa* (Amsterdam: InterAcademy Council, 2004).
68. G. Conway, *The Doubly Green Revolution: Food for All in the 21st Century* (New York: Penguin Books, 1997); M. S. Swaminathan, "An Evergreen Revolution," *Crop Science* 46, no. 5 (2006): 2293–303.
69. MDG Centre, *Africa's Green Revolution.*
70. The estimate of 96 percent is based on 2019 data (100 percent minus the land area equipped for irrigation as a percentage of total arable land area). See https://www.fao.org/faostat/en/#data/RL. FAO defines arable land as land under temporary crops (double-cropped areas are counted once), temporary meadows for mowing or for pasture, land under market or kitchen gardens, and land temporarily fallow. Perennial crops and land abandoned as a result of shifting cultivation are excluded.
71. Pedro A. Sanchez, "Soil Fertility and Hunger in Africa," *Science* 295, no. 5562 (2002): 2019–20.
72. There is no formal definition of what area of land defines a smallholder farmer. Douglas Gollin explains in a working paper for the International Institute for Environment and Development (IIED) that "most sub-Saharan African households depend on smallholder agriculture for their livelihoods, and most agriculture in Africa, in turn, is carried out by smallholder households." He noted that evidence from numerous household surveys indicated that the median size of a crop farm in most countries is probably 1–2 hectares. Douglas Gollin, *Smallholder Agriculture in Africa: An Overview and Implications for Policy*, IIED Working Paper (London: IIED, 2014), 5, http://pubs.iied.org/14640IIED.
73. UN Millennium Project Task Force on Hunger, *Halving Hunger*; P. A. Sanchez and M. S. Swaminathan, "Cutting World Hunger in Half," *Science* 307, no. 5708 (2005): 357–59.
74. United Nations, "Secretary-General Calls for 'Uniquely African Green Revolution.' "
75. Michael Morris et al., *Fertilizer Use in African Agriculture: Lessons Learned and Good Practice Guidelines* (Washington, D.C.: World Bank, 2007).
76. Carl K. Eicher and Derek Byerlee, "Accelerating Maize Production: Synthesis," in *Africa's Emerging Maize Revolution*, ed. Derek Byerlee and Carl K. Eicher (London: Lynne Rienner, 1977), 247–62.
77. Sasakawa Africa Association, *Take It to the Farmer: The Sasakawa Experience in Africa* (Tokyo: Sasakawa Africa Association, 2015).
78. FAOSTAT, http://www.fao.org/faostat/en/#data/QCL.
79. "Alliance for a Green Revolution in Africa," Tropical Agriculture Platform, accessed August 14, 2021, https://tapipedia.org/content/alliance-green-revolution-africa.
80. World Bank, *World Development Report 2008: Agriculture for Development* (Washington, D.C.: World Bank, 2007), https://openknowledge.worldbank.org/handle/10986/5990.
81. "L'Aquila Joint Statement on Global Food Security," L'Aquila Food Security Initiative, L'Aquila Summit, July 9, 2009, http://www.g7.utoronto.ca/summit/2009laquila/2009-food.html; "G20 Leaders Statement: The Pittsburgh Summit, September 24–25, 2009, Pittsburgh," G20 Research Group, University of Toronto, para. 39, http://www.g20.utoronto.ca/2009/2009communique0925.html.
82. "Our Impact," GAFSP, accessed August 10, 2021, https://www.gafspfund.org/index.php/impact.

83. African Union Commission and NEPAD Agency, *Malabo Declaration on Accelerated Agricultural Growth and Transformation for Shared Prosperity and Improved Livelihoods* (Addis Ababa, Ethiopia: AUC and Midrand, South Africa: NEPAD, 2016), https://www.nepad.org/caadp/publication/malabo-declaration-accelerated-agricultural-growth.
84. African Development Bank, *Feed Africa* (Abidjan, Côte d'Ivoire: AfDB, 2019), https://www.afdb.org/en/documents/document/feed-africa-brochure-89189.
85. Glenn Denning and Hanson Nyantakyi-Frimpong, "Is the New Green Revolution Approach the Best Way to Address Hunger in Africa?," in *Debating African Issues: Conversations Under the Palaver Tree*, ed. William G. Moseley and Kefa M. Otiso (London: Taylor & Francis, 2022), 151–63.
86. Unless otherwise cited, all data in this section was sourced from FAOSTAT, http://www.fao.org/faostat/en/#data/QCL.
87. Glenn Denning et al., "Input Subsidies to Improve Smallholder Maize Productivity in Malawi: Toward an African Green Revolution," *PLoS Biology* 7, no. 1 (2009): e1000023.
88. R. L. Nielsen, "Historical Corn Grain Yields in the U.S," updated August 2021, https://www.agry.purdue.edu/ext/corn/news/timeless/yieldtrends.html. See fig. 1, with data from the U.S. Department of Agriculture.
89. United States Department of Agriculture, National Agricultural Statistics Service, "Quick Stats," https://quickstats.nass.usda.gov/results/7B832B80-F468-398F-8A28-F9C76A50551B?pivot=short_desc.
90. Denning et al., "Input Subsidies"; Generose Nziguheba et al., "The African Green Revolution: Results from the Millennium Villages Project," *Advances in Agronomy*, 109 (2010): 75–115.
91. Keijiro Otsuka and Rie Muraoka, "Green Revolution for Sub-Saharan Africa: Past Failures and Future Prospects," *Journal of African Economies* 26, suppl. 1 (August 2017): i73–i98.
92. Thomas S. Jayne and Pedro A. Sanchez, "Agricultural Productivity Must Improve in Sub-Saharan Africa," *Science* 372, no. 6546 (2021): 1045–47.
93. Evenson and Gollin, "Impact of the Green Revolution."
94. Denning, "Is the New Green Revolution Approach the Best Way to Address Hunger in Africa?"
95. Pedro A. Sanchez, "En Route to Plentiful Food Production in Africa," *Nature Plants* 1 (2015): 14014.
96. FAO et al., *The State of Food Security and Nutrition in the World 2021: Transforming Food Systems for Food Security, Improved Nutrition and Affordable Healthy Diets for All* (Rome: FAO, 2021); United Nations, Department of Economic and Social Affairs, Population Division, *World Population Prospects 2019*, vol. 1: *Comprehensive Tables* (ST/ESA/SER.A/426) (New York: United Nations, 2019).
97. FAO et al., *The State of Food Security.*
98. The African Union's Agenda 2063 is "Africa's blueprint and master plan for transforming Africa." African heads of state and government signed the 50th Anniversary Solemn Declaration during the Golden Jubilee celebrations of the formation of the Organisation of African Unity/African Union in May 2013. "Agenda 2063: The Africa We Want," African Union, accessed August 14, 2021, https://au.int/en/agenda2063/overview.

3. SOIL AND LAND

1. Firsthand experience from that period is masterfully captured by Ken Burns's documentary *The Dust Bowl* (2012). See https://www.pbs.org/show/dust-bowl/.
2. Franklin D. Roosevelt, *Letter to All State Governors on a Uniform Soil Conservation Law*, February 26, 1937, online by Gerhard Peters and John T. Woolley, American Presidency Project, https://www.presidency.ucsb.edu/node/209378.
3. Norman Myers et al., "Biodiversity Hotspots for Conservation Priorities," *Nature* 403 (2000): 853–58.

4. P. Randrianarijaona, "The Erosion of Madagascar," *Ambio* 12 (1983): 308–11. The Malagasy Republic was the official name of Madagascar from 1960 to 1975.
5. R. Cox et al., "Erosion Rates and Sediment Sources in Madagascar Inferred from ^{10}Be Analysis of Lavaka, Slope, and River Sediment," *Journal of Geology* 117, no. 4 (2009): 363–76.
6. Randrianarijaona, "The Erosion of Madagascar."
7. Randrianarijaona, "The Erosion of Madagascar"; Mark Kaufman, "Grim Space Photo Shows the Horror of Demolishing Forests," Mashable, July 13, 2018, https://mashable.com/article/nasa-space-photo-madagascar/.
8. C. Fontanier et al., "Are Deep-Sea Ecosystems Surrounding Madagascar Threatened by Land-Use or Climate Change?," *Deep Sea Research Part I: Oceanographic Research Papers* 131 (2018): 93–100.
9. "Madagascar Overview," World Bank, last updated July 31, 2020, https://www.worldbank.org/en/country/madagascar/overview.
10. Ghislain Vieilledent et al., "Combining Global Tree Cover Loss Data with Historical National Forest Cover Maps to Look at Six Decades of Deforestation and Forest Fragmentation in Madagascar," *Biological Conservation* 222 (2018): 189–97, doi.org/10.1016/j.biocon.2018.04.008.
11. Bart Minten and Christopher B. Barrett, "Agricultural Technology, Productivity, and Poverty in Madagascar," *World Development* 36, no. 5 (2008): 797–822.
12. "Country Nutrition Profiles: Madagascar," Global Nutrition Report, accessed August 14, 2021, https://globalnutritionreport.org/resources/nutrition-profiles/africa/eastern-africa/madagascar/.
13. Harold Van Es, "A New Definition of Soil," *CSA News* 62, no. 10 (2017): 20–21. As described in this paper, the SSSA board approved this new definition of soil in August 2017 to be "inclusive, direct, precise, and succinct" and "to be understandable to scientists and professionals." Moreover, for the first time, the SSSA definition acknowledged the existence of soils on any planet, not only Earth.
14. P. H. Nye and D. J. Greenland, "The Soil Under Shifting Cultivation," *Soil Science* 92, no. 5 (1961): 354.
15. See Peter Gregory's comprehensive memoir on Greenland published by the British Royal Society in January 2019: "Dennis James Greenland. 13 June 1930–23 December 2012," *Biographical Memoirs of Fellows of the Royal Society* 66 (2019): 225–41, https://royalsocietypublishing.org/doi/10.1098/rsbm.2018.0030.
16. Pedro A. Sanchez, *Properties and Management of Soils in the Tropics*, 2nd ed. (Cambridge: Cambridge University Press, 2019).
17. USDA Natural Resources Conservation Service, *Soil Taxonomy, a Basic System of Soil Classification for Making and Interpreting Soil Surveys*, 2nd ed. (Washington, D.C.: USDA, 1999), https://www.nrcs.usda.gov/Internet/FSE_DOCUMENTS/nrcs142p2_051232.pdf; International Union of Soil Sciences (IUSS) Working Group WRB, *World Reference Base for Soil Resources 2014, Update 2015 International Soil Classification System for Naming Soils and Creating Legends for Soil Maps*, World Soil Resources Reports no. 106 (Rome: FAO, 2015), http://www.fao.org/3/i3794en/I3794en.pdf.
18. Sanchez, *Properties and Management of Soils*, 97.
19. J. L. Heller, "The Nomenclature of Soils, or What's in a Name?," *Soil Science Society of America Journal* 27, no. 2 (1963): 216–20.
20. T. J. Smyth and P. A. Sanchez, "Effects of Lime, Silicate, and Phosphorus Applications to an Oxisol on Phosphorus Sorption and Ion Retention," *Soil Science Society of America Journal* 44, no. 3 (1980): 500–505; W. J. Goedert, "Management of the Cerrado Soils of Brazil: A Review," *Journal of Soil Science* 34, no. 3 (1983): 405–28.
21. See WWF, "Cerrado-Threats," Critical Ecosystem Partnership Fund, accessed September 5, 2021, https://www.cepf.net/our-work/biodiversity-hotspots/cerrado/threats.
22. Hari Eswaran et al., "An Assessment of the Soil Resources of Africa in Relation to Productivity," *Geoderma* 77, no. 1 (1997): 1–18.
23. Sanchez, *Properties and Management of Soils*, "Functional Capability Classification," 120–33.

24. Sanchez, 310.
25. "Glossary of Soil Science Terms," Soil Science Society of America, accessed August 14, 2021, https://www.soils.org/publications/soils-glossary#.
26. John W. Doran and Michael R. Zeiss, "Soil Health and Sustainability: Managing the Biotic Component of Soil Quality," *Applied Soil Ecology* 15, no. 1 (2000): 3–11.
27. Sanchez, *Properties and Management of Soils*, 310 (table 12.1).
28. Sanchez, 311–12, describes these three principles in greater detail.
29. Justus Liebig, *Die Organische Chemie in Ihrer Anwendung auf Agricultur und Physiologie* (Braunschweig: F. Vieweg und Sohn, 1840). German agronomist Carl Sprengel (1787–1859) is widely credited for first formulating the conceptual basis for the law of the minimum. Sprengel's publications on the topic predated von Liebig's publication of 1840, though it is unclear whether Liebig was aware of Sprengel's work. The Association of German Agricultural Experimental and Research Stations created the Sprengel–Liebig Medal, recognizing Sprengel as a cofounder of agricultural chemistry. Rienk R. van der Ploeg, Wolfgang Böhm, and Mary Beth Kirkham, "On the Origin of the Theory of Mineral Nutrition of Plants and the Law of the Minimum," *Soil Science Society of America Journal* 63, no. 5 (1999): 1055–62.
30. Sanchez, *Properties and Management of Soils*, 311 (fig. 12.2).
31. T. K. Hartz and G. J. Hochmuth, "Fertility Management of Drip-Irrigated Vegetables," *HortTechnology* 6, no. 3 (1996): 168–72.
32. Sanchez, *Properties and Management of Soils*, 326.
33. E. John Russell, "Rothamsted and Its Experiment Station," *Agricultural History* 16, no. 4 (1942): 161–83.
34. See Thomas Hager, *The Alchemy of Air* (New York: Broadway Books, 2008), for a well-researched and readable account of the work of Haber and Bosch.
35. Kathryn Harkup, "Chlorine: The Gas of War Crimes," *Guardian*, September 16, 2016, https://www.theguardian.com/science/blog/2016/sep/16/chlorine-the-gas-of-war-crimes.
36. Johan Rockström et al., "Planetary Boundaries: Exploring the Safe Operating Space for Humanity," *Ecology and Society* 14, no. 2 (2009): 32.
37. Vaclav Smil, "Global Population and the Nitrogen Cycle," *Scientific American* 277, no. 1 (1997): 76–81.
38. Will Steffen et al., "Planetary Boundaries: Guiding Human Development on a Changing Planet," *Science* 347, no. 6223 (2015): 1259855.
39. Peter M. Vitousek et al., "Human Alteration of the Global Nitrogen Cycle: Sources and Consequences," *Ecological Applications* 7, no. 3 (1997): 737–50; Jan Willem Erisman et al., *Nitrogen: Too Much of a Vital Resource*, Science Brief (Zeist: WWF Netherlands, 2015).
40. Sanchez, *Properties and Management of Soils*, 358–59.
41. "Slow-Release Nitrogen Fertilizers Measure Up," International Fertilizer Development Center (IFDC), June 3, 2019, https://ifdc.org/2019/06/03/slow-release-nitrogen-fertilizers-measure-up/.
42. Pedro A. Sanchez, "Science in Agroforestry," *Agroforestry Systems* 30, no. 1–2 (1995): 5–55; Oluyede Clifford Ajayi et al., "Agricultural Success from Africa: The Case of Fertilizer Tree Systems in Southern Africa (Malawi, Tanzania, Mozambique, Zambia and Zimbabwe)," *International Journal of Agricultural Sustainability* 9, no. 1 (2011): 129–36; Dennis Philip Garrity et al., "Evergreen Agriculture: A Robust Approach to Sustainable Food Security in Africa," *Food Security* 2, no. 3 (2010): 197–214.
43. The caveats and limitation on legume use in farming systems draw on my personal experience and on Bernard Vanlauwe and Ken E. Giller, "Popular Myths Around Soil Fertility Management in Sub-Saharan Africa," *Agriculture, Ecosystems and Environment* 116, no. 1–2 (2006): 34–46.
44. Deep-rooted agroforestry species have shown capacity to extract nitrogen and sulfur that has been leached into the subsoils beyond the root zones of annual crops. Some subsoils in the tropics have significant anion exchange capacity that enables the accumulation of negatively charged nitrate and sulfate ions. Sanchez, *Properties and Management of Soils*, 206–7.
45. "The History of Rothamsted Research," Rothamsted Research, accessed August 14, 2021, https://www.rothamsted.ac.uk/history-and-heritage.

46. Sanchez, *Properties and Management of Soils*, 371.
47. Pedro A. Sánchez and José G. Salinas, "Low-Input Technology for Managing Oxisols and Ultisols in Tropical America," *Advances in Agronomy* 34 (1981): 279–406.
48. Sanchez, *Properties and Management of Soils*, 372.
49. U.S. Geological Survey, Mineral Commodity Summaries, *Phosphate Rock* (Washington, D.C.: U.S. Geological Survey, January 2020), https://pubs.usgs.gov/periodicals/mcs2020/mcs2020-phosphate.pdf.
50. Sanchez, *Properties and Management of Soils*, 374.
51. Sanchez was quoted in Renee Cho, "Phosphorus: Essential to Life—Are We Running Out?," Columbia Climate School, April 1, 2013, https://blogs.ei.columbia.edu/2013/04/01/phosphorus-essential-to-life-are-we-running-out/.
52. See "Glossary of Soil Science Terms," Soil Science Society of America, accessed August 14, 2021, https://www.soils.org/files/publications/soils-glossary/figure-1.pdf for the textural triangle.
53. For a more complete description of the ribboning method for determining soil texture, see New South Wales Department of Primary Industries, *Determining Soil Texture Using the Ribboning Technique*, Primefact 1363 (NSW, Australia: DPI, 2014), https://www.dpi.nsw.gov.au/__data/assets/pdf_file/0005/164615/determining_soil_texture_using_-ribboning_technique.pdf.
54. Charles C. Mann, "Our Good Earth: The Future Rests on the Soil Beneath Our Feet. Can We Save It?," *National Geographic* 214, no. 3 (2008): 80–106.
55. Sanchez, *Properties and Management of Soils*, 143–46.
56. "Soil Biology," U.S. Department of Agriculture Natural Resources Conservation Service, accessed August 14, 2021, https://www.nrcs.usda.gov/wps/portal/nrcs/main/soils/health/biology/.
57. Sanchez, *Properties and Management of Soils*, 260.
58. SOM also includes soil organic nitrogen (SON), soil organic phosphorus (SOP), soil organic sulfur (SOS), and more. Soil organic carbon (SOC) is typically measured in the soil and multiplied by two to estimate SOM. Sanchez, 260.
59. Sanchez, 278.
60. Estimates based on cattle manure quality data from the TSBF database cited by Cheryl A. Palm, Herbert K. Murwira, and Simon E. Carter, "Organic Matter Management: From Science to Practice," table 1, in *Soil Fertility Research for Maize-Based Farming Systems in Malawi and Zimbabwe*, ed. S. R. Waddington et al. (Harare: CIMMYT, 1998), 21–27.
61. Sanchez, *Properties and Management of Soils*, 300–301.
62. Bernard Vanlauwe et al. "Integrated Soil Fertility Management: Operational Definition and Consequences for Implementation and Dissemination," *Outlook on Agriculture* 39, no. 1 (2010): 17–24.
63. Verena Seufert, Navin Ramankutty, and Jonathan A. Foley, "Comparing the Yields of Organic and Conventional Agriculture," *Nature* 485, no. 7397 (2012): 229–32.
64. Seufert, Ramankutty, and Foley.
65. Biovision Foundation for Ecological Development and IPES-Food, *Money Flows: What Is Holding Back Investment in Agroecological Research for Africa?* (Zurich: Biovision Foundation for Ecological Development and Brussels: International Panel of Experts on Sustainable Food Systems, 2020), 107, 105, 19.
66. Camila Bonilla Cedrez et al., "Spatial Variation in Fertilizer Prices in Sub-Saharan Africa," *PloS One* 15, no. 1 (2020): e0227764.
67. "Fertilizer Consumption (Kilograms per Hectare of Arable Land)," World Bank, Data, citing the most recent data from 2018, https://data.worldbank.org/indicator/AG.CON.FERT.ZS.
68. Pedro A. Sanchez, "Soil Fertility and Hunger in Africa," *Science* 295, no. 5562 (2002): 2019–20.
69. Sanchez, *Properties and Management of Soils*, 301.
70. "*Critias* by Plato, Written 360 BCE, Translated by Benjamin Jowett," Internet Classics Archive, accessed August 14, 2021, http://classics.mit.edu/Plato/critias.html, para. 8.
71. "*Critias* by Plato," para. 9.
72. "Erosion," USDA, Natural Resources Conservation Service, accessed August 14, 2021, https://www.nrcs.usda.gov/wps/portal/nrcs/main/national/landuse/crops/erosion/.

73. David R. Montgomery, "Soil Erosion and Agricultural Sustainability," *Proceedings of the National Academy of Sciences* 104, no. 33 (2007): 13268–72.
74. Montgomery, figs. 1 and 2.
75. Montgomery, 13271.
76. "RUSLE Factors," Michigan State University, accessed September 5, 2021, http://www.iwr.msu.edu/rusle/factors.htm.
77. Howard G. Buffett, "Conservation: Reaping the Benefits of No-Tillage Farming," *Nature* 484, no. 7395 (2012): 455.
78. Pasquale Borrelli et al., "An Assessment of the Global Impact of 21st Century Land Use Change on Soil Erosion," *Nature Communications* 8, no. 1 (2017): 1–13.
79. Bernard Vanlauwe et al., "A Fourth Principle Is Required to Define Conservation Agriculture in Sub-Saharan Africa: The Appropriate Use of Fertilizer to Enhance Crop Productivity," *Field Crops Research* 155 (2014): 10–13.
80. K. E. Giller, "No Silver Bullets for African Soil Problems," *Nature* 485, no. 41 (2012), doi.org/10.1038/485041c.
81. "Erosion," USDA, Natural Resources Conservation Service.
82. Michael C. Duniway et al., "Wind Erosion and Dust from US Drylands: A Review of Causes, Consequences, and Solutions in a Changing World," *Ecosphere* 10, no. 3 (2019): e02650.
83. DeAnn Presley and John Tatarko, *Principles of Wind Erosion and Its Control* (Manhattan: Kansas State University, September 2009), https://www.ars.usda.gov/ARSUserFiles/30200525/419PrinciplesofWinderosionanditscontrol.pdf; Gary Tibke, "5. Basic Principles of Wind Erosion Control," *Agriculture, Ecosystems and Environment* 22–23 (1988): 103–22; B. Ben Salem, "Prevention and Control of Wind Erosion in Arid Regions," *Unasylva NO. 164 Watershed Mangement* 42, no. 1 (1991), http://www.fao.org/3/u1510e/u1510e07.htm.
84. Rattan Lal, "Soil Erosion and Land Degradation: The Global Risks," in *Advances in Soil Science*, ed. R. Lal and B. A. Stewart (New York: Springer, 1990), 129–72.

4. WATER RESOURCES

1. FAO, *Water for Sustainable Food and Agriculture: A Report Produced for the G20 Presidency of Germany* (Rome: FAO, 2017), http://www.fao.org/3/i7959e/i7959e.pdf.
2. R. Hofste et al., *Aqueduct 3.0: Updated Decision-Relevant Global Water Risk Indicators*, Technical Note (Washington, D.C.: World Resources Institute, 2019), https://www.wri.org/publication/aqueduct-30.
3. "Aqueduct 3.0 Country Rankings," World Resources Institute, accessed September 5, 2021, https://www.wri.org/data/aqueduct-30-country-rankings.
4. Tianyi Luo, Robert Young, and Paul Reig, *Aqueduct Projected Water Stress Country Rankings*, Technical Note (Washington, D.C.: World Resources Institute, 2015), www.wri.org/publication/aqueduct-projected-water-stress-country-rankings.
5. Elizabeth Whitman, "A Land Without Water: The Scramble to Stop Jordan from Running Dry," *Nature* 573 (2019): 20–23.
6. Jim Yoon et al., "A Coupled Human–Natural System Analysis of Freshwater Security Under Climate and Population Change," *Proceedings of the National Academy of Sciences* 118, no. 14 (2021): e2020431118,doi.org/10.1073/pnas.2020431118.
7. Malin Falkenmark and Johan Rockström, "The New Blue and Green Water Paradigm: Breaking New Ground for Water Resources Planning and Management," *Journal of Water Resources Planning and Management* 132, no. 3 (2006): 129–32. The concept of green and blue water was first published in Malin Falkenmark and Johan Rockström, *Balancing Water for Humans and Nature: The New Approach in Ecohydrology* (London: Earthscan, 2004).

8. FAO, *The State of the World's Land and Water Resources for Food and Agriculture: Managing Systems at Risk* (Rome: FAO and London: Earthscan, 2011).
9. FAOSTAT, https://www.fao.org/faostat/en/#data/RL, accessed October 13, 2021. The estimate of 96 percent was based on the area equipped for irrigation as a percentage of arable land area, using 2019 data.
10. Theib Oweis, Ahmed Hachum, and Jacob Kijne, *Water Harvesting and Supplemental Irrigation for Improved Water Use Efficiency in Dry Areas*, SWIM Paper 7 (Colombo, Sri Lanka: International Water Management Institute, 1999), 2.
11. It should be recognized that water "losses" through percolation and cross-surface flow are only on-site losses to the crop. Those same water resources may ultimately find use elsewhere through the withdrawal from rivers, dams, and aquifers.
12. Johan Rockström and Malin Falkenmark, "Agriculture: Increase Water Harvesting in Africa," *Nature* 519, no. 7543 (2015): 283.
13. Walter K. Anderson, David Stephens, and Kadambot HM Siddique, "Dryland Agriculture in Australia: Experiences and Innovations," in *Innovations in Dryland Agriculture*, ed. M. Farooq and K. Siddique (Cham, Switzerland: Springer, 2016), 299–319.
14. James W. Hansen et al., "Innovations in Climate Risk Management: Protecting and Building Rural Livelihoods in a Variable and Changing Climate," *Journal of Semi-Arid Tropical Agricultural Research* 4, no. 1 (2007).
15. Hansen et al.
16. Johan Rockström, Mats Lannerstad, and Malin Falkenmark, "Assessing the Water Challenge of a New Green Revolution in Developing Countries," *Proceedings of the National Academy of Sciences* 104, no. 15 (2007): 6253–60.
17. *Culavamsa*, part 1, chap. 68, "The Improvement of His Own Kingdom," trans. Wilhelm Geiger (London: Pali Text Society, 1929), 277, https://archive.org/details/in.ernet.dli.2015.188749/page/n321/mode/2up.
18. World Water Assessment Programme, *The United Nations World Water Development Report 3: Water in a Changing World* (Paris: UNESCO and London: Earthscan, 2009). Some publications refer to 25 percent of cropland being irrigated. The estimate here refers to the global harvested area and takes into account that some croplands produce more than one crop per year.
19. FAOSTAT, https://www.fao.org/faostat/en/#data/RL, accessed October 13, 2021. Calculations were based on the area equipped for irrigation as a percentage of arable land area, using 2019 data.
20. Raj Gupta and Ashok Seth, "A Review of Resource Conserving Technologies for Sustainable Management of the Rice–Wheat Cropping Systems of the Indo-Gangetic Plains (IGP)," *Crop Protection* 26, no. 3 (2007): 436–47.
21. A. M. MacDonald et al., "Groundwater Quality and Depletion in the Indo-Gangetic Basin Mapped from In Situ Observations," *Nature Geoscience* 9, no. 10 (2016): 762–66.
22. A. M. MacDonald et al., "Groundwater Resources in the Indo-Gangetic Basin: Resilience to Climate Change and Abstraction," data made live September 29, 2015 (Nottingham, UK: British Geological Survey [OR/15/047], unpublished), http://nora.nerc.ac.uk/id/eprint/511898.
23. MacDonald et al., "Groundwater Quality."
24. Philip Micklin, "The Aral Sea Disaster," *Annual Review of Earth and Planetary Sciences* 35 (2007): 47–72.
25. Malabo Montpellier Panel, *Water-Wise: Smart Irrigation Strategies for Africa* (Dakar, Senegal: International Food Policy Research Institute and Malabo Montpellier Panel, 2018).
26. AQUASTAT, http://www.fao.org/aquastat/statistics/query/index.html?lang=en, accessed January 3, 2022. This estimate was based on a comparison of area equipped for irrigation in 2003–2007 with 2013–2017. For Sudan and South Sudan, missing data in 2003–2007 was substituted with data from 2008–2012.
27. T. P. Higginbottom et al., "Performance of Large-Scale Irrigation Projects in Sub-Saharan Africa," *Nature Sustainability* 4 (2021): 501–8, doi.org/10.1038/s41893-020-00670-7.

28. Cool evening temperatures reduce the nighttime losses associated with plant respiration, which reverses the photosynthesis process by breaking down carbohydrates to produce carbon dioxide and water. Reduction in nighttime respiration contributes to very high yields in temperate zones.
29. B. A. M. Bouman and T. P. Tuong, "Field Water Management to Save Water and Increase Its Productivity in Irrigated Lowland Rice," *Agricultural Water Management* 1615, no. 1 (2000): 20.
30. J. A. Allan, "Virtual Water: A Strategic Resource Global Solutions to Regional Deficits," *Groundwater* 36, no. 4 (1998): 545–46.
31. Arjen Y. Hoekstra, "Water Footprint Assessment: Evolvement of a New Research Field," *Water Resources Management* 31, no. 10 (2017): 3061–81.
32. Pedro A. Sánchez, "Tripling Crop Yields in Tropical Africa," *Nature Geoscience* 3, no. 5 (2010): 299–300.
33. "Water, Food and Energy," UN Water, accessed August 15, 2021, https://www.unwater.org/water-facts/water-food-and-energy/.
34. David Molden, ed., *Water for Food Water for Life: A Comprehensive Assessment of Water Management in Agriculture* (London: Routledge, 2007). A shorter, more science-focused report of the study was published as Charlotte De Fraiture, David Molden, and Dennis Wichelns, "Investing in Water for Food, Ecosystems, and Livelihoods: An Overview of the Comprehensive Assessment of Water Management in Agriculture," *Agricultural Water Management* 97, no. 4 (2010): 495–501.
35. Molden, *Water for Food*, 1.
36. Molden, 1–4.

5. SEEDS OF LIFE

1. "International Rice Genebank," IRRI, accessed August 15, 2021, https://www.irri.org/international-rice-genebank. The term *accessions* is often used to describe distinct seed samples held in a gene bank. It avoids the mistake of assuming that every accession is a genetically distinct variety. I will use *accessions* and *samples* interchangeably on the understanding that some varieties may be duplicated as different accessions to a collection.
2. Ewen Callaway, "World's Largest Rice Gene Bank Secures Permanent Funding," *Nature*, News, October 12, 2018, doi.org/10.1038/d41586-018-07029-1. In October 2018 the Crop Trust, an NGO based in Bonn, Germany, announced funding of US$1.4 million a year "in perpetuity" to support the International Rice Genebank, ensuring long-term support for the conservation of rice *germplasm* (a broad term used in the scientific community to include all living material from which plants or other living organisms can be grown).
3. *Improved varieties* is a label generally attached to crossbred varieties that bring some kind of trait improvement. Typically, the term is used from the researchers' perspective, though whether it leads to an improvement in yield, quality, or some other trait ultimately depends on its performance as perceived by the producer and/or consumer.
4. Robert Flint Chandler, *An Adventure in Applied Science: A History of the International Rice Research Institute* (Los Baños, Philippines: IRRI, 1982).
5. Gene Hettel, *The IRRI Pioneer Interviews* (Los Baños, Philippines: IRRI, 2010), 2, http://books.irri.org/Pioneer_Interviews.pdf. The interview first appeared in *Rice Today* 7, no. 1 (January–March 2008).
6. The Mendelian ratio is the 3:1 ratio of phenotypes shown by the second generation of offspring from parents differing in respect to a single character. "Mendelian Ratio," *Merriam-Webster.com Dictionary*, accessed August 15, 2021, https://www.merriam-webster.com/dictionary/Mendelian%20ratio.
7. Don Puckridge, *The Burning of the Rice: A Cambodian Success Story* (Hartwell, Victoria, Australia: Temple House, 2004), 163.
8. Sam Fujisaka, *Rice Agroecosystems, Farmer Management, and Social Organization in Kampuchea: A Preliminary Assessment and Research Recommendations*, IRRI Research Paper Series, no. 136 (Los

Baños, Philippines: IRRI, 1988); Richard P. Lando and Solieng Mak, *Rainfed Lowland Rice in Cambodia: A Baseline Survey*, IRRI Research Paper Series, no. 152 (Los Baños, Philippines: IRRI, 1994).

9. Sam Fujisaka, "Rainfed Lowland Rice: Building Research on Farmer Practice and Technical Knowledge," *Agriculture, Ecosystems and Environment* 33, no. 1 (1990): 57–74.
10. Glenn Denning, "Fostering International Collaboration for Food Security and Sustainable Development: A Personal Perspective of M. S. Swaminathan's Vision, Impact and Legacy for Humanity," *Current Science* (2015): 447–55.
11. Harry Nesbitt was IRRI's team leader in Cambodia from 1988 to 2000. His remarkable contributions are documented in Puckridge, *Burning of the Rice*, and Brad Collis, "A Rice God Named Harry," Coretext, accessed August 15, 2021, http://stories.coretext.com.au/a-rice-god-named-harry/.
12. Ahmed Amri, "ICARDA Genebank: Collecting, Conserving, and Using Dryland Agrobiodiversity," ICARDA, https://www.icarda.org/research/innovations/icarda-genebank-collecting-conserving-and-using-dryland-agrobiodiversity.
13. S. Bhattacharya, "Syrian Seed Bank Gets New Home Away from War," *Nature* 538 (2016): 16–17, doi.org/10.1038/538016a.
14. "International Treaty on Plant Genetic Resources for Food and Agriculture: About Us," FAO, accessed August 15, 2021, http://www.fao.org/plant-treaty/overview/en/.
15. "International Treaty on Plant Genetic Resources for Food and Agriculture: Texts of the Treaty: Official Versions," FAO, accessed August 15, 2021, http://www.fao.org/plant-treaty/overview/texts-treaty/en/.
16. "Svalbard Global Seed Vault," Crop Trust, accessed August 15, 2020, https://www.croptrust.org/our-work/svalbard-global-seed-vault/.
17. Mauricio R. Bellon et al., "In Situ Conservation—Harnessing Natural and Human-Derived Evolutionary Forces to Ensure Future Crop Adaptation," *Evolutionary Applications* 10, no. 10 (2017): 965–77.
18. Mauricio R. Bellon, Elisabetta Gotor, and Francesco Caracciolo, "Conserving Landraces and Improving Livelihoods: How to Assess the Success of On-Farm Conservation Projects?," *International Journal of Agricultural Sustainability* 13, no. 2 (2015): 167–82.
19. Bellon et al., "In Situ Conservation," 973.
20. Bellon et al., 973.
21. Jared Diamond, "Evolution, Consequences and Future of Plant and Animal Domestication," *Nature* 418, no. 6898 (2002): 700–707.
22. Michael D. Purugganan and Dorian Q. Fuller, "The Nature of Selection During Plant Domestication," *Nature* 457, no. 7231 (2009): 843–48.
23. Duncan A. Vaughan, Bao-Rong Lu, and Norihiko Tomooka, "The Evolving Story of Rice Evolution," *Plant Science* 174, no. 4 (2008): 394–408.
24. "1996: Beachell and Khush," World Food Prize Foundation, accessed August 15, 2021, https://www.worldfoodprize.org/en/laureates/19871999_laureates/1996_beachell_and_khush/.
25. IRRI, *IR36, the World's Most Popular Rice* (Los Baños, Philippines: IRRI, undated), http://books.irri.org/IR36.pdf.
26. David J. Mackill, and Gurdev S. Khush, "IR64: A High-Quality and High-Yielding Mega Variety," *Rice* 11, no. 1 (2018): 18, doi.org/10.1186/s12284-018-0208-3.
27. Mahesh Ghimiray, "An Analysis of Rice Varietal Improvement and Adoption Rate by Farmers in Bhutan," *Journal of Renewable Natural Resources of Bhutan* 10, no. 1 (2012): 13–24.
28. Mahesh Ghimiray and Ronnie Vernooy, "The Importance and Challenges of Crop Germplasm Interdependence: The Case of Bhutan," *Food Security* 9, no. 2 (2017): 301–10.
29. "Hybrid," *Merriam-Webster.com Dictionary*, accessed August 15, 2021, https://www.merriam-webster.com/dictionary/hybrid.
30. *Genotype* identifies the underlying genetic makeup of an organism. *Phenotype* describes the observable features of an organism. Genotype influences phenotype, but phenotype is also influenced by environment.

31. In this book I use the term *maize* consistently to describe corn (*Zea mays*) while acknowledging that the term *corn* is almost always used to describe maize in the United States.
32. USDA, National Agricultural Statistics Service, *Crop Production Historical Track Records* (Washington, D.C.: USDA, April 2018) https://www.nass.usda.gov/Publications/Todays_Reports/reports/croptr18.pdf.
33. USDA, National Agricultural Statistics Service, *Crop Production 2019 Summary* (Washington, D.C.: USDA, January 2020).
34. Donald N. Duvick, "Biotechnology in the 1930s: The Development of Hybrid Maize," *Nature Reviews Genetics* 2, no. 1 (2001): 69–74. Duvick provides a clear and concise history of hybrid maize in the United States.
35. Joseph Rusike and Carl K. Eicher, "Institutional Innovations in the Maize Seed Industry," in *Africa's Emerging Maize Revolution*, ed. Derek Byerlee and Carl K. Eicher (London: Lynne Rienner, 1977), 173–91.
36. Melinda Smale and T. S. Jayne, *Maize Breeding in East and Southern Africa, 1900–2000*, 2020 Vision Focus 12, no. 4 (Washington, D.C.: International Food Policy Research Institute, 2004).
37. Glenn Denning et al., "Input Subsidies to Improve Smallholder Maize Productivity in Malawi: Toward an African Green Revolution," *PLoS Biol* 7, no. 1 (2009): e1000023.
38. Charles Darwin, *The Effects of Cross and Self Fertilisation in the Vegetable Kingdom* (New York: Appleton, 1885).
39. Shellen X. Wu, "Yuan Longping (1930–2021)," *Nature* 595, no. 7865 (2021): 26; "2004: Jones and Yuan," World Food Prize Foundation, accessed August 15, 2021, https://www.worldfoodprize.org/en/laureates/20002009_laureates/2004_jones_and_yuan/.
40. Dennis Normile, "Crossing Rice Strains to Keep Asia's Rice Bowls Brimming," *Science* 283, no. 5400 (1999): 313.
41. "Hybrid Rice," International Rice Research Institute, accessed August 15, 2021, https://www.irri.org/hybrid-rice.
42. "Xi Focus: Xi Confers Highest State Honors on Individuals Ahead of National Day," *Xinhuanet*, September 29, 2019, http://www.xinhuanet.com/english/2019-09/29/c_138433793.htm.
43. "Yuan Longping, Plant Scientist Who Helped Curb Famine, Dies at 90," *New York Times*, May 23, 2021, https://www.nytimes.com/2021/05/23/world/asia/yuan-longping-dead.html.
44. Charles Darwin, *On the Origin of Species by Means of Natural Selection, or the Preservation of Favoured Races in the Struggle for Life* (London: John Murray, 1859).
45. Ilona Miko, "Gregor Mendel and the Principles of Inheritance," *Nature Education* 1, no. 1 (2008): 134.
46. Michal Simunek, Uwe Hoßfeld, and Volker Wissemann, "'Rediscovery' Revised—the Cooperation of Erich and Armin von Tschermak-Seysenegg in the Context of the 'Rediscovery' of Mendel's Laws in 1899–1901," *Plant Biology* 13, no. 6 (2011): 835–41.
47. Duvick, "Biotechnology in the 1930s."
48. Anthony J. F. Griffiths, "Recombinant DNA, Genetic Engineering," *Britannica*, accessed August 15, 2021, https://www.britannica.com/science/recombinant-DNA-technology.
49. R. L. Hellmich and K. A. Hellmich, "Use and Impact of Bt Maize," *Nature Education Knowledge* 3, no. 10 (2012): 4.
50. Krisy Gashler, "Bt Eggplant Improving Lives in Bangladesh," *Cornell Chronicle*, July 16, 2018, https://news.cornell.edu/stories/2018/07/bt-eggplant-improving-lives-bangladesh.
51. R. S. Telem et al., "Cisgenics—a Sustainable Approach for Crop Improvement," *Current Genomics* 14, no. 7 (2013): 468–76.
52. H. J. Schouten, F. A. Krens, and E. Jacobsen, "Cisgenic Plants Are Similar to Traditionally Bred Plants: International Regulations for Genetically Modified Organisms Should Be Altered to Exempt Cisgenesis," *EMBO Reports* 7, no. 8 (2016): 750–53, doi.org/10.1038/sj.embor.7400769.
53. Sarah Garland, "USDA Definitions Around Biotechnology Need Some Crucial Updating," *Slate*, March 22, 2021, https://slate.com/technology/2021/03/usda-coordinated-framework-regulation-biotechnology-gene-editing.html.

54. For example, Pamela Ronald, "Moving Beyond Pro/Con Debates Over Genetically Engineered Crops," *The Conversation*, June 2, 2016. https://theconversation.com/moving-beyond-pro-con-debates-over-genetically-engineered-crops-59564.
55. ISAAA, *Global Status of Commercialized Biotech/GM Crops: 2018*, ISAAA Brief no. 54 (Ithaca, N.Y.: ISAAA, 2018). ISAAA was founded in 1991 and describes itself as "a not-for-profit international organization that shares the benefits of new bioscience technologies to key stakeholders, particularly resource-poor farmers in developing countries, through knowledge sharing, support to capacity building initiatives, and partnerships." "About ISAAA," International Service for the Acquisition of Agri-biotech Applications, accessed August 15, 2021, https://www.isaaa.org/.
56. All statistics in this paragraph are sourced from ISAAA, *Global Status*.
57. Jonas Kathage and Matin Qaim. "Economic Impacts and Impact Dynamics of Bt (Bacillus thuringiensis) Cotton in India," *Proceedings of the National Academy of Sciences* 109, no. 29 (2012): 11652–56.
58. N. J. Taylor et al., "The VIRCA Project: Virus Resistant Cassava for Africa," *GM Crops Food* 3, no. 2 (2012): 93–103.
59. "Water Efficient Maize for Africa (WEMA)," International Maize and Wheat Improvement Center (CIMMYT), accessed August 15, 2021, https://www.cimmyt.org/projects/water-efficient-maize-for-africa-wema/; Sylvester Oikeh et al., "The Water Efficient Maize for Africa Project as an Example of a Public–Private Partnership," in *Convergence of Food Security, Energy Security and Sustainable Agriculture*, ed. David D. Songstad, Jerry L. Hatfield, and Dwight T. Tomes (Berlin: Springer, 2014), 317–29.
60. "Major Breakthrough for Farmers and Scientists as Nigerian Biotech Body Approves Commercial Release of Genetically Modified Cowpea," International Institute of Tropical Agriculture (IITA), news release, February 9, 2019, https://www.iita.org/news-item/major-breakthrough-for-farmers-and-scientists-as-nigerian-biotech-body-approves-commercial-release-of-genetically-modified-cowpea/.
61. Anthony M. Shelton et al., "Impact of Bt Brinjal Cultivation in the Market Value Chain in Five Districts of Bangladesh," *Frontiers in Bioengineering and Biotechnology* 8 (2020): 498.
62. "Golden Rice," IRRI, accessed August 15, 2021, https://www.irri.org/golden-rice.
63. "Vitamin A Deficiency," World Health Organization, accessed August 15, 2021, https://www.who.int/data/nutrition/nlis/info/vitamin-a-deficiency.
64. Jacqueline A. Paine et al., "Improving the Nutritional Value of Golden Rice Through Increased Pro-Vitamin A Content," *Nature Biotechnology* 23, no. 4 (2005): 482.
65. Kai Kupferschmidt, "Activists Destroy 'Golden Rice' Field Trial," *Science*, August 9 (corrected August 13), 2013, https://www.sciencemag.org/news/2013/08/activists-destroy-golden-rice-field-trial.
66. J. Madeleine Nash, "This Rice Could Save a Million Kids a Year," *Time*, July 31, 2000, http://content.time.com/time/magazine/article/0,9171,997586,00.html.
67. Michael Pollan, "The Way We Live Now: The Great Yellow Hype," *New York Times*, March 4, 2001, https://michaelpollan.com/articles-archive/the-way-we-live-now-the-great-yellow-hype/.
68. B. Owens, "Golden Rice Is Safe to Eat, Says FDA," *Nature Biotechnology* 36 (2018): 559–60, doi.org/10.1038/nbt0718-559a.
69. "Philippines Approves Golden Rice for Direct Use as Food and Feed, or for Processing," IRRI, December 18, 2019, https://www.irri.org/news-and-events/news/philippines-approves-golden-rice-direct-use-food-and-feed-or-processing.
70. "Philippines Becomes First Country to Approve Nutrient-Enriched 'Golden Rice' for Planting," IRRI, July 23, 2021, https://www.irri.org/news-and-events/news/philippines-becomes-first-country-approve-nutrient-enriched-golden-rice.
71. "Laureates Letter Supporting Precision Agriculture (GMOs)," Support Precision Agriculture, June 29, 2016, https://www.supportprecisionagriculture.org/nobel-laureate-gmo-letter_rjr.html.
72. "Nobel Laureates Sign Letter on Greenpeace 'Golden' Rice Position—Statement," Greenpeace International, June 30, 2016, https://www.greenpeace.org/international/press-release/6866/nobel-laureates-sign-letter-on-greenpeace-golden-rice-position-statement/.

73. "#Food," Greenpeace International, accessed August 15, 2021, https://www.greenpeace.org/international/tag/food/.
74. Judith L. Fridovich-Keil, "Gene Editing, Genetics," *Britannica*, accessed August 15, 2021, https://www.britannica.com/science/gene-editing.
75. Sarah Garland, "EU Policy Must Change to Reflect the Potential of Gene Editing for Addressing Climate Change," *Global Food Security* 28 (2021): 100496.
76. "How Does Genome Editing Work?," National Human Genome Research Institute, accessed August 15, 2021, https://www.genome.gov/about-genomics/policy-issues/Genome-Editing/How-genome-editing-works.
77. Martin Jinek et al., "A Programmable Dual-RNA–Guided DNA Endonuclease in Adaptive Bacterial Immunity," *Science* 337, no. 6096 (2012): 816–21. CRISPR-Cas9 is short for Clustered Regularly Interspaced Short Palindromic Repeats and CRISPR-associated protein 9.
78. "The Nobel Prize in Chemistry 2020," Nobel Prize, news release, October 7, 2020, https://www.nobelprize.org/prizes/chemistry/2020/press-release/.
79. Merlin Crossley, "What Is CRISPR Gene Editing, and How Does It Work?," *The Conversation*, February 1, 2018, https://theconversation.com/what-is-crispr-gene-editing-and-how-does-it-work-84591.
80. Emily Waltz, "With CRISPR and Machine Learning, Startups Fast-Track Crops to Consume Less, Produce More," *Nature Biotechnology* 37, no. 11 (2019): 1251.
81. Emily Waltz, "Gene-Edited CRISPR Mushroom Escapes US Regulation," *Nature* 532, no. 7599 (2016): 293.
82. Swati Tyagi et al., "CRISPR-Cas9 System: A Genome-Editing Tool with Endless Possibilities," *Journal of Biotechnology* 319, no. 1 (2020): 36–53.
83. A. Zsögön et al., "De Novo Domestication of Wild Tomato Using Genome Editing," *Nature Biotechnology* 36 (2018): 1211–16, doi.org/10.1038/nbt.4272.
84. "Super-Tomato Shows What Plant Scientists Can Do," *Nature* 562, no. 8 (2018).
85. Florian Hahn and Vladimir Nekrasov, "CRISPR/Cas Precision: Do We Need to Worry About Off-targeting in Plants?" *Plant Cell Reports* 38, no. 4 (2019): 437–41.
86. Syed Shan-e-Ali Zaidi et al., "New Plant Breeding Technologies for Food Security," *Science* 363, no. 6434 (2019): 1390–91.

6. CLIMATE CHANGE

1. While acknowledging that some domestication began earlier than 12,000 BP (before present), current evidence indicates that most domestication occurred in the past ten to twelve thousand years.
2. Anthony McMichael, *Climate Change and the Health of Nations: Famines, Fevers, and the Fate of Populations* (Oxford: Oxford University Press, 2017), 1–21.
3. Paul J. Crutzen and Eugene F. Stoermer, "The Anthropocene," *Global Change Newsletter*, 41 (2000): 17–18. Also see Paul J. Crutzen, "Geology of Mankind," in *Paul J. Crutzen: A Pioneer on Atmospheric Chemistry and Climate Change in the Anthropocene*, ed. P. Crutzen and H. Brauch, SpringerBriefs on Pioneers in Science and Practice, vol 50 (Cham, Switzerland: Springer, 2016), doi.org/10.1007/978-3-319-27460-7_10.
4. *Rice Today, IRRI*, "Coping with Climate Change," *Rice Today* (July–September 2007): 10–15.
5. R. Wassmann et al., "Characterization of Methane Emissions from Rice Fields in Asia. I. Comparison Among Field Sites in Five Countries," *Nutrient Cycling in Agroecosystems* 58, no. 1 (2000): 1–12.
6. S. Peng et al., eds., *Climate Change and Rice* (Berlin: Springer-Verlag and Manila: IRRI, 1995), ix.
7. IPCC, "Summary for Policymakers," in *Climate Change 2021: The Physical Science Basis*, contribution of Working Group I to the Sixth Assessment Report of the Intergovernmental Panel on Climate Change, ed. V. Masson-Delmotte et al. (Cambridge: Cambridge University Press, in press).

8. IPCC, "Summary for Policymakers," 5, 17.
9. IPCC, *Climate Change and Land: An IPCC Special Report on Climate Change, Desertification, Land Degradation, Sustainable Land Management, Food Security, and Greenhouse Gas Fluxes in Terrestrial Ecosystems*, ed. P. R. Shukla et al. (Geneva: IPCC, 2019), chap. 5, https://www.ipcc.ch/srccl/chapter/chapter-5/.
10. "Understanding Global Warming Potentials," United States Environmental Protection Agency, accessed August 23, 2021, https://www.epa.gov/ghgemissions/understanding-global-warming-potentials.
11. IPCC, *Climate Change and Land.*
12. Hanqin Tian et al., "A Comprehensive Quantification of Global Nitrous Oxide Sources and Sinks," *Nature* 586, no. 7828 (2020): 248–56.
13. IPCC, *Climate Change and Land.*
14. Francesco N. Tubiello et al., "Greenhouse Gas Emissions from Food Systems: Building the Evidence Base," *Environmental Research Letters* 16, no. 6 (2021): 065007.
15. FAO et al., *The State of Food Security and Nutrition in the World 2018: Building Climate Resilience for Food Security and Nutrition* (Rome: FAO, 2018).
16. Bin Wang et al., "Historical Change of El Niño Properties Sheds Light on Future Changes of Extreme El Niño," *Proceedings of the National Academy of Sciences* 116, no. 45 (2019): 22512–17. An El Niño occurs when sea surface temperatures in the central and eastern tropical Pacific Ocean become substantially warmer than average, causing a shift in atmospheric circulation and related weather patterns in various parts of the world. For additional information about El Niño and the opposing climate pattern of La Niña (together known as the El Niño–Southern Oscillation or ENSO), see "What Are El Niño and La Niña?," NOAA National Ocean Service, accessed August 23, 2021, https://oceanservice.noaa.gov/facts/ninonina.html.
17. FAO et al., *State of Food Security 2018.*
18. Carl-Friedrich Schleussner et al., "Differential Climate Impacts for Policy-Relevant Limits to Global Warming: The Case of 1.5°C and 2°C," *Earth System Dynamics* 7, no. 2 (2016): 327–51.
19. Carl-Friedrich Schleussner et al., "Crop Productivity Changes in 1.5°C and 2°C Worlds Under Climate Sensitivity Uncertainty," *Environmental Research Letters* 13, no. 6 (2018): 064007.
20. IPCC, *Climate Change and Land.*
21. Louise Fox and Thomas S. Jayne, "Unpacking the Misconceptions About Africa's Food Imports," Brookings Institution, December 14, 2020, https://www.brookings.edu/blog/africa-in-focus/2020/12/14/unpacking-the-misconceptions-about-africas-food-imports/.
22. Jikun Huang et al., *Global Food Security and Market Stability: The Role and Concerns of Large Net Food Importers and Exporters*, T20 Argentina: Food Security and Sustainable Development Task Force Brief, https://www.g20-insights.org/policy_briefs/global-food-security-and-market-stability-the-role-and-concerns-of-large-net-food-importers-and-exporters/.
23. FAO et al., *State of Food Security 2018.*
24. FAO et al., 82.
25. Joshua E. Cinner et al., "Vulnerability of Coastal Communities to Key Impacts of Climate Change on Coral Reef Fisheries," *Global Environmental Change* 22, no. 1 (2012): 12–20.
26. FAO et al., *State of Food Security 2018.*
27. This coping mechanism is well documented by researchers in Malawi and elsewhere. For example: Anne Conroy et al., *Poverty, AIDS and Hunger: Breaking the Poverty Trap in Malawi* (London: Palgrave Macmillan, 2006).
28. IPCC, "Annex I: Glossary," ed. J. B. R. Matthews, in *Global Warming of 1.5°C. An IPCC Special Report on the Impacts of Global Warming of 1.5°C Above Pre-Industrial Levels and Related Global Greenhouse Gas Emission Pathways, in the Context of Strengthening the Global Response to the Threat of Climate Change, Sustainable Development, and Efforts to Eradicate Poverty*, ed. V. Masson-Delmotte et al. (Geneva: IPCC, 2018), https://www.ipcc.ch/sr15/chapter/glossary/.
29. I. R. Noble et al. "Adaptation Needs and Options," in *Climate Change 2014: Impacts, Adaptation, and Vulnerability. Part A: Global and Sectoral Aspects. Contribution of Working Group II to the Fifth*

Assessment Report of the Intergovernmental Panel on Climate Change, ed. C. B. Field et al. (Cambridge: Cambridge University Press, 2014), 836.

30. "What Is Food System Resilience?," Resilience of the UK Food System in a Global Context, accessed August 23, 2021, https://www.foodsystemresilienceuk.org/what-is-food-system-resilience/. Resilience is described in terms of three Rs: (1) robustness—"the ability of the food system to resist disruptions to current outcomes by preventing shocks and stress impacting food system activities"; (2) recovery—"the ability of the food system to return to prior outcomes following disruption (*bounce back*)"; and (3) reorientation—"the ability of the food system to deliver acceptable alternative outcomes before or following disruption (*bounce forward*)."
31. M. Fisher et al., "Drought Tolerant Maize for Farmer Adaptation to Drought in Sub-Saharan Africa: Determinants of Adoption in Eastern and Southern Africa," *Climatic Change* 133 (2015): 283–99.
32. Kenong Xu et al., "Sub1A Is an Ethylene-Response-Factor-Like Gene That Confers Submergence Tolerance to Rice," *Nature* 442, no. 7103 (2006): 705–8.
33. Asian Development Bank, Country Partnership Strategy, *Cambodia, 2019–2023—Inclusive Pathways to a Competitive Economy* (Manila: ADB, October 2019), https://www.adb.org/sites/default/files/institutional-document/534691/cps-cam-2019-2023.pdf.
34. Barry J. Barnett and Olivier Mahul, "Weather Index Insurance for Agriculture and Rural Areas in Lower-Income Countries," *American Journal of Agricultural Economics* 89, no. 5 (2007): 1241–47.
35. Ho Huu Loc et al., "How the Saline Water Intrusion Has Reshaped the Agricultural Landscape of the Vietnamese Mekong Delta, a Review," *Science of the Total Environment* 794 (2021): 148651.
36. Daniel Maxwell, "Famine Early Warning and Information Systems in Conflict Settings: Challenges for Humanitarian Metrics and Response," November 2019, LSE and UKaid, http://eprints.lse.ac.uk/102836/1/Maxwell_famine_early_warning_and_information_systems_published.pdf.
37. See FEWS NET website, https://fews.net/.
38. Chris Hewitt, Simon Mason, and David Walland, "The Global Framework for Climate Services," *Nature Climate Change* 2, no. 12 (2012): 831–32.
39. International Institute for Climate and Society, "Seasonal Climate Forecasts," https://iri.columbia.edu/our-expertise/climate/forecasts/#Seasonal_Climate_Forecasts.
40. Gerald A. Meehl et al., "How Much More Global Warming and Sea Level Rise?," *Science* 307, no. 5716 (2005): 1769–72.
41. "Definition of Terms Used Within the DDC Pages, Glossary L–M," Data Distribution Centre, Intergovernmental Panel on Climate Change, accessed August 23, 2021, https://www.ipcc-data.org/guidelines/pages/glossary/glossary_lm.html.
42. Paris Agreement, December 12, 2015, U.S.T. (United States Treaties and Other International Agreements) I-54113, https://unfccc.int/sites/default/files/english_paris_agreement.pdf.
43. United Nations Framework Convention on Climate Change, *Nationally Determined Contributions Under the Paris Agreement: Synthesis Report by the Secretariat* (Bonn, Germany: UNFCC, February 26, 2021), https://unfccc.int/process-and-meetings/the-paris-agreement/nationally-determined-contributions-ndcs/nationally-determined-contributions-ndcs/ndc-synthesis-report.
44. Stephanie Roe et al., "Contribution of the Land Sector to a 1.5°C World," *Nature Climate Change* 9, no. 11 (2019): 817–28.
45. William F. Laurance, Jeffrey Sayer, and Kenneth G. Cassman, "Agricultural Expansion and Its Impacts on Tropical Nature," *Trends in Ecology and Evolution* 29, no. 2 (2014): 107–16.
46. Tubiello et al., "Greenhouse Gas Emissions."
47. Tubiello et al.
48. A. De Stefano and M. G. Jacobson "Soil Carbon Sequestration in Agroforestry Systems: A Meta-Analysis," *Agroforestry Systems* 92 (2018): 285–99, doi.org/10.1007/s10457-017-0147-9.
49. Bram Govaerts et al., "Conservation Agriculture and Soil Carbon Sequestration: Between Myth and Farmer Reality," *Critical Reviews in Plant Sciences* 28, no. 3 (2009): 97–122.

50. Ken E. Giller et al., "Conservation Agriculture and Smallholder Farming in Africa: The Heretics' View," *Field Crops Research* 114, no. 1 (2009): 23–34.
51. Roe et al., "Contribution of the Land Sector."
52. Roe et al.
53. Mario Herrero et al., "Greenhouse Gas Mitigation Potentials in the Livestock Sector," *Nature Climate Change* 6, no. 5 (2016): 452–61.
54. K. Crumpler et al., *Assessing the Role of Agriculture and Land Use in Nationally Determined Contributions*, Environment and Natural Resources Management Working Paper no. 76 (Rome: FAO, 2019).
55. Alexandre Meybeck and Vincent Gitz, *"Climate-Smart" Agriculture: Policies, Practices and Financing for Food Security, Adaptation and Mitigation* (Rome: FAO, 2010), ii.
56. FAO, *Climate-Smart Agriculture: Case Studies 2018: Successful Approaches from Different Regions* (Rome: FAO, 2018); "Climate Smart Agriculture Sourcebook," FAO, accessed August 23, 2021, http://www.fao.org/climate-smart-agriculture-sourcebook/en/. First published in 2013, the CSA Sourcebook has undergone a complete revision accessible on a new digital platform as the second edition, 2017.
57. R. Strohmaier et al., *The Agriculture Sectors in the Intended Nationally Determined Contributions: Analysis*, Environment and Natural Resources Management Working Paper no. 62 (Rome: FAO, 2016).
58. J. Bayala et al., *Towards Developing Scalable Climate-Smart Village Models: Approach and Lessons Learnt from Pilot Research in West Africa*, ICRAF (World Agroforestry) Occasional Paper no. 25 (Nairobi: World Agroforestry Centre, 2016).
59. Rhys Bucknall-Williams, "How the CCAFS Legacy Lives On," CGIAR-CCAFS, December 16, 2021, https://ccafs.cgiar.org/news/how-ccafs-legacy-lives.
60. World Bank, International Finance Corporation (IFC), and Multilateral Investment Guarantee Agency (MIGA), *World Bank Group Climate Change Action Plan 2016–2020* (Washington, D.C.: World Bank, 2016).
61. "Climate-Smart Agriculture."
62. "Livestock and Dairy Development Project (Bangladesh)," World Bank, accessed August 23, 2021, https://projects.worldbank.org/en/projects-operations/project-detail/P161246.
63. Leslie Lipper et al., "Climate-Smart Agriculture for Food Security," *Nature Climate Change* 4, no. 12 (2014): 1068–72.
64. S. J. Vermeulen, J. R. Porter, and E. Bennetzen, "Climate-Smart Food Systems, Plenary Presentation," in *Proceedings of the 3rd Climate Smart Agriculture Global Science Conference* (Montpellier, France, March 16–18, 2015); Stephen Whitfield, Andrew Juan Challinor, and Robert M. Rees, "Frontiers in Climate Smart Food Systems: Outlining the Research Space," *Frontiers in Sustainable Food Systems* 2 (2018): 2.

7. HUMAN NUTRITION

1. Jessica Fanzo, *Can Fixing Dinner Fix the Planet?* (Baltimore: Johns Hopkins University Press, 2021).
2. Barry M. Popkin, Linda S. Adair, and Shu Wen Ng, "Global Nutrition Transition and the Pandemic of Obesity in Developing Countries," *Nutrition Reviews* 70, no. 1 (2012): 3–21, doi.org/10.1111/j.1753-4887.2011.00456.x.
3. See preface, nn. 8 and 9, for supporting assumptions about consumption.
4. Jim Mann and A. Stewart Truswell, eds., *Essentials of Human Nutrition*, 5th ed. (Oxford: Oxford University Press, 2017), 4.
5. A. Altemimi et al., "Phytochemicals: Extraction, Isolation, and Identification of Bioactive Compounds from Plant Extracts," *Plants* 6, no. 4 (2017): 42, doi.org/10.3390/plants6040042.

6. Walter Willett et al., "Food in the Anthropocene: The EAT–Lancet Commission on Healthy Diets from Sustainable Food Systems," *Lancet* 393, no. 10170 (2019): 447–92.
7. Regan L. Bailey, Keith P. West Jr., and Robert E. Black, "The Epidemiology of Global Micronutrient Deficiencies," *Annals of Nutrition and Metabolism* 66, suppl. 2 (2015): 22–33.
8. William Gardner and Nicholas Kassebaum, "Global, Regional, and National Prevalence of Anemia and Its Causes in 204 Countries and Territories, 1990–2019," *Current Developments in Nutrition* 4, suppl. 2 (June 2020): 830, doi.org/10.1093/cdn/nzaa053_035.
9. "Vitamin A Deficiency," World Health Organization, accessed August 24, 2021, https://www.who.int/data/nutrition/nlis/info/vitamin-a-deficiency.
10. Bailey, West, and Black, "Epidemiology of Global."
11. Uma Lele et al., *Measuring Food and Nutrition Security: An Independent Technical Assessment and User's Guide for Existing Indicators*, Measuring Food and Nutrition Security Technical Working Group (Rome: Food Security Information Network, 2016), 56, https://sites.tufts.edu/willmasters/files/2016/06/FSIN-TWG_UsersGuide_12June2016.pdf.
12. Also known as body mass index (BMI).
13. FAO, International Fund for Agricultural Development, and World Food Programme, *The State of Food Insecurity in the World 2015: Meeting the 2015 International Hunger Targets: Taking Stock of Uneven Progress* (Rome: FAO, 2015). See annex 2 for a description of the PoU methodology.
14. The other indicator for MDG Target 1.C was "prevalence of underweight children under five years of age." See "Official List of MDG Indicators," United Nations Statistics Division, January 15, 2008, http://mdgs.un.org/unsd/mdg/Host.aspx?Content=Indicators/OfficialList.htm.
15. FAO et al., *The State of Food Security and Nutrition in the World 2019: Safeguarding Against Economic Slowdowns and Downturns* (Rome: FAO, 2019).
16. Lele et al., *Measuring Food.*
17. For example, see Kathleen Beegle et al., "More to Do on Measuring Hunger," World Bank Blogs, March 19, 2014, https://blogs.worldbank.org/developmenttalk/more-do-measuring-hunger.
18. FAO et al., *The State of Food Security and Nutrition in the World 2020: Transforming Food Systems for Affordable Healthy Diets* (Rome: FAO, 2020).
19. UN Millennium Project Task Force on Hunger, *Halving Hunger: It Can Be Done* (London: Earthscan, 2005).
20. Hunger Task Force, *Report to the Government of Ireland* (Limerick, Ireland: Irish Aid, 2008), https://www.irishaid.ie/media/irishaid/allwebsitemedia/20newsandpublications/publicationpdfsenglish/hunger-task-force.pdf.
21. Carolina Milhorance, Marcel Bursztyn, and Eric Sabourin, "The Politics of the Internationalisation of Brazil's 'Zero Hunger' Instruments," *Food Security* 11, no. 2 (2019): 447–60.
22. UNICEF, WHO, and International Bank for Reconstruction and Development/The World Bank, *Levels and Trends in Child Malnutrition: Key Findings of the 2021 Edition of the Joint Child Malnutrition Estimates* (Geneva: World Health Organization, 2021), 14. The subregions are in accordance with UN definitions—Oceania data excludes Australia and New Zealand.
23. Jef L. Leroy et al., "Linear Growth Deficit Continues to Accumulate Beyond the First 1000 Days in Low- and Middle-Income Countries: Global Evidence from 51 National Surveys," *Journal of Nutrition* 144, no. 9 (2014): 1460–66.
24. UNICEF, *Strategy for Improved Nutrition of Children and Women in Developing Countries*, UNICEF Policy Review (New York: UNICEF, 1990).
25. Maureen M. Black et al., "Early Childhood Development Coming of Age: Science Through the Life Course," *Lancet* 389, no. 10064 (2017): 77–90.
26. K. D. Bruce and M. A. Hanson, "The Developmental Origins, Mechanisms, and Implications of Metabolic Syndrome," *Journal of Nutrition* 140 (2010): 648–52, doi.org/10.3945/jn.109.111179.
27. Sue Horton and Richard H. Steckel, "Malnutrition: Global Economic Losses Attributable to Malnutrition 1900–2000 and Projections to 2050," in *How Much Have Global Problems Cost the Earth? A Scorecard from 1900 to 2050*, ed. B. Lomborg (Cambridge: Cambridge University Press, 2013), 247–72.

28. E. Galasso et al., *The Economic Costs of Stunting and How to Reduce Them*, Policy Research Note (PRN) 17/5 (Washington, D.C.: World Bank Group, 2017). *Income penalty* refers to the per capita income reduction that a country incurs for not having eliminated stunting when today's workers were children.
29. Zulfiqar A. Bhutta et al., "How Countries Can Reduce Child Stunting at Scale: Lessons from Exemplar Countries," *American Journal of Clinical Nutrition* 112, suppl. 2 (2020): 894S–904S.
30. Jennifer Bryce et al., "Maternal and Child Undernutrition: Effective Action at National Level," *Lancet* 371, no. 9611 (2008): 510–26.
31. See Roger Thurow, *The First Thousand Days: A Crucial Time for Mothers and Children—and the World* (New York: PublicAffairs, US, 2016).
32. F. Branca et al., "Nutrition and Health in Women, Children, and Adolescent Girls," *BMJ* 351 (2015): h4173, doi.org/10.1136/bmj.h4173.
33. Branca et al.
34. WHO, *WHA Global Nutrition Targets 2025: Wasting Policy Brief* (Washington, D.C.: WHO, 2014), https://apps.who.int/iris/bitstream/handle/10665/149023/WHO_NMH_NHD_14.8_eng.pdf.
35. UNICEF et al., *Levels and Trends in Child Malnutrition 2021.*
36. WHO, *WHA Global Nutrition Targets 2025.*
37. Christine M. McDonald et al., "The Effect of Multiple Anthropometric Deficits on Child Mortality: Meta-Analysis of Individual Data in 10 Prospective Studies from Developing Countries," *American Journal of Clinical Nutrition* 97, no. 4 (2013): 896–901, doi.org/10.3945/ajcn.112.047639; Z. A. Bhutta et al., "What Works? Interventions for Maternal and Child Undernutrition and Survival," *Lancet* 371, no. 9610 (2008): 417–40.
38. WHO, *WHA Global Nutrition Targets 2025.*
39. UNICEF et al., *Levels and Trends in Child Malnutrition 2021,* 14.
40. WHO, *WHA Global Nutrition Targets 2025.*
41. "Nutrition and Care for Children with Wasting," UNICEF, accessed August 24, 2021, https://www.unicef.org/nutrition/index_sam.html.
42. WHO, *WHA Global Nutrition Targets 2025.*
43. "Community-Based Management of Acute Malnutrition: Technical Guidance Brief," U.S. Agency for International Development (USAID), accessed April 2, 2022, https://www.usaid.gov/sites/default/files/documents/1864/CMAM-technical-guidance-brief-508-revFeb2017.pdf.
44. S. Collins, "Treating Severe Acute Malnutrition Seriously," *Archives of Disease in Childhood* 92 no. 5 (2007): 453–61, doi.org/10.1136/adc.2006.098327.
45. USAID, "Community-Based Management of Acute Malnutrition."
46. See UNICEF, "Nutrition and Care for Children with Wasting" for details.
47. "Water, Sanitation and Hygiene (WASH)," UNICEF, accessed August 24, 2021, https://www.unicef.org/wash.
48. "Rome Declaration on World Food Security," World Food Summit, November 13–17, 1996, http://www.fao.org/3/w3613e/w3613e00.htm.
49. "Voices of the Hungry: The Food Insecurity Experience Scale," FAO, accessed August 25, 2021, http://www.fao.org/in-action/voices-of-the-hungry/fies/en/.
50. FAO, "Voices of the Hungry."
51. See FAO, *The Food Insecurity Experience Scale: Measuring Food Insecurity Through People's Experiences*, I7835EN/1/09.17 (Rome: FAO, n.d.), http://www.fao.org/3/a-i7835e.pdf.
52. "Food Insecurity Experience Scale (FIES)," International Dietary Data Expansion Project, accessed August 24, 2021, https://inddex.nutrition.tufts.edu/data4diets/indicator/food-insecurity-experience-scale-fies?back=/data4diets/indicators.
53. Anna Herforth et al., "A Global Review of Food-Based Dietary Guidelines," *Advances in Nutrition* 10, no. 4 (2019): 590–605.
54. C. Peter Timmer, Walter P. Falcon, and Scott R. Pearson, *Food Policy Analysis* (Baltimore: Johns Hopkins University Press for the World Bank, 1983).

55. M. T. Ruel, "Operationalizing Dietary Diversity: A Review of Measurement Issues and Research Priorities," *Journal of Nutrition* 133, no. 11, suppl. 2 (2003): 3911S–26S.
56. G. L. Kennedy et al., "Dietary Diversity Score Is a Useful Indicator of Micronutrient Intake in Non-Breast-Feeding Filipino Children," *Journal of Nutrition* 137, no. 2 (2007): 472–77.
57. "Data4Diets: Building Blocks for Diet-Related Food Security Analysis," International Dietary Data Expansion Project, accessed August 21, 2021, https://inddex.nutrition.tufts.edu/data4diets.
58. Indira Bose et al., "The 'Fill the Nutrient Gap' Analysis: An Approach to Strengthen Nutrition Situation Analysis and Decision Making Towards Multisectoral Policies and Systems Change," *Maternal and Child Nutrition* 15, no. 3 (2019): e12793. WFP's technical partners included the University of California, Davis, International Food Policy Research Institute (IFPRI), Epicentre, Harvard University, Mahidol University, Save the Children UK, and United Nations Children's Fund.
59. Bose et al., 1.
60. Mariko Kawabata et al., "Food Security and Nutrition Challenges in Tajikistan: Opportunities for a Systems Approach," *Food Policy* 96 (2020): 101872.
61. Edward A. Gargan, "Nyerere Steps Down but Keeps His Hand In," *New York Times*, November 3, 1985, https://www.nytimes.com/1985/11/03/weekinreview/nyerere-steps-down-but-keeps-his-hand-in.html.
62. Roger Nord et al., *Tanzania: The Story of an African Transition*, no. 09/02 (Washington, D.C.: International Monetary Fund, 2009). *Ujamaa* is a Kiswahili word meaning brotherhood or extended family. Julius Nyerere advanced ujamaa socialism in Tanzania as a uniquely African approach to collective action. See Julius K. Nyerere, "Ujamaa: The Basis of African Socialism," *Journal of Pan African Studies* 1, no. 1 (1987): 4–11.
63. "Country Nutrition Profiles: United Republic of Tanzania," Global Nutrition Report, accessed August 24, 2021, https://globalnutritionreport.org/resources/nutrition-profiles/africa/eastern-africa/united-republic-tanzania/#profile.
64. Kedir Y. Ahmed et al., "Factors Associated with Underweight, Overweight, and Obesity in Reproductive Age Tanzanian Women," *PloS One* 15, no. 8 (2020): e0237720.
65. B. M. Popkin, "The Nutrition Transition in Low-Income Countries: An Emerging Crisis," *Nutrition Reviews* 52, no. 9 (1994): 285–98.
66. "Assessing Your Weight and Health Risk," NIH National Heart, Lung, and Blood Institute, accessed August 24, 2021, https://www.nhlbi.nih.gov/health/educational/lose_wt/risk.htm.
67. "Obesity and Overweight," WHO, June 9, 2021, https://www.who.int/news-room/fact-sheets/detail/obesity-and-overweight.
68. NCDs are nontransmissible diseases and are commonly referred to as chronic diseases.
69. Global Nutrition Report, "Country Nutrition Profiles: United Republic of Tanzania."
70. WHO, "Obesity and Overweight."
71. Renata Micha et al., *2021 Global Nutrition Report: The State of Global Nutrition* (Bristol, UK: Development Initiatives, 2021), https://globalnutritionreport.org/reports/2021-global-nutrition-report/.
72. Between 2005 and 2019 the total population of overweight (including obese) adults rose from 1.33 billion to 2.2 billion, an increase of 65 percent. Assuming a modest 15 percent increase from 2019 to 2025, the number rises to more than 2.5 billion adults. The 1.33 billion estimate for 2005 is taken from T. Kelly et al., "Global Burden of Obesity in 2005 and Projections to 2030," *International Journal of Obesity* 32, no. 9 (September 2008):1431–37.
73. Meera Shekar and Barry Popkin, eds., *Obesity: Health and Economic Consequences of an Impending Global Challenge*, Human Development Perspectives series (Washington, D.C.: World Bank, 2020), https://openknowledge.worldbank.org/handle/10986/32383.
74. See Centers for Disease Control and Prevention, "Science Brief: Evidence Used to Update the List of Underlying Medical Conditions That Increase a Person's Risk of Severe Illness from COVID-19," accessed September 2, 2021, https://www.cdc.gov/coronavirus/2019-ncov/science/science-briefs/underlying-evidence-table.html.

75. Shekar and Popkin, *Obesity*.
76. SDG indicator 2.2.2: Prevalence of malnutrition (weight for height > +2 or < −2 standard deviation from the median of the WHO Child Growth Standards) among children under five years of age, by type (wasting and overweight). See "Goal 2: Zero Hunger," UNICEF, accessed August 24, 2021, https://data.unicef.org/sdgs/goal-2-zero-hunger/.
77. Corinna Hawkes et al., "Double-Duty Actions: Seizing Programme and Policy Opportunities to Address Malnutrition in All Its Forms," *Lancet* 395, no. 10218 (2020): 142–55. Erratum in *Lancet* 395, no. 10221 (February 1, 2020): 338, PMID: 31852603.
78. WHO, *The Double Burden of Malnutrition*, Policy brief (Geneva: WHO, 2017).
79. Francesco Branca et al., "A New Nutrition Manifesto for a New Nutrition Reality," *Lancet* 395, no. 10217 (2020): 8–10.
80. Bruce and Hanson, "Developmental Origins."

8. FOOD SYSTEMS TRANSFORMATION

1. An adaptation of definitions used by P. J. Ericksen, "Conceptualizing Food Systems for Global Environmental Change Research," *Global Environmental Change* 18 (2008): 234–45; and High Level Panel of Experts, *Nutrition and Food Systems*, A report by the High Level Panel of Experts on Food Security and Nutrition (HLPE) (Rome: Committee on World Food Security, 2017), http://www.fao.org/3/i7846e/i7846e.pdf. The HLPE is a science–policy interface of the UN Committee on World Food Security (CFS) that was created in October 2009 as an essential element of the CFS reform.
2. Glenn Denning and Jessica Fanzo, "Ten Forces Shaping the Global Food System," in *Good Nutrition: Perspectives for the 21st Century*, ed. M. Eggersdorfer et al. (Basel, Switzerland: Karger, 2016).
3. Anthony McMichael, *Climate Change and the Health of Nations: Famines, Fevers, and the Fate of Populations* (Oxford: Oxford University Press, 2017).
4. P. H. Nye and D. J. Greenland, *The Soil Under Shifting Cultivation*, Technical Communication 51 (Farnham Royal, UK: Commonwealth Bureau of Soils, Commonwealth Agricultural Bureau, 1960).
5. Cheryl Palm et al., eds., *Slash-and-Burn Agriculture: The Search for Alternatives* (New York: Columbia University Press, 2005).
6. For a more nuanced assessment of the dynamics and controversies around shifting cultivation in the Southeast Asian context, see Ole Mertz et al., "Swidden Change in Southeast Asia: Understanding Causes and Consequences," *Human Ecology* 37, no. 3 (2009): 259–64.
7. Palm et al., *Slash-and-Burn Agriculture*.
8. Pedro A. Sanchez, *Properties and Management of Soils in the Tropics*, 2nd ed. (Cambridge: Cambridge University Press, 2019), 464–79.
9. Dimithri Devinda Jayagoda, "Community-Based Mangrove Forest Management in Association with Sustainable Tourism in Puerto Princesa City of the Philippines," *International Journal of Sustainable Future for Human Security* 3, no. 2 (2015): 23–30.
10. Pieter A. Zuidema and René G. A. Boot, "Demography of the Brazil Nut Tree (*Bertholletia excelsa*) in the Bolivian Amazon: Impact of Seed Extraction on Recruitment and Population Dynamics," *Journal of Tropical Ecology* 18, no. 1 (2002): 1–31.
11. Luciana Porter-Bolland et al., "Community Managed Forests and Forest Protected Areas: An Assessment of their Conservation Effectiveness across the Tropics," *Forest Ecology and Management* 268 (2012): 6–17.
12. Götz Schroth, Celia A. Harvey, and Grégoire Vincent, "Complex Agroforests: Their Structure, Diversity, and Potential Role in Landscape Conservation," in *Agroforestry and Biodiversity Conservation in Tropical Landscapes*, ed. Götz Schroth et al. (Washington, D.C.: Island Press, 2004), 227–60.
13. Sanchez, *Properties and Management of Soils*.

14. Dennis J. Greenland, ed., *The Sustainability of Rice Farming* (Wallingford, UK: CABI, 1997).
15. Malleshaiah SharathKumar, Ep Heuvelink, and Leo F. M. Marcelis, "Vertical Farming: Moving from Genetic to Environmental Modification," *Trends in Plant Science* 25, no. 8 (2020): 724–27.
16. C. Peter Timmer, *Food Security and Scarcity* (Philadelphia: University of Pennsylvania Press, 2015), 2.
17. Timmer, 2–3.
18. C. Peter Timmer, "The Structural Transformation and the Changing Role of Agriculture in Economic Development," Wendt Lecture, October 30, 2007, American Enterprise Institute, Washington, D.C.
19. See Timmer, *Food Security*, chap. 4, for a thorough analysis of structural transformation as it relates to food security.
20. "Meet Farm Aid's Current Grantees," Farm Aid, accessed August 24, 2021, https://www.farmaid.org/our-work/grants/.
21. "Farming and Farm Income," Economic Research Service, U.S. Department of Agriculture, accessed August 24, 2021, https://www.ers.usda.gov/data-products/ag-and-food-statistics-charting-the-essentials/farming-and-farm-income/.
22. Carolyn Dimitri, Anne Effland, and Neilson Conklin, *The 20th Century Transformation of US Agriculture and Farm Policy*, Economic Information Bulletin Number 3 (Washington, D.C.: USDA, June 2005).
23. "Ag and Food Sectors and the Economy," Economic Research Service, USDA, accessed February 23, 2022, https://www.ers.usda.gov/data-products/ag-and-food-statistics-charting-the-essentials/ag-and-food-sectors-and-the-economy/.
24. Statistics are from USDA's Economic Research Service, "Ag and Food Statistics: Charting the Essentials," https://www.ers.usda.gov/data-products/ag-and-food-statistics-charting-the-essentials.
25. Peter Costello, interview by Emma Simkin, *AM*, Australian Broadcasting Commission (ABC), July 26, 2001, https://www.abc.net.au/am/stories/s335651.htm.
26. Timmer notes that while structural transformation may initially lead to a spurt in population growth, a new "modern" equilibrium of low birth rates and low death rates is subsequently reached. Timmer, *Food Security*, 108.
27. Timmer, 79, 78.
28. As described in Hendrik Samuel Houthakker, "An International Comparison of Household Expenditure Patterns, Commemorating the Centenary of Engel's Law," *Econometrica* 25 (1957): 532–51.
29. Bruce F. Johnston and John W. Mellor. "The Role of Agriculture in Economic Development," *American Economic Review* 51, no. 4 (1961): 566.
30. See Steven Haggblade, Peter B. R. Hazell, and Thomas Reardon, eds., *Transforming the Rural Nonfarm Economy: Opportunities and Threats in the Developing World,* International Food Policy Research Institute Issue Brief 58 (Baltimore: Johns Hopkins University Press for the IFPRI, 2007), for a comprehensive review and synthesis of research on the rural nonfarm economy in the developing world.
31. World Bank, *World Development Report 2008: Agriculture for Development* (Washington, D.C.: World Bank, 2007), 26, https://openknowledge.worldbank.org/handle/10986/5990.
32. C. Peter Timmer, "The Role of Agriculture in 'Catching Up': A Gerschenkronian Perspective," in *Diverse Development Paths and Structural Transformation in the Escape from Poverty*, ed. Martin Andersson and Tobias Axelsson (Oxford: Oxford University Press, 2016), 69–80, doi:10.1093/acprof:oso/9780198737407.003.0004; C. Peter Timmer, "Agriculture and Economic Development," in *Handbook of Agricultural Economics*, ed. B. L. Gardner and G. C. Rausser, 2A (Amsterdam: Elsevier/North Holland, 2002), 1487–1546.
33. See several papers on the rise of supermarkets in developing countries authored by Thomas Reardon and colleagues, e.g., Thomas Reardon et al., "The Rise of Supermarkets in Africa, Asia, and Latin America," *American Journal of Agricultural Economics* 85, no. 5 (2003): 1140–46; Thomas Reardon and Julio A. Berdegué, "The Rapid Rise of Supermarkets in Latin America: Challenges

and Opportunities for Development," *Development Policy Review* 20, no. 4 (2002): 371–88; and Thomas Reardon, C. Peter Timmer, and Bart Minten, "Supermarket Revolution in Asia and Emerging Development Strategies to Include Small Farmers," *Proceedings of the National Academy of Sciences* 109, no. 31 (2012): 12332–37.

34. "Rome Declaration on World Food Security," World Food Summit, November 13–17, 1996, http://www.fao.org/3/w3613e/w3613e00.htm.
35. World Commission on Environment and Development, *Our Common Future* (New York: United Nations, 1987), https://sustainabledevelopment.un.org/content/documents/5987our-common-future.pdf. Also known as the Brundtland Report.
36. United Nations, *The Future We Want: Outcome Document of the United Nations Conference on Sustainable Development, Rio de Janeiro, Brazil, 20–22 June 2012* (New York: United Nations, 2012).
37. David Baulcombe et al., *Reaping the Benefits: Science and the Sustainable Intensification of Global Agriculture* (London: The Royal Society, 2009).
38. N. Mahon et al., "Towards a Broad-Based and Holistic Framework of Sustainable Intensification Indicators," *Land Use Policy* 77 (2018): 576–97; Johan Rockström et al., "Sustainable Intensification of Agriculture for Human Prosperity and Global Sustainability," *Ambio* 46, no. 1 (2017): 4–17.
39. FAO, *The State of Food and Agriculture 2019: Moving Forward on Food Loss and Waste Reduction* (Rome: FAO, 2019); United Nations Environment Programme, *Food Waste Index Report 2021* (Nairobi: UNEP, 2021).

9. SUSTAINABLE INTENSIFICATION

1. "GNI per Capita, Atlas Method (Current US$)—Malawi," World Bank, accessed August 24, 2021, https://data.worldbank.org/indicator/NY.GNP.PCAP.CD?locations=MW; UNDP, *Human Development Report 2005: International Cooperation at a Crossroads—Aid, Trade and Security in an Unequal World* (New York: UNDP, 2005), http://hdr.undp.org/sites/default/files/reports/266/hdr05_complete.pdf.
2. World Bank, *Malawi—Poverty and Vulnerability Assessment: Investing in Our Future* (Washington, D.C.: World Bank, December 2007), 12, fig. 1.7.
3. "Prevalence of Stunting, Height for Age (% of Children Under 5)—Malawi," World Bank, accessed August 24, 2021, https://data.worldbank.org/indicator/SH.STA.STNT.ZS?locations=MW. Estimates were 52.5 (2004) and 53.1 (2006).
4. World Bank, *Malawi—Poverty and Vulnerability Assessment*, 20, table 1.6.
5. Famine Early Warning Systems Network, monthly reports (2005–2007), https://fews.net/southern-africa/malawi, cited in G. Denning et al., "Input Subsidies to Improve Smallholder Maize Productivity in Malawi: Toward an African Green Revolution," *PLoS Biol* 7 no. 1 (2009): e1000023, doi.org/10.1371/journal.pbio.1000023.
6. Two critical assessments of Malawi's Starter Pack program are Sarah Levy, ed., *Starter Packs: A Strategy to Fight Hunger in Developing Countries? Lessons from the Malawi Experience 1998–2003* (Wallingford, UK: CABI, 2005); and Jane Harrigan, "Food Insecurity, Poverty and the Malawian Starter Pack: Fresh Start or False Start?," *Food Policy* 33, no. 3 (2008): 237–49.
7. The following details of Farm Input Subsidy Programme are drawn from Denning et al., "Input Subsidies." For description and assessments of the FISP, see that article and Andrew Dorward and Ephraim Chirwa, "The Malawi Agricultural Input Subsidy Programme: 2005/06 to 2008/09," *International Journal of Agricultural Sustainability* 9, no. 1 (2011): 232–47.
8. Glenn Denning and Jeffrey Sachs, "How the Rich World Can Help Africa Help Itself," *Financial Times*, May 29, 2007.
9. President Bingu wa Mutharika, "Capitalizing on Opportunity," opening plenary session, World Economic Forum on Africa, June 4, 2008, Cape Town, South Africa, http://www.youtube.com

/watch?v=rmUjw1J35V4. See a segment of President Mutharika's speech from 34:02 to 35:10 minutes.

10. Based on FAOSTAT, http://www.fao.org/faostat/en/#data/QCL.
11. Based on a comparison of the FAO estimate for 2021 (FAO, *GIEWS Country Brief, Malawi* [Rome: FAO, February 8, 2022], https://reliefweb.int/sites/reliefweb.int/files/resources/MWI_21.pdf) and the average production of 2002–2004 (FAOSTAT, http://www.fao.org/faostat/en/#data/QCL).
12. FAO, *GIEWS Country Brief, Malawi.*
13. Denning et al., "Input Subsidies."
14. Committee on World Food Security, *Global Strategic Framework for Food Security and Nutrition*, third version (Rome: CFS, 2014), http://www.fao.org/fileadmin/templates/cfs/Docs1314/GSF/GSF_Version_3_EN.pdf.
15. Merrill Bennett, *The World's Food: A Study of the Interrelations of World Populations, National Diets, and Food Potentials* (New York: Harper, 1954).
16. FAO, *Global Food Losses and Food Waste: Extent, Causes, and Prevention* (Rome: FAO, 2011).
17. N. Alexandratos and J. Bruinsma, *World Agriculture Towards 2030/2050: The 2012 Revision*, ESA Working Paper no. 12–03 (Rome: FAO, 2012).
18. FAO, *Global Agriculture Towards 2050*, High-Level Expert Forum—How to Feed the World in 2050, October 12–13 (Rome: FAO, 2009), https://reliefweb.int/sites/reliefweb.int/files/resources/E6441A901ECFFDDDC125763C0034ED32-Full_Report.pdf.
19. The report is David Tilman et al., "Global Food Demand and the Sustainable Intensification of Agriculture," *Proceedings of the National Academy of Sciences* 108, no. 50 (2011): 20260–64. The quote is from Gordon Conway, *One Billion Hungry: Can We Feed the World?* (Ithaca, N.Y.: Cornell University Press, 2012).
20. Mitchell C. Hunter et al., "Agriculture in 2050: Recalibrating Targets for Sustainable Intensification," *BioScience* 67, no. 4 (2017): 386–91.
21. Based on a comparison of the average of 2002–2004 and 2017–2019. FAOSTAT, http://www.fao.org/faostat/en/#data/QCL.
22. Based on Derek Byerlee, James Stevenson, and Nelson Villoria, "Does Intensification Slow Crop Land Expansion or Encourage Deforestation?," *Global Food Security* 3, no. 2 (2014): 92–98. Calculations were adapted for a different time period using FAOSTAT data.
23. Byerlee, Stevenson, and Villoria. Calculations were adapted for a different time period using FAOSTAT data.
24. Jesse H. Ausubel, Iddo K. Wernick, and Paul E. Waggoner, "Peak Farmland and the Prospect for Land Sparing," *Population and Development Review* 38 (2013): 221–42, http://www.jstor.org/stable/23655296.
25. Byerlee, Stevenson, and Villoria, "Does Intensification Slow."
26. United Nations Environment Programme, *Becoming #GenerationRestoration: Ecosystem Restoration for People, Nature and Climate* (Nairobi: UNEP, 2021).
27. Navin Ramankutty et al., "Farming the Planet: 1. Geographic Distribution of Global Agricultural Lands in the Year 2000," *Global Biogeochemical Cycles* 22, no. 1 (2008): GB1003.
28. H. K. Gibbs and J. Meghan Salmon, "Mapping the World's Degraded Lands," *Applied Geography* 57 (2015): 12–21.
29. Ruth S. DeFries, Jonathan A. Foley, and Gregory P. Asner, "Land-Use Choices: Balancing Human Needs and Ecosystem Function," *Frontiers in Ecology and the Environment* 2, no. 5 (June 2004): 249–57.
30. Jules N. Pretty, "The Sustainable Intensification of Agriculture," *Natural Resources Forum* 21, no. 4 (1997): 247–56.
31. There have been several influential published reviews and perspectives on sustainable intensification in addition to Pretty's. Most prominent among them are Tilman et al., "Global Food Demand"; T. Garnett et al., "Sustainable Intensification in Agriculture: Premises and Policies," *Science* 341 (2013): 33–34; J. Pretty and Z. P. Bharucha., "Sustainable Intensification in Agricultural Systems,"

Annals of Botany 114 (2014): 1571–96; H. C. J. Godfray and T. Garnett, "Food Security and Sustainable Intensification," *Philosophical Transactions of the Royal Society B* 369 (2014): 20120273.

32. Johan Rockström et al., "Sustainable Intensification of Agriculture for Human Prosperity and Global Sustainability," *Ambio* 46, no. 1 (2017): 7.
33. United Nations, *Transforming Our World: The 2030 Agenda for Sustainable Development* (New York: UN Division for Sustainable Development Goals, 2015), https://sustainabledevelopment.un.org/post2015/transformingourworld/publication.
34. Timothy Reeves, "Sustainable Intensification of Agriculture—for Food and Nutritional Security," Farrer Memorial Oration, August 2019, 19–20, https://www.dpi.nsw.gov.au/__data/assets/pdf_file/0008/1147967/Farrer-memorial-oration-2019-reeves.pdf. The Farrer Memorial Medal is awarded annually to commemorate William James Farrer, Australia's leading wheat breeder, in recognition of distinguished service in agricultural science and contribution to Australia's cropping industries. The awardee delivers the Farrer Memorial Oration.
35. Reeves, "Sustainable Intensification of Agriculture," 13–16.
36. Xiao-Tang Ju et al., "Reducing Environmental Risk by Improving N Management in Intensive Chinese Agricultural Systems," *Proceedings of the National Academy of Sciences* 106, no. 9 (2009): 3041–46.
37. Reeves, "Sustainable Intensification of Agriculture," 1.
38. K. G. Cassman and P. Grassini, "A Global Perspective on Sustainable Intensification Research," *Nature Sustainability* 3 (2020): 262–68, doi.org/10.1038/s41893-020-0507-8.
39. Cassman and Grassini, 267.

10. MARKET INFRASTRUCTURE

1. *Masagana* is a Filipino word meaning "bountiful," while 99 represented the number of 50 kg sacks of rice that farmers could obtain from a hectare of land in the program, corresponding to a yield of almost 5 t/ha.
2. Arthur T. Mosher, *Getting Agriculture Moving: Essentials for Development and Modernization* (New York: Praeger, 1966), 141.
3. Hubert Zandstra et al., *Cáqueza: Living Rural Development* (Ottawa: International Development Research Centre, 1979).
4. Zandstra et al., 228.
5. G. L. Denning, "An Australian-Assisted Credit Scheme for Smallholder Upland Crop Production in the Southern Philippines," *Agricultural Administration* 10, no. 2 (1982): 123–43.
6. Barry J. Barnett and Olivier Mahul, "Weather Index Insurance for Agriculture and Rural Areas in Lower-Income Countries," *American Journal of Agricultural Economics* 89, no. 5 (2007): 1241–47.
7. H. Greatrex et al., *Scaling Up Index Insurance for Smallholder Farmers: Recent Evidence and Insights*, CCAFS Report No. 14 (Copenhagen: CGIAR Research Program on Climate Change, Agriculture, and Food Security [CCAFS], 2015).
8. Gerardo P. Sicat, *The Economic Legacy of Marcos,* Discussion Paper no. 2011–11 (Manila: University of the Philippines School of Economics, 2011), https://econ.upd.edu.ph/dp/index.php/dp/article/viewFile/679/144.
9. Mosher, *Getting Agriculture Moving*, chap. 9.
10. Thomas P. Tomich, Peter Kilby, and Bruce F. Johnston, *Transforming Agrarian Economies: Opportunities Seized, Opportunities Missed* (Ithaca, N.Y.: Cornell University Press, 1995), 172, http://www.jstor.org/stable/10.7591/j.ctv1nhms5.
11. Montague Yudelman, "Agriculture in Integrated Rural Development: The Experience of the World Bank," *Food Policy* 1, no. 5 (1976): 367–81; Vernon W. Ruttan, "Integrated Rural Development Programmes: A Historical Perspective," *World Development* 12, no. 4 (1984): 393–401.

12. Kenneth Quinn, "Roads and Rice: How Innovation and Infrastructure Can Feed the World," *Guardian* Poverty Matters blog, June 21, 2011, https://www.theguardian.com/global-development/poverty-matters/2011/jun/21/roads-rice-to-feed-world.
13. UN Millennium Project Task Force on Hunger, *Halving Hunger: It Can Be Done* (London: Earthscan, 2005).
14. Agro-dealers are decentralized, (generally) small-scale suppliers of agricultural inputs like seeds and fertilizer. The Alliance for a Green Revolution in Africa (AGRA) has supported the expanded reach of agro-dealers in Africa. See "Input Distributions," AGRA, accessed August 28, 2021, https://agra.org/input-distributions/.
15. United Nations, "Secretary-General Calls for 'Uniquely African Green Revolution' in 21st Century, to End Continent's Plague of Hunger, in Addis Ababa Remarks," news release SG/SM/9405-AFR/988, July 6, 2004, https://www.un.org/press/en/2004/sgsm9405.doc.htm.
16. Claudia N. Berg et al., *Transport Policies and Development*, Policy Research Working Paper no. 7366. (Washington, D.C.: World Bank, 2015).
17. FAO, *Dynamic Development, Shifting Demographics, Changing Diets* (Bangkok: FAO, 2018), 96, http://www.fao.org/3/I8499EN/i8499en.pdf.
18. Asian Development Bank, *Strategy 2030: Achieving a Prosperous, Inclusive, Resilient, and Sustainable Asia and the Pacific* (Manila: ADB, 2018), https://www.adb.org/documents/strategy-2030-prosperous-inclusive-resilient-sustainable-asia-pacific.
19. Yiping Huang, "Understanding China's Belt and Road Initiative: Motivation, Framework and Assessment," *China Economic Review* 40 (2016): 314–21.
20. United Nations, *Transforming Our World: The 2030 Agenda for Sustainable Development* (New York: UN Division for Sustainable Development Goals, 2015), https://sustainabledevelopment.un.org/post2015/transformingourworld/publication.
21. Tomich, Kilby, and Johnston, *Transforming Agrarian Economies*, 172.
22. The controversial role of the PADAP roads in providing military access and fueling corruption was featured in an ABC (Australia) documentary program *Four Corners* in 1983: "Aiding or Abetting," https://www.abc.net.au/4corners/aiding-or-abetting—1983/2832022.
23. William F. Laurance et al., "A Global Strategy for Road Building," *Nature* 513, no. 7517 (2014): 229–32.
24. John Dulac, *Global Land Transport Infrastructure Requirements: Estimating Road and Railway Infrastructure Capacity and Costs to 2050* (Paris: International Energy Agency, 2013).
25. Laurance et al. "A Global Strategy."
26. Derek Byerlee, James Stevenson, and Nelson Villoria, "Does Intensification Slow Crop Land Expansion or Encourage Deforestation?," *Global Food Security* 3, no. 2 (2014): 92–98.
27. Impetus to incorporate road safety into infrastructure investments was given a boost with the inclusion of two explicit SDG targets on road safety: Target 3.6, "By 2020, halve the number of global deaths and injuries from road traffic accidents"; and Target 11.2, "By 2030, provide access to safe, affordable, accessible and sustainable transport systems for all, improving road safety, notably by expanding public transport, with special attention to the needs of those in vulnerable situations, women, children, persons with disabilities and older persons."
28. Pedro Sanchez et al., "The African Millennium Villages," *Proceedings of the National Academy of Sciences* 104, no. 43 (2007): 16775–80.
29. Pedro A. Sanchez, Glenn L. Denning, and Generose Nziguheba, "The African Green Revolution Moves Forward," *Food Security* 1, no. 1 (2009): 37–44.
30. Shira Mitchell et al., "The Millennium Villages Project: A Retrospective, Observational, Endline Evaluation," *Lancet Global Health* 6, no. 5 (2018): e500–e513, doi.org/10.1016/s2214-109x(18)30065-2.
31. Bill Gates, "Prepared Remarks," 2009 World Food Prize Symposium, October 15, 2009, https://www.gatesfoundation.org/ideas/speeches/2009/10/bill-gates-2009-world-food-prize-symposium.
32. African Development Bank, "AfDB's High 5s: A Game Changer in Africa's Development Discourse," accessed August 28, 2021, https://www.afdb.org/en/high5s. AfDB president Akinwumi

Adesina was World Food Prize laureate in 2017 in recognition of his efforts to transform African agriculture.

33. Raisuddin Ahmed et al., "The Policy and Institutional Environment Affecting the Rural Nonfarm Economy," in *Transforming the Rural Nonfarm Economy: Opportunities and Threats in the Developing World*, ed. Steven Haggblade, Peter B. R. Hazell, and Thomas Reardon (Baltimore: Johns Hopkins University Press, 2007), 237–53.
34. FAO, *Dynamic Development*, 96–97.
35. African Development Bank, *The New Deal on Energy for Africa: A Transformative Partnership to Light Up and Power Africa by 2025* (Abidjan, Côte d'Ivoire: AfDB, 2018), https://www.afdb.org/fileadmin/uploads/afdb/Documents/Generic-Documents/Brochure_New_Deal_2-En.pdf.
36. Estimate cited by the executive director of the International Energy Agency, Fatih Birol. Reported in Addis Getachew Tadesse, "600 Million Africans Go Without Electricity: IEA Chief," *Anadolu Agency*, June 12, 2019, https://www.aa.com.tr/en/africa/600-million-africans-go-without-electricity-iea-chief/1502097.
37. World Bank, *World Development Report 2016: Digital Dividends* (Washington, D.C.: World Bank, 2016), 90–92, https://www.worldbank.org/en/publication/wdr2016.
38. World Bank, *World Development Report 2020: Trading for Development in the Age of Global Value Chains* (Washington, D.C.: World Bank, 2020), chap. 6. https://www.worldbank.org/en/publication/wdr2020
39. For further information on the blockchain concept and its potential applications in agricultural value chains, see *ICT Update* 88 (September 2018), https://ictupdate.cta.int/en/issues/88-blockchain.
40. Haire, Benjamin, "Blockchain: Revolutionising the Agriculture Industry," *Agribusiness Bulletin*, October 2018, https://www2.deloitte.com/au/en/pages/consumer-business/articles/blockchain-revolutionising-agriculture-industry.html.
41. Lesly Goh, "How AgriTech Is Transforming Traditional Agriculture in Emerging Markets," in *Breakthrough: The Promise of Frontier Technologies for Sustainable Development*, ed. Homi Kharas, John W. McArthur, and Izumi Ohno (Washington, D. C.: Brookings Institution Press, 2022), 125–48.
42. World Bank, *World Development Report 2016*, 90–92.
43. Feed the Future is described as "the U.S. Government's global hunger & food security initiative." See "About," Feed the Future, accessed August 28, 2021, https://www.feedthefuture.gov/about/.
44. "About Us," TechnoServe, accessed August 28, 2021, https://www.technoserve.org/about-us/.
45. See "Enabling the Business of Agriculture," World Bank, accessed August 28, 2021, https://eba.worldbank.org/en/eba.
46. For example, World Bank, *Enabling the Business of Agriculture 2019* (Washington D.C.: World Bank, 2019), https://openknowledge.worldbank.org/bitstream/handle/10986/31804/9781464813870.pdf.

11. POSTHARVEST STEWARDSHIP

1. Based on a comparison of the Food and Agriculture Organization estimate for 2021, FAO, *GIEWS Country Brief, Malawi* (Rome: FAO, March 25, 2021), https://reliefweb.int/report/malawi/giews-country-brief-malawi-25-march-2021), and the average production of 2002–2004 (FAOSTAT, http://www.fao.org/faostat/en/#data/QCL).
2. CABI, "Invasive Species Compendium. *Prostephanus truncatus* (Larger Grain Borer)," accessed May 22, 2022, https://www.cabi.org/isc/datasheet/44524.
3. N. A. Phiri and G. Otieno, *Managing Pests of Stored Maize in Kenya, Malawi and Tanzania, Nairobi (*Kenya: CABI Africa, 2008).
4. Phiri and Otieno, 48–59.

5. Jenny Gustavsson et al., *Global Food Losses and Food Waste: Extent, Causes and Prevention* (Rome: FAO, 2011).
6. FAO, *The State of Food and Agriculture 2019: Moving Forward on Food Loss and Waste Reduction* (Rome: FAO, 2019).
7. UNEP, *Food Waste Index Report 2021* (Nairobi: UNEP, 2021). The apparent discrepancy between the total (17 percent) and the three components (11, 5, and 2 percent, respectively) is due to rounding.
8. C. Mbow et al., "Food Security," in *Climate Change and Land: An IPCC Special Report on Climate Change, Desertification, Land Degradation, Sustainable Land Management, Food Security, and Greenhouse Gas Fluxes in Terrestrial Ecosystems*, ed. P. R. Shukla et al. (Geneva: IPCC, 2019), sec. 5.5.2.
9. "Goal 12: Ensure sustainable consumption and production patterns," United Nations, accessed August 28, 2021, https://www.un.org/sustainabledevelopment/sustainable-consumption-production/.
10. Champions 12.3, accessed August 28, 2021, https://champions123.org/.
11. For example, see Julian Parfitt, Mark Barthel, and Sarah Macnaughton, "Food Waste Within Food Supply Chains: Quantification and Potential for Change to 2050," *Philosophical Transactions of the Royal Society B: Biological Sciences* 365, no. 1554 (2010): 3065–81; High Level Panel of Experts, *Food Losses and Waste in the Context of Sustainable Food Systems: A Report by the High Level Panel of Experts on Food Security and Nutrition of the Committee on World Food Security* (Rome: Committee on World Food Security, 2014).
12. FAO, *State of Food and Agriculture 2019*, 5.
13. UNEP, *Food Waste Index Report 2021*, 21.
14. FAO, *State of Food and Agriculture 2019*. In particular, see chaps. 1 and 2 on the design and application of the Food Loss Index methodology.
15. UNEP, *Food Waste Index Report 2021*, Executive Summary, 7–19.
16. FAO, *State of Food and Agriculture 2019*, 5.
17. Tadele Tefera, "Post-Harvest Losses in African Maize in the Face of Increasing Food Shortage," *Food Security* 4 (2012): 267–77, doi.org/10.1007/s12571-012-0182-3.
18. Hippolyte Affognon et al., "Unpacking Postharvest Losses in Sub-Saharan Africa: A Meta-Analysis," *World Development* 66 (2015): 49–68.
19. FAO, *State of Food and Agriculture 2019*, 27.
20. FAO, 27.
21. M. d. J. da Costa et al. "Household Food Insecurity in Timor-Leste," *Food Security* 5 (2013): 83–94, doi.org/10.1007/s12571-012-0228-6.
22. World Health Organization, *Mycotoxins*, May 9, 2018, https://www.who.int/news-room/fact-sheets/detail/mycotoxins.
23. Pooja Soni et al., "Functional Biology and Molecular Mechanisms of Host-Pathogen Interactions for Aflatoxin Contamination in Groundnut (*Arachis hypogaea* L.) and Maize (*Zea mays* L.)," *Frontiers in Microbiology* 11 (2020): 227.
24. Kerstin Hell and Charity Mutegi, "Aflatoxin Control and Prevention Strategies in Key Crops of Sub-Saharan Africa," *African Journal of Microbiology Research* 5, no. 5 (2011): 459–66.
25. Francisco Sánchez-Bayo et al., "The Aflatoxin QuicktestTM—a Practical Tool for Ensuring Safety in Agricultural Produce," in *Poisoning—from Specific Toxic Agents to Novel Rapid and Simplified Techniques for Analysis*, ed. Ntambwe Malangu (Rijeka, Croatia: Intech, 2017), 193–208.
26. Affognon et al., "Unpacking Postharvest Losses."
27. Megan Sheahan and Christopher B. Barrett, "Food Loss and Waste in Sub-Saharan Africa," *Food Policy* 70 (2017): 1–12.
28. Larry L. Murdock et al., "Death by Desiccation: Effects of Hermetic Storage on Cowpea Bruchids," *Journal of Stored Products Research* 49 (2012): 166–70.
29. Warehouse receipt systems are a mechanism whereby a farmer deposits a commodity in a certified warehouse in return for a warehouse receipt. The farmer can then apply for short-term credit from a participating bank or other financial institution using the warehouse receipt as security for a loan, thus increasing access to finance for farmers. The system means that the farmer is not forced to sell

immediately after harvest when prices are normally lowest. The system may also allow more professional grain storage technology and infrastructure. See "How Warehouse Receipts Can Improve Lives," International Finance Corporation, May 13, 2015, https://www.ifc.org/wps/wcm/connect/news_ext_content/ifc_external_corporate_site/news+and+events/news/how+warehouse+receipts+can+improve+lives; Jonathan Coulter and Gideon Onumah, "The Role of Warehouse Receipt Systems in Enhanced Commodity Marketing and Rural Livelihoods in Africa," *Food Policy* 27, no. 4 (2002): 319–37.

30. Sheahan and Barrett, "Food Loss and Waste," 9.
31. UNEP, *Food Waste Index Report 2021*, 75.
32. WRAP, *Labelling Guidance: Best Practice on Food Date Labelling and Storage Advice* (Banbury, UK: WRAP, updated November 2019), 17–18, https://wrap.org.uk/sites/default/files/2020-07/WRAP-Food-labelling-guidance.pdf.
33. United States Environmental Protection Agency, *Advancing Sustainable Materials Management: 2018 Fact Sheet. Assessing Trends in Materials Generation and Management in the United States, December 2020* (Washington, D.C.: EPA, 2020), https://www.epa.gov/sites/default/files/2021-01/documents/2018_ff_fact_sheet_dec_2020_fnl_508.pdf.
34. UNEP, *Food Waste Index Report 2021*, 8.
35. Comparison based on the estimate in Gustavsson et al., *Global Food Losses*, of 1.3 billion metric tons per year "lost or wasted globally" (p. 4) and distributions in fig. 2 (p. 5).
36. UNEP, *Food Waste Index Report 2021*, 8.
37. WRAP, "Labelling Guidance."
38. See "The Courtauld Commitment 2030," WRAP, accessed August 28, 2021, https://wrap.org.uk/taking-action/food-drink/initiatives/courtauld-commitment.
39. WRAP, *The Courtauld Commitment Annual Report* (Banbury, UK: WRAP, 2021), 4, https://wrap.org.uk/sites/default/files/2021-07/WRAP-Courtauld-Commitment-Annual-Report-2021.pdf.
40. See Love Food Hate Waste, accessed August 28, 2021, https://www.lovefoodhatewaste.com/.
41. Karin Schanes, Karin Dobernig, and Burcu Gözet, "Food Waste Matters—a Systematic Review of Household Food Waste Practices and Their Policy Implications," *Journal of Cleaner Production* 182 (2018): 978–91.
42. Walter Willett et al., "Food in the Anthropocene: The EAT–Lancet Commission on Healthy Diets from Sustainable Food Systems," *Lancet* 393, no. 10170 (2019): 447–92; Katie Flanagan, Kai Robertson, and Craig Hanson, *Reducing Food Loss and Waste: Setting a Global Action Agenda* (Washington, D.C.: World Resources Institute, 2019); Mbow et al., "Food Security"; FAO, *State of Food and Agriculture* 2019; UNEP, *Food Waste Index Report 2021*.
43. C. Hanson et al., *Reducing Food Loss and Waste: Ten Interventions to Scale Impact* (Washington, D.C.: World Resources Institute, 2019).
44. See Global Agriculture and Food Security Program, accessed August 28, 2021, https://www.gafspfund.org/.
45. Emma Kennedy, "Can Kitchen Tech Reduce Excessive Food Waste?," *CNN Business*, July 11, 2018, https://money.cnn.com/2018/07/11/technology/startups/smarterware-fridgecam-food-waste/index.html.
46. Grant Gerlock, "Why The Arctic Apple Means You May Be Seeing More GMOs at the Store," *NPR: The Salt*, February 1, 2017, https://www.npr.org/sections/thesalt/2017/02/01/512633781/why-the-arctic-apple-means-you-may-be-seeing-more-gmos-at-the-store.

12. HEALTHY DIETS

1. Government of Timor-Leste, "Prime Minister Launched the SDP During the Timor-Leste and Development Partners Meeting," news release, July 13, 2011, http://timor-leste.gov.tl/?p=5337&lang=en.

2. Xanana Gusmão, "Address by His Excellency the Prime Minister Kay Rala Xanana Gusmão at the Timor-Leste and Development Partners Meeting, Dili, July 12, 2011," Democratic Republic of Timor-Leste, http://timor-leste.gov.tl/wp-content/uploads/2011/07/Launch-of-the-SDP_TLDPM_12.07.11.pdf.
3. Democratic Republic of Timor-Leste (RDTL), *Timor-Leste Strategic Development Plan 2011–2030* (Dili, Timor Leste: RDTL, 2011), 8, 11.
4. RDTL, 116.
5. RDTL, 38.
6. National Statistics Directorate, Timor-Leste, Ministry of Finance, Timor-Leste, and ICF Macro, *Timor-Leste Demographic and Health Survey 2009–10* (Dili, Timor-Leste: NSD and ICF Macro, 2010), http://dhsprogram.com/pubs/pdf/FR235/FR235.pdf. These DHS reports are among the most valuable resources for obtaining an objective assessment of the health of population in developing countries and for tracking progress. With the support of USAID, the DHS program has undertaken more than four hundred surveys on population, health, HIV, and nutrition in more than ninety countries.
7. Stunting (moderate and severe) in Afghanistan was 52 percent in 2011, according to Central Statistics Organisation and United Nations Children's Fund, *Afghanistan Multiple Indicator Cluster Survey 2010–2011: Final Report* (Kabul: CSO and UNICEF, 2012). The corresponding statistic in Yemen was 47 percent (2011) and in Burundi, 58 percent (2010). See "Prevalence of Stunting, Height for Age (% of Children Under 5)—Afghanistan, Yemen, Rep., Burundi, Timor-Leste," World Bank, accessed August 28, 2021, https://data.worldbank.org/indicator/SH.STA.STNT.ZS?end=2017&locations=AF-YE-BI-TL&start=1997.
8. RDTL, *Timor-Leste Strategic Development Plan*, 216.
9. Jessica Fanzo, "Stunting: A Country's Lasting Burden," *Journal of International Affairs* 67, no. 2 (2014): 165–73.
10. J. Fanzo et al., *A Nutrition Sensitive Agriculture Strategy*, for the Ministry of Agriculture and Fisheries of Timor-Leste and Fini ba Moris (Seeds of Life) (Dili, Timor-Leste, March 2013).
11. Democratic Republic of Timor-Leste Ministry of Health, "Timor-Leste Food and Nutrition Survey," final report (2020), 36, https://www.unicef.org/timorleste/reports/timor-leste-food-and-nutrition-survey.
12. Democratic Republic of Timor-Leste Ministry of Health, 47. Minimum acceptable diet is a composite based on minimum dietary diversity and minimum meal frequency. See "Minimum Dietary Diversity, Minimum Meal Frequency and Minimum Acceptable Diet," in Trevor N. Croft et al., *Guide to DHS Statistics* (Rockville, Md.: ICF, 2018), https://dhsprogram.com/data/Guide-to-DHS-Statistics/Minimum_Dietary_Diversity_Minimum_Meal_Frequency_and_Minimum_Acceptable_Diet.htm.
13. Jessica Fanzo et al., *Timor-Leste Strategic Review: Progress and Success in Achieving the Sustainable Development Goal 2* (Dili, Timor-Leste: Centre of Studies for Peace and Development (CEPAD) and Baltimore: Johns Hopkins University, May 2017), https://docs.wfp.org/api/documents/WFP-0000015583/download/; G. Bonis-Profumo, R. McLaren, and J. Fanzo, "Ravaged Landscapes and Climate Vulnerability: The Challenge in Achieving Food Security and Nutrition in Post-Conflict Timor-Leste," *Advances in Food Security and Sustainability* 4 (2019): 97–132.
14. Since 2012 I have included a teaching case in my Global Food Systems course at Columbia University: Maria Wang, "Grow Your Own? Rice Self-sufficiency in Timor-Leste," SIPA-12-001.0 (New York: Columbia University School of International and Public Affairs, 2012), http://ccnmtl.columbia.edu/projects/caseconsortium/casestudies/84/casestudy/www/layout/case_id_84.html. The case explores the challenges of implementing a rice production policy in a country with multiple, sometimes conflicting, needs.
15. They are also known as *nutrition-specific* interventions, as distinct from *nutrition-sensitive* interventions (see chapter 7). Zulfiqar A. Bhutta et al., "Evidence-Based Interventions for Improvement of Maternal and Child Nutrition: What Can Be Done and at What Cost?," *Lancet* 382, no. 9890 (2013): 452–77, 469, table 5.

16. See chapter 9 for an elaboration of this topic.
17. Bhutta et al., "Evidence-Based Interventions," 472. 469. Note from the last footnote on table 5 that the lives saved estimate for acute malnutrition is based on SAM only.
18. Bhutta et al., 452.
19. M. T. Ruel and C. E. Levin, "Discussion Paper 92. Assessing the Potential for Food-Based Strategies to Reduce Vitamin A and Iron Deficiencies: A Review of Recent Evidence," *Food and Nutrition Bulletin* 22, no. 1 (2001): 94.
20. Ruel and Levin, 94.
21. R. S. Gibson, "Strategies for Preventing Multi-Micronutrient Deficiencies: A Review of Experiences with Food-Based Approaches in Developing Countries," in *Combating Micronutrient Deficiencies: Food-Based Approaches*, ed. Brian Thompson and Leslie Amoroso, (Wallingford, UK: CABI, 2011), 7–27.
22. Global Alliance for Improved Nutrition, *Food Fortification: The Unfinished Agenda*, Briefing Paper (Geneva: GAIN, 2018), https://www.gainhealth.org/sites/default/files/publications/documents/food-fortification-unfinished-agenda-2018.pdf.
23. C. Turner et al., "Concepts and Critical Perspectives for Food Environment Research: A Global Framework with Implications for Action in Low- and Middle-Income Countries," *Global Food Security* 18 (2018): 95.
24. The framework was developed by Jessica Fanzo (Johns Hopkins University), Ahmed Raza (FAO), and Elizabeth Fox (Johns Hopkins University) in collaboration with Saul Morris (GAIN), Nita Dalmiya (UNICEF), Roland Kupka (UNICEF), Arnold Timmer (GAIN), and Joyce Greene (GAIN). The original figure graphic was developed by Nona Reuter (UNICEF). See UNICEF Office of Research–Innocenti, *Food Systems for Children and Adolescents: Working Together to Secure Nutritious Diets* (New York: UNICEF, GAIN, 2018), https://www.unicef.org/media/94086/file/Food-systems-brochure.pdf.
25. High Level Panel of Experts, *Nutrition and Food Systems*, a report by the High Level Panel of Experts on Food Security and Nutrition (HLPE) (Rome: Committee on World Food Security, 2017), http://www.fao.org/3/i7846e/i7846e.pdf.
26. Craig Hadley et al., "Gender Bias in the Food Insecurity Experience of Ethiopian Adolescents," *Social Science and Medicine* 66, no. 2 (2008): 427–38.
27. C. Peter Timmer, *Food Security and Scarcity: Why Ending Hunger Is So Hard* (Philadelphia: University of Pennsylvania Press, 2015), 169.
28. Anna Herforth et al., "A Global Review of Food-Based Dietary Guidelines," *Advances in Nutrition* 10, no. 4 (2019): 590–605.
29. Walter Willett and Patrick J. Skerrett, *Eat, Drink, and Be Healthy: The Harvard Medical School Guide to Healthy Eating* (New York: Simon and Schuster, 2017).
30. B. M. Popkin, "Global Nutrition Dynamics: The World Is Shifting Rapidly Toward a Diet Linked with Noncommunicable Diseases," *American Journal of Clinical Nutrition* 84, no. 2 (2006): 289–98.
31. Democratic Republic of Timor-Leste Ministry of Health, "Timor-Leste Food and Nutrition Survey," 62.
32. Thomas Reardon et al., "The Processed Food Revolution in African Food Systems and the Double Burden of Malnutrition," *Global Food Security* 28 (2021): 100466.
33. FAO, *Dynamic Development, Shifting Demographics, Changing Diets* (Bangkok: FAO, 2018), http://www.fao.org/3/I8499EN/i8499en.pdf.
34. EAT–*Lancet* Commission, *Healthy Diets from Sustainable Food Systems*, Summary Report (Oslo: EAT, 2019), 5, https://eatforum.org/content/uploads/2019/07/EAT-Lancet_Commission_Summary_Report.pdf.
35. EAT–*Lancet* Commission, 5.
36. Walter Willett et al., "Food in the Anthropocene: The EAT–Lancet Commission on Healthy Diets from Sustainable Food Systems," *Lancet* 393, no. 10170 (2019): 447–92; EAT–*Lancet* Commission, *Healthy Diets*. EAT is a nonprofit organization founded by the Stordalen Foundation, Stockholm

Resilience Centre, and the Wellcome Trust to catalyze and support food system transformation. Gunhild A. Stordalen is the founder and executive chair.

37. Willett et al., 459.

38. I. Darnton-Hill, C. Nishida, and W. P. T. James, "A Life Course Approach to Diet, Nutrition and the Prevention of Chronic Diseases," *Public Health Nutrition* 7 (2004): 101–21.

39. See "Obesity and Overweight," World Health Organization, June 9, 2021, https://www.who.int/news-room/fact-sheets/detail/obesity-and-overweight.

40. J. Rockström et al., "Planetary Boundaries: Exploring the Safe Operating Space for Humanity," *Ecology and Society* 14, no. 2 (2009): 32.

41. Walter Willett, personal communication in email to Jessica Fanzo, copied to the author and Marco Springmann, dated August 11, 2019.

42. K. Hirvonen et al., "Affordability of the EAT–*Lancet* Reference Diet: A Global Analysis," *Lancet Global Health* 8, no. 1 (2020): e59–e66.

43. Willett et al., "Food in the Anthropocene," 451, table 1.

44. Renata Micha et al., *2021 Global Nutrition Report: The State of Global Nutrition* (Bristol, UK: Development Initiatives, 2021), 14, 25, https://globalnutritionreport.org/reports/2021-global-nutrition-report/.

45. NCD Risk Factor Collaboration (NCD-RisC), "Rising Rural Body-Mass Index is the Main Driver of the Global Obesity Epidemic in Adults," *Nature* 569 (2019): 260–64, doi.org/10.1038/s41586-019-1171-x. The exception to this trend of a narrowing rural–urban gap in BMI was sub-Saharan Africa, where women continue to perform manual tasks for farming and household activities and where poor infrastructure and poverty have slowed the spread of unhealthy, processed food. I anticipate that the rural–urban gap in BMI will close in the coming years.

46. NCD Risk Factor Collaboration, "Rising Rural Body-Mass Index."

47. Willett et al., "Food in the Anthropocene"; Meera Shekar and Barry Popkin, eds., *Obesity: Health and Economic Consequences of an Impending Global Challenge*, Human Development Perspectives series (Washington, D.C.: World Bank, 2020), https://openknowledge.worldbank.org/handle/10986/32383; and Micha et al., *2021 Global Nutrition Report*.

48. "A Look at Coca-Cola's Advertising Expenses," Investopedia, updated August 10, 2021, https://www.investopedia.com/articles/markets/081315/look-cocacolas-advertising-expenses.asp.

49. The figures are based on a comparison of the following two reports: Statistical Agency under the President of the Republic of Tajikistan (SA), Tajikistan Ministry of Health (MOH), and ICF International, *Tajikistan Demographic and Health Survey 2012* (Dushanbe, Tajikistan: SA, MOH and Calverton, Md.: ICF International, 2013), http://dhsprogram.com/pubs/pdf/FR279/FR279.pdf; SA, MOH, and ICF International, *Tajikistan Demographic and Health Survey 2017* (Dushanbe, Tajikistan: SA, MOH and Rockville, Md.: ICF International, 2018), http://dhsprogram.com/pubs/pdf/FR341/FR341.pdf.

50. Republic of Tajikistan, *National Development Strategy of the Republic of Tajikistan for the Period Up to 2030* (Dushanbe: Republic of Tajikistan, 2016), https://nafaka.tj/images/zakoni/new/strategiya_2030_en.pdf. Also see the National Voluntary Review presented in 2017 to the UN High Level Political Forum: "Tajikistan Voluntary National Review 2017," United Nations Sustainable Development Goals Knowledge Platform, https://sustainabledevelopment.un.org/memberstates/tajikistan.

51. Innovation Development Centre, Republic of Tajikistan, *Country Strategic Review: Food Security and Nutrition* (Dushanbe, Tajikistan: Innovation Development Centre, 2018), https://www.wfp.org/publications/tajikistan-zero-hunger-strategic-review-2018; Mariko Kawabata et al., "Food Security and Nutrition Challenges in Tajikistan: Opportunities for a Systems Approach," *Food Policy* 96 (2020): 101872.

52. C. Hawkes et al., "Double-Duty Actions: Seizing Programme and Policy Opportunities to Address Malnutrition in All Its Forms," *Lancet*, 395, no. 10218 (2020): 142–55.

53. Hawkes et al.

54. Government of Timor-Leste, *Report on the Implementation of the Sustainable Development Goals: From Ashes to Reconciliation, Reconstruction and Sustainable Development*, Voluntary National Review of Timor-Leste 2019 (Dili: Government of Timor-Leste, 2019), 12, https://sustainabledevelopment.un.org/content/documents/23417TimorLeste_VNR_2019_FINAL.pdf.
55. Gianna Bonis-Profumo, Natasha Stacey, and Julie Brimblecombe, "Maternal Diets Matter for Children's Dietary Quality: Seasonal Dietary Diversity and Animal-Source Foods Consumption in Rural Timor-Leste," *Maternal and Child Nutrition* 17, no. 1 (2021): e13071.
56. The Women's Empowerment in Agriculture Index (WEAI) is a multidimensional index that measures the empowerment, agency, and inclusion of women in the agricultural sector. For details, see Sabina Alkire et al., "The Women's Empowerment in Agriculture Index," *World Development* 52 (2013): 71–91.
57. Gianna Bonis-Profumo, Natasha Stacey, and Julie Brimblecombe, "Measuring Women's Empowerment in Agriculture, Food Production, and Child and Maternal Dietary Diversity in Timor-Leste," *Food Policy* 102 (2021): 102102.
58. Democratic Republic of Timor-Leste Ministry of Health, "Timor-Leste Food and Nutrition Survey" (2020), 58.
59. Bonis-Profumo, McLaren, and Fanzo, "Ravaged Landscapes."

13. SOCIAL PROTECTION

1. Nicholas Kulish, "'Never Seen Anything Like It': Cars Line Up for Miles at Food Banks," *New York Times*, April 8, 2020, https://www.nytimes.com/2020/04/08/business/economy/coronavirus-food-banks.html.
2. Karen Allen, "Hunger Grips in Malawi Maize Crisis," *BBC News*, October 7, 2005, http://news.bbc.co.uk/2/hi/africa/4318974.stm.
3. From the definition of food security agreed at the World Food Summit in 1996. "Rome Declaration on World Food Security," World Food Summit, November 13–17, 1996, n. 1, http://www.fao.org/3/w3613e/w3613e00.htm, Plan of Action.
4. Harold Alderman, *Leveraging Social Protection Programs for Improved Nutrition: Summary of Evidence Prepared for the Global Forum on Nutrition-Sensitive Social Protection Programs, 2015*, Global Forum on Nutrition-Sensitive Social Protection Programs (Washington, D.C.: World Bank, 2016).
5. "Universal Declaration of Human Rights," United Nations, accessed August 11, 2021, https://www.un.org/en/about-us/universal-declaration-of-human-rights.
6. "International Covenant on Economic, Social and Cultural Rights," United Nations Human Rights Office of the High Commissioner, accessed August 11, 2021, https://www.ohchr.org/en/professionalinterest/pages/cescr.aspx.
7. World Food Summit, "Rome Declaration."
8. Harold Alderman, Ugo Gentilini, and Ruslan Yemtsov, "Preface," in *The 1.5 Billion People Question: Food, Vouchers, or Cash Transfers?*, ed. Harold Alderman, Ugo Gentilini, and Ruslan Yemtsov (Washington, D.C.: World Bank, 2018), xiii.
9. Jef L. Leroy et al., "Cash and In-Kind Transfers Lead to Excess Weight Gain in a Population of Women with a High Prevalence of Overweight in Rural Mexico," *Journal of Nutrition*, 143, no. 3 (March 2013): 378–83, doi.org/10.3945/jn.112.167627.
10. Shrayana Bhattacharya, Vanita Leah Falcao, and Raghav Puri, "The Public Distribution System in India: Policy Evolution and Program Delivery Trends," in Alderman, Gentilini, and Yemtsov, *The 1.5 Billion People Question*, 43–105.
11. U.S. Department of Agriculture, *Food Distribution Program on Indian Reservations* (Washington, D.C.: USDA, January 2020), https://www.fns.usda.gov/fdpir/fdpir-fact-sheet.

12. D. A. P. Bundy et al., *Re-Imagining School Feeding: A High-Return Investment in Human Capital and Local Economies* (Washington, D.C.: World Bank, 2018).
13. WFP, *Making the Mid-day Meals Functional Following School Reopening, COVID-19 Response* (New Delhi: WFP, 2020), https://www.wfp.org/publications/wfp-india-making-mid-day-meals-functional-following-school-re-opening-covid-19.
14. WFP, *The Impact of School Feeding Programmes* (Rome: WFP, 2019), https://docs.wfp.org/api/documents/WFP-0000102338/download/.
15. WFP, *WFP School Feeding Programmes in 2018* (Rome: WFP, 2019), https://docs.wfp.org/api/documents/WFP-0000107178/download/.
16. WFP, "WFP and Australia Provide Funds to Help 80,000 School Children Receive Take-Home Food Rations Amidst School Closures," news release, May 14, 2020, https://www.wfp.org/news/wfp-and-australia-provide-funds-help-80000-school-children-receive-take-home-food-rations.
17. "School Meal Trends & Stats," School Nutrition Association, accessed August 11, 2021, https://schoolnutrition.org/aboutschoolmeals/schoolmealtrendsstats/. The website cites USDA FY 2018 data.
18. Kalle Hirvonen and John F. Hoddinott, *Payment Modality Preferences: Evidence from Ethiopia's Productive Safety Net Programme*, ESSP Working Paper 125 (Washington, D.C.: International Food Policy Research Institute and Addis Ababa: Ethiopian Development Research Institute, 2018).
19. "Rationing of Food and Clothing During the Second World War," Australian War Memorial, accessed August 29, 2021, https://www.awm.gov.au/articles/encyclopedia/homefront/rationing.
20. Ganga Tilakaratna and Chinthani Sooriyamudali, "Food-Based Social Assistance Programs in Sri Lanka: Evolution and Transition to Cash Transfers," in Alderman, Gentilini, and Yemtsov, *The 1.5 Billion People Question*, 151–71.
21. "The WFP Food Basket," WFP, accessed August 11, 2021, https://www.wfp.org/wfp-food-basket.
22. Harold Alderman, Ugo Gentilini, and Ruslan Yemtsov, "The Evolution of Food as Social Assistance: An Overview," in Alderman, Gentilini, and Yemtsov, *The 1.5 Billion People Question*, 4.
23. Victor Oliveira et al., "Evolution and Implementation of the Supplemental Nutrition Assistance Program in the United States," in Alderman, Gentilini, and Yemtsov, *The 1.5 Billion People Question*, 209–63; Alderman, Gentilini, and Yemtsov, "The Evolution of Food."
24. Laura Tiehen, "Taking a Closer Look at Supplemental Nutrition Assistance Program (SNAP) Participation and Expenditures," Amber Waves, U.S. Department of Agriculture, Economic Research Service, August 3, 2020, https://www.ers.usda.gov/amber-waves/2020/august/taking-a-closer-look-at-supplemental-nutrition-assistance-program-snap-participation-and-expenditures/.
25. Peter Timmer, Hastuti, and Sudarno Sumarto, "Evolution and Implementation of the RASTRA Program in Indonesia," in Alderman, Gentilini, and Yemtsov, *The 1.5 Billion People Question*, 265–310.
26. Alderman, Gentilini, and Yemtsov, "The Evolution of Food" 6.
27. Alderman, *Leveraging*.
28. Lia C. H. Fernald, Paul J. Gertler, and Xiaohui Hou, "Cash Component of Conditional Cash Transfer Program Is Associated with Higher Body Mass Index and Blood Pressure in Adults," *Journal of Nutrition* 138, no. 11 (2008): 2250–57.
29. Per Pinstrup-Andersen, ed., *Food Subsidies in Developing Countries: Costs, Benefits, and Policy Options* (Baltimore: Johns Hopkins University Press, 1988).
30. Robert, T. Jensen and Nolan H. Miller, "Do Consumer Price Subsidies Really Improve Nutrition?," *Review of Economics and Statistics* 93, no. 4 (2011): 1205–23, doi.org/10.1162/REST_a_00118.
31. Alderman, Gentilini, and Yemtsov, "The Evolution of Food," 12.
32. Kofi Annan, "Letter to Head of Donor States, August 8, 2005," in United Nations, *Malawi 2005 Flash Appeal September 2005–March 2006* (New York: Office for the Coordination of Humanitarian Affairs, 2005), 14, https://www.humanitarianresponse.info/sites/www.humanitarianresponse.info/files/documents/files/flash_2005_malawi.pdf.

33. Glenn Denning et al., "Input Subsidies to Improve Smallholder Maize Productivity in Malawi: Toward an African Green Revolution," *PLoS Biology* 7, no. 1 (2009): e1000023, doi.org/10.1371/journal.pbio.1000023.
34. United Nations, *Malawi 2005 Flash Appeal*, 1.
35. "Malawi 2005," Office for the Coordination of Humanitarian Affairs Services, accessed August 11, 2021, https://fts.unocha.org/appeals/186/flows. The table shows a breakdown of funds mobilized.
36. "UN Humanitarian Chief: After COVID-19, It's in Everyone's Interest to Help the World's Poorest Countries," Office for the Coordination of Humanitarian Affairs, May 4, 2020, https://www.unocha.org/story/un-humanitarian-chief-after-covid-19-it%E2%80%99s-everyone%E2%80%99s-interest-help-worlds-poorest-countries. This web page reproduces an op-ed that was published in the *Guardian*.
37. Nyasha Tirivayi, Marco Knowles, and Benjamin Davis, "The Interaction Between Social Protection and Agriculture: A Review of Evidence," *Global Food Security* 10 (2016): 52–62.
38. FAO, *Strengthening Coherence Between Agriculture and Social Protection to Combat Poverty and Hunger in Africa: Framework for Analysis and Action* (Rome: FAO, 2016), http://www.fao.org/3/i5386e/i5386e.pdf.
39. Daniel O. Gilligan, "Safety Nets for Agriculture and Nutrition," in *Agriculture for Improved Nutrition: Seizing the Momentum*, ed. Shenggen Fan, Sivan Yosef and Rajul Pandya-Lorch (Wallingford, UK: International Food Policy Research Institute and CABI, 2019), 104–12.
40. Gilligan, 109.
41. Jemimah Njuki "Women Are the Nucleus of Our Global Food System—but Their Value Is Undermined by the Patriarchy," *Independent*, July 12, 2021, https://www.independent.co.uk/voices/food-systems-sustainable-poverty-hunger-patriarchal-system-b1879840.html.
42. UN Millennium Project Task Force on Hunger, *Halving Hunger: It Can Be Done* (London: Earthscan, 2005).
43. Pedro Sanchez et al., "The African Millennium Villages," *Proceedings of the National Academy of Sciences* 104, no. 43 (2007): 16775–80.
44. CFS, *Connecting Smallholders to Markets: Policy Recommendations* (Rome: CFS, 2015), http://www.fao.org/3/a-bq853e.pdf.
45. WFP, "Smallholder Support," WFP Year in Review 2018: The Last Resort, the First Defence, accessed August 28, 2021, https://publications.wfp.org/2018/en/annual-report/section2.html#Smallholder_support.
46. WFP, "Cash Assistance," WFP Year in Review," accessed August 28, 2021, https://publications.wfp.org/2018/en/annual-report/section2.html#Cash_assistance.
47. Kaori Kitaoka, "The National School Meal Program in Brazil: A Literature Review," *Japanese Journal of Nutrition and Dietetics* 76, suppl. (2018): S115–S125.
48. Corinna Hawkes et al., "How to Engage Across Sectors: Lessons from Agriculture and Nutrition in the Brazilian School Feeding Program," *Revista de Saúde Pública* 50 (2016): 47, doi.org/10.1590/S1518-8787.2016050006506. *Quilombolas* are descendants of Afro-Brazilian slaves who escaped from slave plantations that existed in Brazil until abolition in 1888.
49. Jenny Aker et al., "Payment Mechanisms and Antipoverty Programs: Evidence from a Mobile Money Cash Transfer Experiment in Niger," *Economic Development and Cultural Change* 65, no. 1 (October 2016): 1–37.

14. COVID-19 AND FOOD SECURITY

1. "WHO Director-General's Opening Remarks at the Media Briefing on COVID-19–11 March 2020," World Health Organization, https://www.who.int/director-general/speeches/detail/who-director-general-s-opening-remarks-at-the-media-briefing-on-covid-19---11-march-2020.

2. "Orange Juice Futures," Investing.com, accessed August 29, 2021, https://www.investing.com/commodities/orange-juice-historical-data. Orange Juice Futures Historical Data, March 19–26, 2020.
3. Justin Harper, "Why Orange Juice Prices Are Soaring on Global Markets," *BBC News*, March 26, 2020, https://www.bbc.com/news/technology-52030133. Harper cites industry analysts.
4. Marvin G. Perez, Michael Hirtzer, and Megan Durisin, "Avocados Are In, Pork Bellies Out in Era of Pandemic Eating," *Bloomberg*, April 26, 2020, https://www.bloomberg.com/news/articles/2020-04-26/avocados-are-in-pork-bellies-out-in-the-era-of-pandemic-eating.
5. FSIN and Global Network Against Food Crises, *Global Report on Food Crises 2020* (Rome: FSIN, 2020), 2, https://docs.wfp.org/api/documents/WFP-0000114546/download/.
6. IPC Global Partners, *Integrated Food Security Phase Classification Technical Manual Version 3.0. Evidence and Standards for Better Food Security and Nutrition Decisions* (Rome: FAO, 2019), 4, http://www.ipcinfo.org/fileadmin/user_upload/ipcinfo/manual/IPC_Technical_Manual_3_Final.pdf.
7. FSIN and Global Network Against Food Crises, *Global Report on Food Crises 2021* (Rome: FSIN, 2021), https://www.fsinplatform.org/sites/default/files/resources/files/GRFC%202021%20050521%20med.pdf.
8. FSIN and Global Network, *Global Report 2020*, 2.
9. FAO et al., *The State of Food Security and Nutrition in the World 2021. Transforming Food Systems for Food Security, Improved Nutrition and Affordable Healthy Diets for All* (Rome: FAO, 2021), 11.
10. In accordance with SOFI 2021, I have used the revised estimate of 650 million undernourished people in 2019. SOFI 2020 originally reported the 2019 number as 690 million, based on a newly applied methodology. FAO et al., *The State of Food Security and Nutrition in the World 2020. Transforming Food Systems for Affordable Healthy Diets* (Rome: FAO, 2020), 11, table 2, http://www.fao.org/3/ca9692en/ca9692en.pdf.
11. FAO et al., *State of Food Security 2021*, 18.
12. FSIN and Global Network, *Global Report 2021*, 10.
13. FAO et al., *State of Food Security 2021*, 22.
14. Melissa Locker, "Tom Hanks and Rita Wilson Had Very Different Reactions to COVID-19," *Time*, July 6, 2020, https://time.com/5863326/tom-hanks-coronavirus-reaction/.
15. "Tracking Coronavirus in New York: Latest Map and Case Count," *New York Times*, accessed August 11, 2021, https://www.nytimes.com/interactive/2021/us/new-york-covid-cases.html.
16. Caroline Hopkins, "Covid-19 Patients Are Flooding New York Hospitals, and the Peak May Be 3 Weeks Away," *Vox*, March 27, 2020, https://www.vox.com/2020/3/27/21197400/new-york-covid-19-hospitals-coronavirus.
17. "New York Governor Cuomo Pleads for 30,000 Ventilators," *CBS News*, March 25, 2020, https://www.cbsnews.com/video/new-york-governor-cuomo-pleads-for-30000-ventilators/.
18. Paul D. Sonenthal et al., "COVID-19 Preparedness in Malawi: A National Facility-Based Critical Care Assessment," *Lancet Global Health* 8, no. 7 (2020): E890–92, doi.org/10.1016/S2214-109X(20)30250-3.
19. John Vidal, "'If It Comes, It Will Overwhelm Us': Malawi Braces for Coronavirus," *Guardian*, April 3, 2020, https://www.theguardian.com/global-development/2020/apr/03/if-it-comes-it-will-overwhelm-us-malawi-braces-for-coronavirus.
20. Wanga Gwede, "Lack of Ventilators in Malawi Will See More Die to Coronavirus," *Nyasa Times*, March 27, 2020, https://www.nyasatimes.com/lack-of-ventilators-in-malawi-will-see-more-die-to-coronavirus/.
21. "Tracking Coronavirus in New York: Latest Map and Case Count," *New York Times*, accessed October 27, 2020, https://www.nytimes.com/interactive/2021/us/new-york-covid-cases.html; "COVID-19 National Information Dashboard," Ministry of Health, Malawi, accessed October 27, 2020, https://covid19.health.gov.mw/.
22. "Malawi," Worldometer (citing as source the Malawi Ministry of Health), accessed July 31, 2021, https://www.worldometers.info/coronavirus/country/malawi/.

23. Lameck Masina, "WHO, US Name Malawi a High Risk COVID-19 Country as Cases Spike," *Voice of America*, July 10, 2021, https://www.voanews.com/covid-19-pandemic/who-us-name-malawi-high-risk-covid-19-country-cases-spike.
24. "In Malawi, Health Workers Bring Hope and Experience to COVID-19 Fight," UNICEF, July 27, 2021, https://www.unicef.org/coronavirus/malawi-health-workers-bring-hope-and-experience-covid-19-fight.
25. Ada Cuevas and Simon Barquera, "COVID-19, Obesity, and Undernutrition: A Major Challenge for Latin American Countries," *Obesity* 28, no. 10 (2020): 1791–92.
26. David P. Richardson and Julie A. Lovegrove, "Nutritional Status of Micronutrients as a Possible and Modifiable Risk Factor for COVID-19: A UK Perspective," *British Journal of Nutrition* 125, no. 6 (2021): 682.
27. Barry M. Popkin et al., "Individuals with Obesity and COVID-19: A Global Perspective on the Epidemiology and Biological Relationships," *Obesity Reviews* 21, no. 11 (2020): e13128.
28. Matthew J. Cummings et al., "Epidemiology, Clinical Course, and Outcomes of Critically Ill Adults with COVID-19 in New York City: A Prospective Cohort Study," *Lancet* 395, no. 10239 (2020): P1763–70.
29. Stephen Devereux, Christophe Béné, and John Hoddinott, "Conceptualising COVID-19's Impacts on Household Food Security," *Food Security* 12, no. 4 (2020): 769–72.
30. Julia Winterflood, "Post-Pandemic, Will Bali Rethink Tourism?," *Diplomat*, June 10, 2020, https://thediplomat.com/2020/06/post-pandemic-will-bali-rethink-tourism/.
31. Pascale Hunt, "'Great Reboot' or Short-Term Saviour? Bali's Seaweed Farming Revival," *The Fish Site*, April 30, 2021, https://thefishsite.com/articles/great-reboot-or-short-term-saviour-balis-seaweed-farming-revival.
32. Lisa Robins et al., *COVID-19 and Food Systems in the Indo-Pacific: An Assessment of Vulnerabilities, Impacts and Opportunities for Action*, Australian Centre for International Agricultural Research Technical Report 96 (Canberra: ACIAR, 2020).
33. "FAO Cereal Supply and Demand Brief," FAO, July 8, 2021, http://www.fao.org/worldfoodsituation/csdb/en/.
34. FAO, "FAO Cereal Supply and Demand Brief." Percentage increase based on real prices. The nominal increase was 32 percent.
35. "Food Security and COVID-19, Brief," World Bank, accessed October 15, 2020, https://www.worldbank.org/en/topic/agriculture/brief/food-security-and-covid-19.
36. "Joint Statement on COVID-19 Impacts on Food Security and Nutrition," World Bank, April 21, 2020, https://www.worldbank.org/en/news/statement/2020/04/21/joint-statement-on-covid-19-impacts-on-food-security-and-nutrition.
37. Lesly Goh, "How AgriTech Is Transforming Traditional Agriculture in Emerging Markets," in *Breakthrough: The Promise of Frontier Technologies for Sustainable Development*, ed. Homi Kharas, John W. McArthur, and Izumi Ohno (Washington, D.C.: Brookings Institution Press, 2022), 125–48.
38. David Yaffe-Bellany and Michael Corker, "Dumped Milk, Smashed Eggs, Plowed Vegetables: Food Waste of the Pandemic," *New York Times*, April 11, 2020, https://www.nytimes.com/2020/04/11/business/coronavirus-destroying-food.html.
39. Brian E. Roe, Kathryn Bender, and Danyi Qi, "The Impact of COVID-19 on Consumer Food Waste," *Applied Economic Perspectives and Policy* 43, no. 1, (2020): 401–11.
40. Roe, Bender, and Qi, "The Impact of COVID-19."
41. R. Aldaco et al., "Food Waste Management During the COVID-19 Outbreak: A Holistic Climate, Economic and Nutritional Approach," *Science of the Total Environment* 742 (2020): 140524.
42. The Standing Together for Nutrition consortium is a multidisciplinary alliance of nutrition, economics, food, and health systems experts working to address the scale and reach of COVID-related nutrition challenges. "Standing Together for Nutrition (STfN)," Global Alliance for Improved Nutrition, July 28, 2020, https://www.gainhealth.org/media/news/standing-together-nutrition-stfn.

43. Derek Headey et al., "Impacts of COVID-19 on Childhood Malnutrition and Nutrition-Related Mortality," *Lancet* 396, no. 10250 (2020): 519–21.
44. Henrietta H. Fore et al., "Child Malnutrition and COVID-19: the Time to Act Is Now," *Lancet* 396, no. 10250 (2020): 517–18.
45. Farah Naja and Rena Hamadeh, "Nutrition Amid the COVID-19 Pandemic: A Multi-Level Framework for Action," *European Journal of Clinical Nutrition* 74 (2020): 1117–21, doi.org/10.1038/s41430-020-0634-3.
46. Eat Well Tasmania, *A Survey About Shopping, Cooking and Eating* (Hobart, Australia: Eat Well Tasmania, 2020), https://www.eatwelltas.org.au/wp-content/uploads/2020/05/EWT_Covid-Survey-Summary_A3.pdf.
47. Naja and Hamadeh, "Nutrition Amid the COVID-19 Pandemic."
48. Kalle Hirvonen, Alan de Brauw, and Gashaw T. Abate, "Food Consumption and Food Security During the COVID-19 Pandemic in Addis Ababa," *American Journal of Agricultural Economics* 103, no. 3 (2021): 772–89.
49. FAO, *Impacts of Coronavirus on Food Security and Nutrition in Asia and the Pacific: Building More Resilient Food Systems* (Bangkok: FAO, June 5, 2020), http://www.fao.org/3/ca9473en/CA9473EN.pdf.
50. World Food Programme, "Futures of 370 Million Children in Jeopardy as School Closures Deprive Them of School Meals—UNICEF and WFP," news release, April 29, 2020, https://www.wfp.org/news/futures-370-million-children-jeopardy-school-closures-deprive-them-school-meals-unicef-and-wfp.
51. "Global Monitoring of School Meals During COVID-19 School Closures," World Food Programme, https://cdn.wfp.org/2020/school-feeding-map/?_ga=2.72314643.1844293323.1604985491-67437893.1603343218.
52. Social Protection Interagency Cooperation Board, *A Joint Statement on the Role of Social Protection in Responding to the COVID-19 Pandemic* (New York: SPIAC-B, 2020), https://socialprotection.org/sites/default/files/publications_files/Joint%20SPIAC-B%20COVID-19%20statement.pdf.
53. I have drawn on a telephone conversation with David Dawe on April 7, 2021, and FAO, *Impacts of Coronavirus on Food Security*, which Dawe wrote.
54. Agricultural Market Information System (AMIS), "Market Monitor," no. 95 (February 2022), AMIS Secretariat, FAO, Rome, http://www.amis-outlook.org/fileadmin/user_upload/amis/docs/Market_monitor/AMIS_Market_Monitor_current.pdf.

15. MORE THAN A MIRACLE

1. Robert Flint Chandler, *An Adventure in Applied Science: A History of the International Rice Research Institute* (Los Baños, Philippines: International Rice Research Institute, 1982), 109.
2. "New CARDI Director Cites IRRI's Role in Making Cambodia a Rice-Exporting Country," CARDI, accessed October 22, 2021, https://www.irri.org/news-and-events/news/new-cardi-director-cites-irri%E2%80%99s-role-making-cambodia-rice-exporting-country.
3. Walter Willett et al., "Food in the Anthropocene: The EAT–Lancet Commission on Healthy Diets from Sustainable Food Systems," *Lancet* 393, no. 10170 (2019): 447, 485.
4. Ryan Teague Beckwith, "Transcript: Read the Speech Pope Francis Gave to the United Nations," *Time*, September 25, 2015/, para. 10, https://time.com/4049905/pope-francis-us-visit-united-nations-speech-transcript.
5. Donella Meadows, "What Would the World Be If There Were No Hunger?," Donella Meadows Archives, December 17, 1987, https://donellameadows.org/archives/what-would-the-world-be-if-there-were-no-hunger/.
6. John Lennon, *Imagine*, third verse, https://genius.com/John-lennon-imagine-lyrics.

7. "Rome Declaration on World Food Security," World Food Summit, November 13–17, 1996, http://www.fao.org/3/w3613e/w3613e00.htm.
8. Ralph D. Christy and Lionel Williamson, eds., *A Century of Service: Land-Grant Colleges and Universities, 1890–1990* (Brunswick, N.J.: Transaction, 1992).
9. Teddy Aljuni, "Bollinger Announces Task Forces on Columbia's Role in Climate Solutions, Pressing Global Problems," *Columbia Spectator*, September 19, 2019, https://www.columbiaspectator.com/news/2019/09/19/bollinger-announces-two-new-task-forces-to-address-climate-change-pressing-human-problems-on-a-global-scale/.
10. "Columbia Climate School," Columbia University, accessed August 29, 2021, https://climate.columbia.edu/.
11. Jessica Fanzo skillfully articulates both the power and the limitations of individuals in choosing their diets in *Can Fixing Dinner Fix the Planet?* (Baltimore: Johns Hopkins University Press, 2021).
12. Global Citizen, accessed August 29, 2021, https://www.globalcitizen.org/en/.
13. Committee on World Food Security, accessed March 16, 2022, http://www.fao.org/cfs/en/.
14. "About the HLPE," Committee on World Food Security, High Level Panel of Experts, accessed March 16, 2022, http://www.fao.org/cfs/cfs-hlpe.
15. CGIAR, "New CGIAR Initiatives Announced to Help Radically Realign Food, Land, and Water Systems," news release, July 26, 2021, https://www.cgiar.org/news-events/news/cgiar-announces-new-portfolio-to-transform-food-land-and-water-systems-in-a-climate-crisis.
16. CGIAR, "One CGIAR," accessed January 24, 2022, https://www.cgiar.org/food-security-impact/one-cgiar/.
17. "About the Summit," United Nations World Food Systems Summit 2021, accessed August 10, 2021, https://www.un.org/en/food-systems-summit/about.
18. Sustainable Development Solutions Network, accessed August 29, 2021, https://www.unsdsn.org/.
19. "Does the Fight Against Hunger Need Its Own IPCC?" *Nature* 595, no. 332 (2021), doi.org/10.1038/d41586-021-01904-0.
20. "'If You Don't Feed People, You Feed Conflict,' UN Chief Tells Security Council," *UN News*, March 11, 2021, https://news.un.org/en/story/2021/03/1087032.
21. "Negotiations on Conflict and Food Security Draft Presidential Statement," Security Council Report, *What's in Blue*, March 18, 2021, https://www.securitycouncilreport.org/whatsinblue/2021/03/negotiations-on-conflict-and-food-security-draft-presidential-statement.php.
22. CGIAR, "New CGIAR Initiatives Announced."
23. CGIAR, "Impact Stories," accessed January 25, 2022, https://www.cgiar.org/food-security-impact/impact-stories/.
24. CGIAR, "Overview: CGIAR—2020 at a Glance," accessed August 29, 2021, https://www.cgiar.org/food-security-impact/finance-reports/dashboard/overview/.
25. CGIAR, "New CGIAR Initiatives Announced."
26. "A Look at Coca-Cola's Advertising Expenses," Investopedia, updated August 10, 2021, https://www.investopedia.com/articles/markets/081315/look-cocacolas-advertising-expenses.asp.
27. Randy C. Ploetz, "Panama Disease: A Classic and Destructive Disease of Banana," *Plant Health Progress* 1, no. 1 (2000): 10, doi.org/10.1094/PHP-2000-1204-01-HM.
28. United Nations, *Transforming Our World: The 2030 Agenda for Sustainable Development* (New York: UN Division for Sustainable Development Goals, 2015), https://sustainabledevelopment.un.org/post2015/transformingourworld/publication.
29. Donella H. Meadows et al., *The Limits to Growth* (New York: Universe Books, 1972), 24. The quote referred to "*his* individual human potential."
30. Dariush Mozaffarian et al., "Role of Government Policy in Nutrition—Barriers to and Opportunities for Healthier Eating," *BMJ* 361 (2018).
31. Jeffrey D. Sachs et al., "Six Transformations to Achieve the Sustainable Development Goals," *Nature Sustainability* 2, no. 9 (2019): 805–14.
32. G. L. Denning, "An Australian-Assisted Credit Scheme for Smallholder Upland Crop Production in the Southern Philippines," *Agricultural Administration* 10, no. 2 (1982): 123–43; Steven Franzel

et al., "Scaling Up the Impact of Agroforestry: Lessons from Three Sites in Africa and Asia," *Agroforestry Systems* 61 (2004): 329–44; Generose Nziguheba et al., "The African Green Revolution: Results from the Millennium Villages Project," *Advances in Agronomy* 109 (2010): 75–115.

16. LEARNING TO LEAD

1. United Nations, *Transforming Our World: The 2030 Agenda for Sustainable Development* (New York: UN Division for Sustainable Development Goals, 2015), para. 4, https://sustainabledevelopment.un.org/post2015/transformingourworld/publication.
2. "Green Revolutionary Norman Borlaug Dies," *NPR*, September 13, 2009, https://www.npr.org/templates/story/story.php?storyId=112791886.
3. C. Peter Timmer, *Food Security and Scarcity: Why Ending Hunger Is So Hard* (Philadelphia: University of Pennsylvania Press, 2015), 177, 161.
4. Timmer, 177.
5. Arthur T. Mosher, *Getting Agriculture Moving: Essentials for Development and Modernization* (New York: Praeger, 1966), 189.
6. Mosher, 189.
7. M. S. Swaminathan, "Combating hunger," *Science* 338, no. 6110 (2012): 1009.
8. Lisa Dreier, David Nabarro, and Jane Nelson, *Systems Leadership for Sustainable Development: Strategies for Achieving Systemic Change* (Cambridge, Mass.: Harvard Kennedy School, September 2019), 4, 5, https://www.hks.harvard.edu/centers/mrcbg/publications/fwp/crisept2019.
9. Ben Ramalingam et al., "Principles to Guide Adaptive Leadership," *Harvard Business Review*, September 11, 2020, https://hbr.org/2020/09/5-principles-to-guide-adaptive-leadership.
10. International Commission on Education for Sustainable Development Practice, *Final Report October 2008* (New York: Earth Institute at Columbia University, 2008), https://www.macfound.org/media/article_pdfs/develcomm-execsumm.pdf.
11. A. Flexner, *Medical Education in the United States and Canada*, Bulletin no. 4 (New York: Carnegie Foundation for the Advancement of Teaching, 1910).
12. International Commission on Education for Sustainable Development Practice, *Final Report*, 2.
13. John Ingram et al., "A Future Workforce of Food-System Analysts," *Nature Food* 1, no. 1 (2020): 9–10.
14. Columbia University, "Fourth Purpose Task Force Report: On Directed Action," Memorandum to President Lee C. Bollinger, December 15, 2020, https://president.columbia.edu/sites/default/files/content/Additional/Fourth%20Purpose%20Task%20Force%20Report.pdf.
15. Uongozi Institute: Institute of African Leadership for Sustainable Development, accessed August 29, 2021, https://uongozi.or.tz/.
16. Alliance for a Green Revolution in Africa, "AGRA and Partners Launch New Centre for African Leaders in Agriculture," Centre for African Leaders in Agriculture, April 27, 2021, https://cala.agra.org/updates/cala-launch/.
17. Personal communication with Dr. Apollos Nwafor, vice president, policy and state capability, AGRA, February 2, 2022.
18. Thomas Robert Malthus, *An Essay on the Principle of Population as It Affects the Future Improvement of Society, with Remarks on the Speculations of Mr Godwin, M. Condorcet, and Other Writers* (London: J. Johnson, 1798), p. 103 in the Echo Library edition of 2006.

BIBLIOGRAPHY

Affognon, Hippolyte, Christopher Mutungi, Pascal Sanginga, and Christian Borgemeister. "Unpacking Postharvest Losses in Sub-Saharan Africa: A Meta-Analysis." *World Development* 66 (2015): 49–68.

African Development Bank. "AfDB's High 5s: A Game Changer in Africa's Development Discourse." Accessed August 28, 2021. https://www.afdb.org/en/high5s.

——. *Feed Africa.* Abidjan, Côte d'Ivoire: AfDB, 2019. https://www.afdb.org/en/documents/document/feed-africa-brochure-89189.

——. *The New Deal on Energy for Africa: A Transformative Partnership to Light Up and Power Africa by 2025.* Abidjan, Côte d'Ivoire: AfDB, 2018. https://www.afdb.org/fileadmin/uploads/afdb/Documents/Generic-Documents/Brochure_New_Deal_2-En.pdf.

African Union. "Agenda 2063: The Africa We Want." Accessed August 14, 2021. https://au.int/en/agenda2063/overview.

African Union Commission and NEPAD Agency. *Malabo Declaration on Accelerated Agricultural Growth and Transformation for Shared Prosperity and Improved Livelihoods.* Addis Ababa, Ethiopia: AUC and Midrand, South Africa: NEPAD, 2016. https://www.nepad.org/caadp/publication/malabo-declaration-accelerated-agricultural-growth.

Agricultural Market Information System (AMIS). "Market Monitor," no. 96 (March 2022). AMIS Secretariat, FAO, Rome. http://www.amis-outlook.org/fileadmin/user_upload/amis/docs/Market_monitor/AMIS_Market_Monitor_current.pdf.

Ahmed, Kedir Y., Abdon G. Rwabilimbo, Solomon Abrha, Andrew Page, Amit Arora, Fentaw Tadese, Tigistu Yemane Beyene, et al. "Factors Associated with Underweight, Overweight, and Obesity in Reproductive Age Tanzanian Women." *PloS One* 15, no. 8 (2020): e0237720.

Ahmed, Raisuddin, Steven Haggblade, Peter B. R. Hazell, Richard L. Meyer, and Thomas Reardon. "The Policy and Institutional Environment Affecting the Rural Nonfarm Economy." In *Transforming the Rural Nonfarm Economy: Opportunities and Threats in the Developing World*, ed. Steven Haggblade, Peter B. R. Hazell, and Thomas Reardon, 237–53. Baltimore: Johns Hopkins University Press, 2007.

Ajayi, Oluyede Clifford, Frank Place, Festus Kehinde Akinnifesi, and Gudeta Weldesemayat Sileshi. "Agricultural Success from Africa: The Case of Fertilizer Tree Systems in Southern Africa. Malawi, Tanzania, Mozambique, Zambia and Zimbabwe)." *International Journal of Agricultural Sustainability* 9, no. 1 (2011): 129–36.

Aker, Jenny C., Rachid Boumnijel, Amanda McClelland, and Niall Tierney. "Payment Mechanisms and Antipoverty Programs: Evidence from a Mobile Money Cash Transfer Experiment in Niger." *Economic Development and Cultural Change* 65, no. 1 (October 2016): 1–37.

Aldaco, R., D. Hoehn, J. Laso, M. Margallo, J. Ruiz-Salmón, J. Cristobal, R. Kahhat, et al. "Food Waste Management During the COVID-19 Outbreak: A Holistic Climate, Economic and Nutritional Approach." *Science of the Total Environment* 742 (2020): 140524.

Alderman, Harold. *Leveraging Social Protection Programs for Improved Nutrition: Summary of Evidence Prepared for the Global Forum on Nutrition-Sensitive Social Protection Programs, 2015*. Global Forum on Nutrition-Sensitive Social Protection Programs. Washington, D.C.: World Bank, 2016.

Alderman, Harold, Ugo Gentilini, and Ruslan Yemtsov. "Preface" and "The Evolution of Food as Social Assistance: An Overview." In *The 1.5 Billion People Question: Food, Vouchers, or Cash Transfers?*, ed. Harold Alderman, Ugo Gentilini, and Ruslan Yemtsov, 1–41. Washington, D.C.: World Bank, 2018.

Alexandratos, N., and J. Bruinsma. *World Agriculture Towards 2030/2050: The 2012 Revision*. ESA Working Paper no. 12–03. Rome: FAO, 2012.

Aljuni, Teddy. "Bollinger Announces Task Forces on Columbia's Role in Climate Solutions, Pressing Global Problems." *Columbia Spectator*, September 19, 2019. https://www.columbiaspectator.com/news/2019/09/19/bollinger-announces-two-new-task-forces-to-address-climate-change-pressing-human-problems-on-a-global-scale/.

Alkire, Sabina, Ruth Meinzen-Dick, Amber Peterman, Agnes Quisumbing, Greg Seymour, and Ana Vaz. "The Women's Empowerment in Agriculture Index." *World Development* 52 (2013): 71–91.

Allan, J. A. "Virtual Water: A Strategic Resource Global Solutions to Regional Deficits." *Groundwater* 36, no. 4 (1998): 545–46.

Allen, Karen. "Hunger Grips in Malawi Maize Crisis." *BBC News*, October 7, 2005. http://news.bbc.co.uk/2/hi/africa/4318974.stm.

Alliance for a Green Revolution in Africa. "AGRA and Partners Launch New Centre for African Leaders in Agriculture." Centre for African Leaders in Agriculture, April 27, 2021. https://cala.agra.org/updates/cala-launch/.

——. "Input Distributions." AGRA. Accessed August 28, 2021. https://agra.org/input-distributions/.

Altemimi, A., N. Lakhssassi, A. Baharlouei, D. G. Watson, and D. A. Lightfoot. "Phytochemicals: Extraction, Isolation, and Identification of Bioactive Compounds from Plant Extracts." *Plants* 6, no. 4 (2017): 42. doi.org/10.3390/plants6040042.

Amri, Ahmed. "ICARDA Genebank: Collecting, Conserving, and Using Dryland Agrobiodiversity." ICARDA. https://www.icarda.org/research/innovations/icarda-genebank-collecting-conserving-and-using-dryland-agrobiodiversity.

Anderson, Walter K., David Stephens, and Kadambot H. M. Siddique. "Dryland Agriculture in Australia: Experiences and Innovations." In *Innovations in Dryland Agriculture*, ed. M. Farooq and K. Siddique, 299–319. Cham, Switzerland: Springer, 2016.

Annan, Kofi. "Letter to Head of Donor States, August 8, 2005." In United Nations, *Malawi 2005 Flash Appeal September 2005—March 2006*. New York: Office for the Coordination of Humanitarian Affairs, 2005. https://www.humanitarianresponse.info/sites/www.humanitarianresponse.info/files/documents/files/flash_2005_malawi.pdf.

Asian Development Bank. *Strategy 2030: Achieving a Prosperous, Inclusive, Resilient, and Sustainable Asia and the Pacific*. Manila: ADB, 2018. https://www.adb.org/documents/strategy-2030-prosperous-inclusive-resilient-sustainable-asia-pacific.

——. Country Partnership Strategy. *Cambodia, 2019–2023—Inclusive Pathways to a Competitive Economy*. Manila: ADB, October 2019. https://www.adb.org/sites/default/files/institutional-document/534691/cps-cam-2019-2023.pdf.

Australian Broadcasting Commission. "Aiding or abetting." *Four Corners*. 1983. https://www.abc.net.au/4corners/aiding-or-abetting—1983/2832022.

Australian War Memorial. "Rationing of Food and Clothing During the Second World War." Accessed August 29, 2021. https://www.awm.gov.au/articles/encyclopedia/homefront/rationing.

Ausubel, Jesse H., Iddo K. Wernick, and Paul E. Waggoner. "Peak Farmland and the Prospect for Land Sparing." *Population and Development Review* 38 (2013): 221–42. http://www.jstor.org/stable/23655296.

Bailey, Regan L., Keith P. West Jr., and Robert E. Black. "The Epidemiology of Global Micronutrient Deficiencies." *Annals of Nutrition and Metabolism* 66, no. Suppl. 2 (2015): 22–33.

Barnett, Barry J., and Olivier Mahul. "Weather Index Insurance for Agriculture and Rural Areas in Lower-Income Countries." *American Journal of Agricultural Economics* 89, no. 5 (2007): 1241–47.

Baulcombe, David, Ian Crute, Bill Davies, Jim Dunwell, Mike Gale, Jonathan Jones, Jules Pretty, et al., *Reaping the Benefits: Science and the Sustainable Intensification of Global Agriculture*. London: Royal Society, 2009.

Bayala J., R. Zougmoré, C. Ky-Dembele, B. A. Bationo, S. Buah, D. Sanogo, J. Somda, A. Tougiani, K. Traoré, and A. Kalinganire. *Towards Developing Scalable Climate-Smart Village Models: Approach and Lessons Learnt from Pilot Research in West Africa*. ICRAF (World Agroforestry) Occasional Paper no. 25. Nairobi: World Agroforestry Centre, 2016.

Beckwith, Ryan Teague. "Transcript: Read the Speech Pope Francis Gave to the United Nations." *Time*, September 25, 2015. https://time.com/4049905/pope-francis-us-visit-united-nations-speech-transcript/.

Beddington, John. *Food, Energy, Water and the Climate: A Perfect Storm of Global Events*. London: Government Office for Science, 2009.

Beegle, Kathleen, Jed Friedman, John Gibson, and Joachim de Weerdt. "More to Do on Measuring Hunger." World Bank Blogs, March 19, 2014. https://blogs.worldbank.org/developmenttalk/more-do-measuring-hunger.

Bellon, Mauricio R., Ehsan Dulloo, Julie Sardos, Imke Thormann, and Jeremy J. Burdon. "In Situ Conservation—Harnessing Natural and Human-Derived Evolutionary Forces to Ensure Future Crop Adaptation." *Evolutionary Applications* 10, no. 10 (2017): 965–77.

Bellon, M. R., E. Gotor, and F. Caracciolo. "Conserving landraces and improving livelihoods: how to assess the success of on-farm conservation projects?" *International Journal of Agricultural Sustainability*, 13, no. 2 (2015): 167–82.

Bennett, Merrill. *The World's Food: A Study of the Interrelations of World Populations, National Diets, and Food Potentials*. New York: Harper, 1954.

Berg, Claudia N., Uwe Deichmann, Yishen Liu, and Harris Selod. *Transport Policies and Development*. Policy Research Working Paper no. 7366. Washington, D.C.: World Bank, 2015.

Bhattacharya, Shrayana. "Syrian Seed Bank Gets New Home Away from War." *Nature* 538 (2016): 16–17. doi.org/10.1038/538016a.

Bhattacharya, Shrayana, Vanita Leah Falcao, and Raghav Puri. "The Public Distribution System in India: Policy Evolution and Program Delivery Trends." In *The 1.5 Billion People Question*, ed. Harold Alderman, Ugo Gentilini, and Ruslan Yemtsov, 43–105. Washington, D.C.: World Bank, 2018.

Bhutta Zulfiqar A., T. Ahmed, R. E. Black, S. Cousens, K. Dewey, E. Giugliani, B. A. Haider, et al.; Maternal and Child Undernutrition Study Group. "What Works? Interventions for Maternal and Child Undernutrition and Survival." *Lancet* 371, no. 9610 (2008): 417–40.

Bhutta, Zulfiqar A., Nadia Akseer, Emily C. Keats, Tyler Vaivada, Shawn Baker, Susan E. Horton, Joanne Katz, et al. "How Countries Can Reduce Child Stunting at Scale: Lessons from Exemplar Countries." *American Journal of Clinical Nutrition* 112, no. Suppl. 2 (2020): 894S–904S.

Bhutta, Zulfiqar A., Jai K. Das, Arjumand Rizvi, Michelle F. Gaffey, Neff Walker, Susan Horton, Patrick Webb, et al. "Evidence-Based Interventions for Improvement of Maternal and Child Nutrition: What Can Be Done and at What Cost?" *Lancet* 382, no. 9890 (2013): 452–77.

Biovision Foundation for Ecological Development and IPES-Food. *Money Flows: What Is Holding Back Investment in Agroecological Research for Africa?* Zurich: Biovision Foundation for Ecological Development and Brussels: International Panel of Experts on Sustainable Food Systems, 2020.

Black, Maureen M., Susan P. Walker, Lia C. H. Fernald, Christopher T. Andersen, Ann M. DiGirolamo, Chunling Lu, Dana C. McCoy, et al. "Early Childhood Development Coming of Age: Science Through the Life Course." *Lancet* 389, no. 10064 (2017): 77–90.

Bonilla Cedrez, Camila, Jordan Chamberlin, Zhe Guo, and Robert J. Hijmans. "Spatial Variation in Fertilizer Prices in Sub-Saharan Africa." *PloS One* 15, no. 1 (2020): e0227764.

Bonis-Profumo, Gianna, R. McLaren, and J. Fanzo. "Ravaged Landscapes and Climate Vulnerability: The Challenge in Achieving Food Security and Nutrition in Post-Conflict Timor-Leste." *Advances in Food Security and Sustainability* 4 (2019): 97–132.

Bonis-Profumo, Gianna, Natasha Stacey, and Julie Brimblecombe. "Maternal Diets Matter for Children's Dietary Quality: Seasonal Dietary Diversity and Animal-Source Foods Consumption in Rural Timor-Leste." *Maternal and Child Nutrition* 17, no. 1 (2021): e13071.

——. "Measuring Women's Empowerment in Agriculture, Food Production, and Child and Maternal Dietary Diversity in Timor-Leste." *Food Policy* 102 (2021): 102102.

Borlaug, Norman. "The Green Revolution, Peace, and Humanity." Nobel Lecture, December 11, 1970. https://www.nobelprize.org/prizes/peace/1970/borlaug/lecture/.

Borrelli, Pasquale, David A. Robinson, Larissa R. Fleischer, Emanuele Lugato, Cristiano Ballabio, Christine Alewell, Katrin Meusburger, et al. "An Assessment of the Global Impact of 21st Century Land Use Change on Soil Erosion." *Nature Communications* 8, no. 1 (2017): 1–13.

Bose, Indira, Giulia Baldi, Lynnda Kiess, and Saskia de Pee. "The 'Fill the Nutrient Gap' Analysis: An Approach to Strengthen Nutrition Situation Analysis and Decision Making Towards Multisectoral Policies and Systems Change." *Maternal and Child Nutrition* 15, no. 3 (2019): e12793.

Bouman, B. A. M., and T. P. Tuong. "Field Water Management to Save Water and Increase Its Productivity in Irrigated Lowland Rice." *Agricultural Water Management* 1615, no. 1 (2000): 1–20.

Branca, Francesco, Alessandro Demaio, Emorn Udomkesmalee, Phillip Baker, Victor M. Aguayo, Simon Barquera, Katie Dain, et al. "A New Nutrition Manifesto for a New Nutrition Reality." *Lancet* 395, no. 10217 (2020): 8–10.

Branca, Francesco, E. Piwoz, W. Schultink, and L. M. Sullivan. "Nutrition and Health in Women, Children, and Adolescent Girls." *BMJ* 351 (2015): h4173. doi.org/10.1136/bmj.h4173.

Bruce, K. D., and Hanson M. A. "The Developmental Origins, Mechanisms, and Implications of Metabolic Syndrome." *Journal of Nutrition* 140 (2010): 648–52. doi.org/10.3945/jn.109.111179.

Bryce, Jennifer, Denise Coitinho, Ian Darnton-Hill, David Pelletier, Per Pinstrup-Andersen, and Maternal and Child Undernutrition Study Group. "Maternal and Child Undernutrition: Effective Action at National Level." *Lancet* 371, no. 9611 (2008): 510–26.

Bucknall-Williams, R. "How the CCAFS Legacy Lives on." CGIAR-CCAFS. December 16, 2021. https://ccafs.cgiar.org/news/how-ccafs-legacy-lives.

Buffett, Howard G. "Conservation: Reaping the Benefits of No-Tillage Farming." *Nature* 484, no. 7395 (2012): 455.

Bundy, D. A. P., N. de Silva, S. Horton, D. T. Jamison, and G. C. Patton. *Re-Imagining School Feeding: A High-Return Investment in Human Capital and Local Economies.* Washington, D.C.: World Bank, 2018.

Burns, Ken. *The Dust Bowl* (documentary). https://www.pbs.org/show/dust-bowl/.

Byerlee, Derek. *The Birth of CIMMYT: Pioneering the Idea and Ideals of International Agricultural Research.* Mexico City: CIMMYT, 2016.

Byerlee, Derek, James Stevenson, and Nelson Villoria. "Does Intensification Slow Crop Land Expansion or Encourage Deforestation?" *Global Food Security* 3, no. 2 (2014): 92–98.

CABI. "Invasive Species Compendium. *Prostephanus truncates* (larger grain borer)." Accessed August 28, 2021. https://www.cabi.org/isc/datasheet/44524.

Callaway, Ewen. "World's Largest Rice Gene Bank Secures Permanent Funding." *Nature News*, October 12, 2018. doi.org/10.1038/d41586-018-07029-1.

Cambodian Agricultural Research and Development Institute (CARDI). "Inauguration." CARDI. December 2, 2011. http://www.cardi.org.kh/?page=detail&menu1=6&menu2=69&menu3=76&ctype=article&id=76&lg=en.

——. "New CARDI Director Cites IRRI's Role in Making Cambodia a Rice-Exporting Country." CARDI. Accessed October 22, 2021. https://www.irri.org/news-and-events/news/new-cardi-director-cites-irri%E2%80%99s -role-making-cambodia-rice-exporting-country.

Carson, Rachel. *Silent Spring.* Boston: Houghton Mifflin, 1962.

Cassman, K. G., and P. Grassini. "A Global Perspective on Sustainable Intensification Research." *Nature Sustainability* 3 (2020): 262–68. doi.org/10.1038/s41893-020-0507-8.

CBS News. "New York Governor Cuomo Pleads for 30,000 Ventilators." March 25, 2020. https://www.cbsnews.com/video/new-york-governor-cuomo-pleads-for-30000-ventilators/.

Centers for Disease Control and Prevention. "Science Brief: Evidence Used to Update the List of Underlying Medical Conditions That Increase a Person's Risk of Severe Illness from COVID-19." Accessed September 2, 2021. https://www.cdc.gov/coronavirus/2019-ncov/science/science-briefs/underlying-evidence-table.html.

Central Statistics Organisation and United Nations Children's Fund (UNICEF). *Afghanistan Multiple Indicator Cluster Survey 2010–2011: Final Report*. Kabul: CSO and UNICEF, 2012.

CGIAR. "Impact Stories." Accessed January 25, 2022. https://www.cgiar.org/food-security-impact/impact-stories/.

——. "One CGIAR." Accessed January 24, 2022. https://www.cgiar.org/food-security-impact/one-cgiar/.

——. "New CGIAR Initiatives Announced to Help Radically Realign Food, Land, and Water Systems." News release, July 26, 2021. https://www.cgiar.org/news-events/news/cgiar-announces-new-portfolio-to-transform-food-land-and-water-systems-in-a-climate-crisis/.

——. "Overview: CGIAR—2020 at a Glance." Accessed August 29, 2021. https://www.cgiar.org/food-security-impact/finance-reports/dashboard/overview/.

Chandler, David P. *Brother Number One: A Political Biography of Pol Pot*. Boulder, Colo.: Westview Press, 1999.

Chandler, Robert Flint. *An Adventure in Applied Science: A History of the International Rice Research Institute*. Los Baños, Philippines: International Rice Research Institute, 1982.

Chandler, Robert W. *War of Ideas: The US Propaganda Campaign in Vietnam*. New York: Routledge, 1981.

Cho, Renée. "Phosphorus: Essential to Life—Are We Running Out?" Columbia Climate School, April 1, 2013. https://blogs.ei.columbia.edu/2013/04/01/phosphorus-essential-to-life-are-we-running-out/.

Christy, Ralph D., and Lionel Williamson, eds. *A Century of Service: Land-Grant Colleges and Universities, 1890–1990*. Brunswick, N.J.: Transaction, 1992.

Cinner, Joshua E., Tim R. McClanahan, Nicholas A. J. Graham, Tim M. Daw, Joseph Maina, Selina M. Stead, Andrew Wamukota, Katrina Brown, and Örjan Bodin. "Vulnerability of Coastal Communities to Key Impacts of Climate Change on Coral Reef Fisheries." *Global Environmental Change* 22, no. 1 (2012): 12–20.

Collins, S. "Treating Severe Acute Malnutrition Seriously." *Archives of Disease in Childhood* 92, no. 5 (2007): 453–61. doi.org/10.1136/adc.2006.098327.

Collis, Brad. "A Rice God named Harry." Coretext. Accessed August 15, 2021. http://stories.coretext.com.au/a-rice-god-named-harry/.

Columbia University. "Columbia Climate School." Accessed 29 August, 2021. https://climate.columbia.edu/.

——. "Fourth Purpose Task Force Report: On Directed Action." Memorandum to President Lee C. Bollinger, December 15, 2020. https://president.columbia.edu/sites/default/files/content/Additional/Fourth%20Purpose%20Task%20Force%20Report.pdf.

Committee on World Food Security. *Connecting Smallholders to Markets: Policy Recommendations*. Rome: CFS, 2015. http://www.fao.org/3/a-bq853e.pdf.

——. *Global Strategic Framework for Food Security and Nutrition (GSF)*. Third version. Rome: CFS, 2014. http://www.fao.org/fileadmin/templates/cfs/Docs1314/GSF/GSF_Version_3_EN.pdf.

——. High Level Panel of Experts. "About the HLPE." Accessed August 29, 2021. http://www.fao.org/cfs/cfs-hlpe.

Conroy, Anne, Malcolm Blackie, Alan Whiteside, Justin Malewezi, and Jeffrey Sachs. *Poverty, AIDS and Hunger: Breaking the Poverty Trap in Malawi*. London: Palgrave Macmillan, 2006.

Conway, Gordon. *The Doubly Green Revolution: Food for All in the 21st Century*. New York: Penguin Books, 1997.

——. *One Billion Hungry: Can We Feed the World?* Ithaca, N.Y.: Cornell University Press, 2012.

Costello, Peter. "Interview with Peter Costello," by Emma Simkin. *AM*, Australian Broadcasting Commission (ABC), July 26, 2001. https://www.abc.net.au/am/stories/s335651.htm.

Coulter, Jonathan, and Gideon Onumah. "The Role of Warehouse Receipt Systems in Enhanced Commodity Marketing and Rural Livelihoods in Africa." *Food Policy* 27, no. 4 (2002): 319–37.

Cox, R., P. Bierman, M. C. Jungers, and A. M. Rakotondrazafy. "Erosion Rates and Sediment Sources in Madagascar Inferred from ^{10}Be Analysis of Lavaka, Slope, and River Sediment." *Journal of Geology* 117, no. 4 (2009): 363–76.

Critical Past. "United States President Lyndon B. Johnson Visiting International Rice Institute in Los Baños and Cam Ranh Bay in Vietnam, October 26, 1966." https://www.criticalpast.com/video/65675039499_Lyndon-B-Johnson_paddy-field_President-decorating-troops_President-Marcos.

Croft, Trevor N., Aileen M. J. Marshall, Courtney K. Allen, et al. "Minimum Dietary Diversity, Minimum Meal Frequency and Minimum Acceptable Diet." In *Guide to DHS Statistics*. Rockville, Md.: ICF, 2018. https://dhsprogram.com/data/Guide-to-DHS-Statistics/Minimum_Dietary_Diversity_Minimum_Meal_Frequency_and_Minimum_Acceptable_Diet.htm.

Cromartie, J. "Modest Improvement in Nonmetro Population Change During the Decade Masks Larger Geographic Shifts." Amber Waves. https://www.ers.usda.gov/amber-waves/2020/july/modest-improvement-in-nonmetro-population-change-during-the-decade-masks-larger-geographic-shifts.

Crookes, William. "Address of the President Before the British Association for the Advancement of Science, Bristol, 1898." *Science* 8, no. 200 (1898): 561–75.

Crop Trust. "Svalbard Global Seed Vault." Accessed August 15, 2020. https://www.croptrust.org/our-work/svalbard-global-seed-vault/.

Crossley, Merlin. "What Is CRISPR Gene Editing, and How Does It Work?" *The Conversation*, January 31, 2018. https://theconversation.com/what-is-crispr-gene-editing-and-how-does-it-work-84591.

Crowley, John, William J. Smyth, and Michael Murphy, eds. *Atlas of the Great Irish Famine, 1845–52*. Cork, Ireland: Cork University Press, 2012.

Crumpler, K., A. Meybeck, S. Federici, M. Salvatore, B. Damen, S. Dasgupta, J. Wolf, and M. Bernoux. *Assessing the Role of Agriculture and Land Use in Nationally Determined Contributions*. Environment and Natural Resources Management Working Paper no. 76. Rome: FAO, 2019.

Crutzen, Paul J. "Geology of Mankind." In *Paul J. Crutzen: A Pioneer on Atmospheric Chemistry and Climate Change in the Anthropocene*, ed. P. Crutzen and H. Brauch. SpringerBriefs on Pioneers in Science and Practice, vol 50. Cham, Switzerland: Springer, 2016. doi.org/10.1007/978-3-319-27460-7_10.

Crutzen, Paul J., and Eugene F. Stoermer. "The Anthropocene." *Global Change Newsletter* 41 (2000): 17–18.

Cuevas, Ada, and Simon Barquera. "COVID-19, Obesity, and Undernutrition: A Major Challenge for Latin American Countries." *Obesity* 28, no. 10 (2020): 1791–92.

Culavamsa. Part 1, chap. 63, "The Improvement of His Own Kingdom," trans. Wilhelm Geiger. London: Pali Text Society, 1929. https://archive.org/details/in.ernet.dli.2015.188749/page/n321/mode/2up.

Cummings, M. J., M. R. Baldwin, D. Abrams, S. D. Jacobson, B. J. Meyer, E. M. Balough, J. G. Aaron, et al. "Epidemiology, Clinical Course, and Outcomes of Critically Ill Adults with COVID-19 in New York City: A Prospective Cohort Study." *Lancet* 395, no. 10239 (2020): P1763–70.

da Costa, M. d. J., M. Lopes, A. Ximenes, A. d. R. Ferreira, L. Spyckerelle, R. Williams, H. Nesbitt, et al. "Household Food Insecurity in Timor-Leste." *Food Security* 5 (2013): 83–94. doi.org/10.1007/s12571-012-0228-6.

Darnton-Hill, I., C. Nishida, and W. P. T. James. "A Life Course Approach to Diet, Nutrition and the Prevention of Chronic Diseases." *Public Health Nutrition* 7 (2004): 101–21.

Darwin, Charles. *The Effects of Cross and Self Fertilisation in the Vegetable Kingdom*. New York: Appleton, 1885.

——. *On the Origin of Species by Means of Natural Selection, or the Preservation of Favoured Races in the Struggle for Life*. London: John Murray, 1859.

De Datta, S. K., A. C. Tauro, and S. N. Balaoing. "Effect of plant type and nitrogen level on the growth characteristics and grain yield of indica rice in the tropics," *Agronomy Journal* 60, no. 6 (1968): 643–47.

De Datta, Surajit K. *Principles and practices of rice production*. New York: John Wiley and Sons, 1981.

De Fraiture, Charlotte, David Molden, and Dennis Wichelns. "Investing in Water for Food, Ecosystems, and Livelihoods: An Overview of the Comprehensive Assessment of Water Management in Agriculture." *Agricultural Water Management* 97, no. 4 (2010): 495–501.

DeFries, Ruth S., Jonathan A. Foley, and Gregory P. Asner. "Land-Use Choices: Balancing Human Needs and Ecosystem Function." *Frontiers in Ecology and the Environment* 2, no. 5 (June 2004): 249–57.

Democratic Republic of Timor-Leste (RDTL). *Timor-Leste Strategic Development Plan 2011–2030*. Dili, Timor-Leste: RDTL, 2011.

——. Ministry of Health. "Timor-Leste Food and Nutrition Survey," final report (2020). https://www.unicef.org/timorleste/reports/timor-leste-food-and-nutrition-survey.

De Stefano, A., and M. G. Jacobson. "Soil Carbon Sequestration in Agroforestry Systems: A Meta-Analysis." *Agroforestry Systems* 92 (2018): 285–99. doi.org/10.1007/s10457-017-0147-9.

Denning, Glenn. "An Australian-Assisted Credit Scheme for Smallholder Upland Crop Production in the Southern Philippines." *Agricultural Administration* 10, no. 2 (1982): 123–43.

——. "Fostering International Collaboration for Food Security and Sustainable Development: A Personal Perspective of M. S. Swaminathan's Vision, Impact and Legacy for Humanity." *Current Science* 109 (2015): 447–55.

Denning, Glenn, and Jessica Fanzo. "Ten Forces Shaping the Global Food System." In *Good Nutrition: Perspectives for the 21st Century*, ed. M. Eggersdorfer et al. Basel, Switzerland: Karger, 2016.

Denning, Glenn, and Hanson Nyantakyi-Frimpong. "Is the New Green Revolution Approach the Best Way to Address Hunger in Africa?" In *Debating African Issues: Conversations Under the Palaver Tree*, ed. William G. Moseley and Kefa M. Otiso, 151–63. London: Taylor & Francis, 2022.

Denning, Glenn, Patrick Kabambe, Pedro Sanchez, Alia Malik, Rafael Flor, Rebbie Harawa, Phelire Nkhoma, et al. "Input Subsidies to Improve Smallholder Maize Productivity in Malawi: Toward an African Green Revolution." *PLoS Biology* 7, no. 1 (2009): e1000023.

Denning, Glenn, and Jeffrey Sachs. "How the Rich World Can Help Africa Help Itself." *Financial Times*, May 29, 2007.

Denning, Glenn, and Vo-Tong Xuan, eds. *Vietnam and IRRI: A Partnership in Rice Research. Proceedings of a Conference Held in Hanoi, Vietnam, 4–7 May, 1994*. Manila: International Rice Research Institute, 1995.

Devereux, Stephen, Christophe Béné, and John Hoddinott. "Conceptualising COVID-19's Impacts on Household Food Security." *Food Security* 12, no. 4 (2020): 769–72.

Diamond, Jared. "Evolution, Consequences and Future of Plant and Animal Domestication." *Nature* 418, no. 6898 (2002): 700–707.

Dimitri, Carolyn, Anne Effland, and Neilson Conklin. *The 20th Century Transformation of US Agriculture and Farm Policy*. Economic Information Bulletin no. 3. Washington, D.C.: U.S. Department of Agriculture, June 2005.

Doran, John W., and Michael R. Zeiss. "Soil Health and Sustainability: Managing the Biotic Component of Soil Quality." *Applied Soil Ecology* 15, no. 1 (2000): 3–11.

Dorward, Andrew, and Ephraim Chirwa. "The Malawi Agricultural Input Subsidy Programme: 2005/06 to 2008/09." *International Journal of Agricultural Sustainability* 9, no. 1 (2011): 232–47.

Dreier, Lisa, David Nabarro, and Jane Nelson. *Systems Leadership for Sustainable Development: Strategies for Achieving Systemic Change*. Cambridge, Mass.: Harvard Kennedy School, September 2019. https://www.hks.harvard.edu/centers/mrcbg/publications/fwp/crisept2019.

Dulac, John. *Global Land Transport Infrastructure Requirements Estimating Road and Railway Infrastructure Capacity and Costs to 2050*. Paris: International Energy Agency, 2013.

Duniway, Michael C., Alix A. Pfennigwerth, Stephen E. Fick, Travis W. Nauman, Jayne Belnap, and Nichole N. Barger. "Wind Erosion and Dust from US Drylands: A Review of Causes, Consequences, and Solutions in a Changing World." *Ecosphere* 10, no. 3 (2019): e02650.

Duvick, Donald N. "Biotechnology in the 1930s: The Development of Hybrid Maize." *Nature Reviews Genetics* 2, no. 1 (2001): 69–74.

——. "The Contribution of Breeding to Yield Advances in Maize (*Zea mays* L.)." *Advances in Agronomy* 86 (2005): 83–145.

Eat Well Tasmania. *A Survey About Shopping, Cooking and Eating*. Hobart, Australia: Eat Well Tasmania, 2020. https://www.eatwelltas.org.au/wp-content/uploads/2020/05/EWT_Covid-Survey-Summary_A3.pdf.

EAT–*Lancet* Commission. *Healthy Diets from Sustainable Food Systems*, Summary Report of the EAT–*Lancet* Commission. Oslo: EAT, 2019. https://eatforum.org/content/uploads/2019/07/EAT-Lancet_Commission_Summary_Report.pdf.

Ehrlich, Paul R. *The Population Bomb*. New York: Ballantine Books, 1968.

Eicher, Carl K., and Derek Byerlee. "Accelerating Maize Production: Synthesis." In *Africa's Emerging Maize Revolution*, ed. Derek Byerlee and Carl K. Eicher, 247–62. London: Lynne Rienner, 1977.

Ericksen, P. J. 2008. Conceptualizing Food Systems for Global Environmental Change Research. *Global Environmental Change* 18: 234–45.

Erisman, Jan Willem, James Galloway, Nancy B. Dise, Albert Bleeker, Bruna Grizzetti, Allison M. Leach, and Wim de Vries. *Nitrogen: Too Much of a Vital Resource*. Science Brief. Zeist: WWF Netherlands, 2015.

Eswaran, Hari, Russell Almaraz, Evert van den Berg, and Paul Reich. "An assessment of the soil resources of Africa in relation to productivity." *Geoderma* 77, no. 1 (1997): 1–18.

Evenson, Robert E., and Douglas Gollin. "Assessing the Impact of the Green Revolution, 1960 to 2000." *Science* 300, no. 5620 (2003): 758–62.

Falkenmark, Malin and Johan Rockström. *Balancing Water for Humans and Nature: The New Approach in Ecohydrology*. London: Earthscan, 2004.

——. "The New Blue and Green Water Paradigm: Breaking New Ground for Water Resources Planning and Management." *Journal of Water Resources Planning and Management* 132, no. 3 (2006): 129–32.

Famine Early Warning Systems Network. Monthly reports (2005–2007). https://fews.net/southern-africa/malawi.

Fanzo, Jessica. *Can Fixing Dinner Fix the Planet?* Baltimore: Johns Hopkins University Press, 2021.

——. "Stunting: A Country's Lasting Burden." *Journal of International Affairs* 67, no. 2 (2014): 165–73.

Fanzo, Jessica, João Boavida, Gianna Bonis-Profumo, Rebecca McLaren, and Claire Davis. *Timor-Leste Strategic Review: Progress and Success in Achieving the Sustainable Development Goal 2*. Dili, Timor-Leste: Centre of Studies for Peace and Development (CEPAD) and Baltimore: Johns Hopkins University, May 2017. https://docs.wfp.org/api/documents/WFP-0000015583/download/.

Fanzo, Jessica, S. Curran, G. Denning, and Seeds of Life Team. *A Nutrition Sensitive Agriculture Strategy*. For the Ministry of Agriculture and Fisheries of Timor-Leste and for Fini ba Moris (Seeds of Life). Dili, March 2013.

Farm Aid. "Meet Farm Aid's Current Grantees." Accessed August 24, 2021. https://www.farmaid.org/our-work/grants/.

Feed the Future. "About." Accessed August 28, 2021. https://www.feedthefuture.gov/about/.

Feehan, John. "The Potato: Root of the Famine." In *Atlas of the Great Irish Famine, 1845–52*, ed. John Crowley, William J. Smyth, and Michael Murphy, 28–37. Cork, Ireland: Cork University Press, 2012.

Fernald, Lia C. H., Paul J. Gertler, and Xiaohui Hou. "Cash Component of Conditional Cash Transfer Program Is Associated with Higher Body Mass Index and Blood Pressure in Adults." *Journal of Nutrition* 138, no. 11 (2008): 2250–57.

Fisher, M., T. Abate, R. W. Lunduka, W. Asnake, Y. Alemayehu, and R. B. Madalu. "Drought Tolerant Maize for Farmer Adaptation to Drought in Sub-Saharan Africa: Determinants of Adoption in Eastern and Southern Africa." *Climatic Change* 133 (2015): 283–99.

Flanagan, Katie, Kai Robertson, and Craig Hanson. *Reducing Food Loss and Waste: Setting a Global Action Agenda*. Washington, D.C.: World Resources Institute, 2019.

Flexner, A. *Medical Education in the United States and Canada*. Bulletin no. 4. New York: Carnegie Foundation for the Advancement of Teaching, 1910.

Fontanier, C., B. Mamo, S. Toucanne, G. Bayon, S. Schmidt, B. Deflandre, B. Dennielou, et al. "Are Deep-Sea Ecosystems Surrounding Madagascar Threatened by Land-Use or Climate Change?" *Deep Sea Research Part I: Oceanographic Research Papers* 131 (2018): 93–100.

Food and Agriculture Organization (FAO). *Climate-Smart Agriculture. Case Studies 2018: Successful Approaches from Different Regions*. Rome: FAO, 2018.

——. "Climate Smart Agriculture Sourcebook." Second ed. FAO, 2017. http://www.fao.org/climate-smart-agriculture-sourcebook/en/.

——. *Dynamic Development, Shifting Demographics, Changing Diets*. Bangkok: FAO, 2018. http://www.fao.org/3/I8499EN/i8499en.pdf.

——. "FAO Cereal Supply and Demand Brief." July 8, 2021. http://www.fao.org/worldfoodsituation/csdb/en/.
——. *The Food Insecurity Experience Scale: Measuring Food Insecurity Through People's Experiences.* I7835EN/1/09.17. Rome: FAO, n.d. http://www.fao.org/3/a-i7835e.pdf.
——. *Food Outlook—Biannual Report on Global Food Markets: June 2020.* Food Outlook, 1. Rome: FAO, 2020. doi.org/10.4060/ca9509en.
——. *GIEWS Country Brief, Malawi.* Rome: FAO, March 25, 2021. https://reliefweb.int/report/malawi/giews-country-brief-malawi-25-march-2021.
——. *Global Agriculture Towards 2050. High-Level Expert Forum—How to Feed the World in 2050.* Rome: FAO, October 12–13, 2009. https://reliefweb.int/sites/reliefweb.int/files/resources/E6441A901ECFFDDDC125763C0034ED32-Full_Report.pdf.
——. *Global Food Losses and Food Waste: Extent, Causes, and Prevention.* Rome: FAO, 2011.
——. *Impacts of Coronavirus on Food Security and Nutrition in Asia and the Pacific: Building More Resilient Food Systems.* Bangkok: FAO, June 5, 2020. http://www.fao.org/3/ca9473en/CA9473EN.pdf.
——. "International Treaty on Plant Genetic Resources for Food and Agriculture: About Us." Accessed August 15, 2021. http://www.fao.org/plant-treaty/overview/en/.
——. "International Treaty on Plant Genetic Resources for Food and Agriculture: Texts of the Treaty: Official Versions." Accessed August 15, 2021. http://www.fao.org/plant-treaty/overview/texts-treaty/en/.
——. *The State of Food and Agriculture 1960.* Rome: FAO, 1960. https://www.un-ilibrary.org/content/books/9789210472760/read.
——. *The State of Food and Agriculture, 2010–11: Women in Agriculture. Closing the Gender Gap for Development.* Rome: FAO, 2011.
——. *The State of Food and Agriculture 2019: Moving Forward on Food Loss and Waste Reduction.* Rome: FAO, 2019.
——. *The State of the World's Land and Water Resources for Food and Agriculture: Managing Systems at Risk.* Rome: FAO and London: Earthscan, 2011.
——. *Strengthening Coherence Between Agriculture and Social Protection to Combat Poverty and Hunger in Africa: Framework for Analysis and Action.* Rome: FAO, 2016. http://www.fao.org/3/i5386e/i5386e.pdf.
——. "Voices of the Hungry: The Food Insecurity Experience Scale." Accessed August 25, 2021. http://www.fao.org/in-action/voices-of-the-hungry/fies/en/.
——. *Water for Sustainable Food and Agriculture: A Report Produced for the G20 Presidency of Germany.* Rome: FAO, 2017. http://www.fao.org/3/i7959e/i7959e.pdf.
——. "World Food Situation." Accessed August 15, 2021. http://www.fao.org/worldfoodsituation/foodpricesindex/en/.
Food and Agriculture Organization, International Fund for Agricultural Development, United Nations Children's Fund, World Food Programme, and World Health Organization. *The State of Food Security and Nutrition in the World 2018: Building Climate Resilience for Food Security and Nutrition.* Rome: FAO, 2018.
——. *The State of Food Security and Nutrition in the World 2019: Safeguarding Against Economic Slowdowns and Downturns.* Rome: FAO, 2019.
——. *The State of Food Security and Nutrition in the World 2020. Transforming Food Systems for Affordable Healthy Diets.* Rome: FAO, 2020. http://www.fao.org/3/ca9692en/ca9692en.pdf.
——. *The State of Food Security and Nutrition in the World 2021. Transforming Food Systems for Food Security, Improved Nutrition and Affordable Healthy Diets for All.* Rome: FAO, 2021.
Food and Agricultural Organization, International Fund for Agricultural Development, and World Food Programme. *The State of Food Insecurity in the World 2015: Meeting the 2015 International Hunger Targets: Taking Stock of Uneven Progress.* Rome: FAO, 2015.
Food Security Information Network (FSIN) and Global Network Against Food Crises. *Global Report on Food Crises 2020.* Rome: FSIN, 2020. https://docs.wfp.org/api/documents/WFP-0000114546/download/.
——. *Global Report on Food Crises 2021.* Rome: FSIN, 2021. https://www.fsinplatform.org/sites/default/files/resources/files/GRFC%202021%20050521%20med.pdf.

Fore, H. H., Q. Dongyu, D. M. Beasley, and T. A. Ghebreyesus. "Child Malnutrition and COVID-19: The Time to Act Is Now." *Lancet* 396, no. 10250 (2020): 517–18.

Fox, Louise, and Thomas S. Jayne. "Unpacking the Misconceptions About Africa's Food Imports." Brookings Institution, December 14, 2020. https://www.brookings.edu/blog/africa-in-focus/2020/12/14/unpacking-the-misconceptions-about-africas-food-imports/.

Franzel, Steven, Glenn L. Denning, Jens-Peter Barnekow Lillesø, and A. R. Mercado. "Scaling Up the Impact of Agroforestry: Lessons from Three Sites in Africa and Asia." *Agroforestry Systems* 61, no. 1 (2004): 329–44.

Fridovich-Keil, Judith L. "Gene Editing, Genetics." *Britannica*. Accessed August 15, 2021. https://www.britannica.com/science/gene-editing.

Fuglie, Keith, and Nicholas Rada, *Resources, Policies, and Agricultural Productivity in Sub-Saharan Africa*. Department of Agriculture, Economic Research Service Economic Research Report no. 145. Washington, D.C.: USDA-ERS, February 2013. dx.doi.org/10.2139/ssrn.2266459.

Fujisaka, Sam. "Rainfed Lowland Rice: Building Research on Farmer Practice and Technical Knowledge." *Agriculture, Ecosystems and Environment* 33, no. 1 (1990): 57–74.

——. *Rice Agroecosystems, Farmer Management, and Social Organization in Kampuchea: A Preliminary Assessment and Research Recommendations*. International Rice Research Institute Research Paper Series no. 136. Los Baños, Philippines: IRRI, 1988.

G7 France Biarritz. *Biarritz Progress Report G7 Development and Development-Related Commitments*. 2019. https://www.diplomatie.gouv.fr/IMG/pdf/rapport-g7-bat-webcle85fc7e.pdf.

G20 Research Group, University of Toronto. "G20 Leaders Statement: The Pittsburgh Summit September 24–25, 2009. Pittsburgh." Last updated November 29, 2011. http://www.g20.utoronto.ca/2009/2009communique0925.html.

Galasso, E., A. Wagstaff, S. Naudeau, and M. Shekar. *The Economic Costs of Stunting and How to Reduce Them*. Policy Research Note 17/5. Washington, D.C.: World Ban Group, 2017.

Gardner, William, and Nicholas Kassebaum. "Global, Regional, and National Prevalence of Anemia and Its Causes in 204 Countries and Territories, 1990–2019." *Current Developments in Nutrition*, 4, no. Supplement_2 (June 2020): 830.doi.org/10.1093/cdn/nzaa053_035.

Gargan, Edward A. "Nyerere Steps Down but Keeps His Hand In." *New York Times*, November 3, 1985. https://www.nytimes.com/1985/11/03/weekinreview/nyerere-steps-down-but-keeps-his-hand-in.html.

Garland, Sarah. "EU Policy Must Change to Reflect the Potential of Gene Editing for Addressing Climate Change." *Global Food Security* 28 (2021): 100496.

——. "USDA Definitions Around Biotechnology Need Some Crucial Updating," *Slate*, March 22, 2021. https://slate.com/technology/2021/03/usda-coordinated-framework-regulation-biotechnology-gene-editing.html.

Garnett, T., M. C. Appleby, A. Balmford, I. J. Bateman, T. G. Benton, P. Bloomer, B. Burlingame, et al. "Sustainable Intensification in Agriculture: Premises and Policies." *Science* 341 (2013): 33–34.

Garrity, Dennis Philip, Festus K. Akinnifesi, Oluyede C. Ajayi, Sileshi G. Weldesemayat, Jeremias G. Mowo, Antoine Kalinganire, Mahamane Larwanou, and Jules Bayala. "Evergreen Agriculture: A Robust Approach to Sustainable Food Security in Africa." *Food Security* 2, no. 3 (2010): 197–214.

Gashler, Krisy. "Bt Eggplant Improving Lives in Bangladesh." *Cornell Chronicle*, July 16, 2018. https://news.cornell.edu/stories/2018/07/bt-eggplant-improving-lives-bangladesh.

Gates, Bill. "Prepared Remarks." 2009 World Food Prize Symposium, October 15, 2009. https://www.gatesfoundation.org/ideas/speeches/2009/10/bill-gates-2009-world-food-prize-symposium.

Gaud, William S. "The Green Revolution: Accomplishments and Apprehensions." Address before the Society for International Development, Shoreham Hotel, Washington D.C., March 8, 1968. http://www.agbioworld.org/biotech-info/topics/borlaug/borlaug-green.html.

Gerlock, Grant. "Why the Arctic Apple Means You May Be Seeing More GMOs at the Store." *NPR: The Salt*, February 1, 2017. https://www.npr.org/sections/thesalt/2017/02/01/512633781/why-the-arctic-apple-means-you-may-be-seeing-more-gmos-at-the-store.

Ghimiray, Mahesh. "An Analysis of Rice Varietal Improvement and Adoption Rate by Farmers in Bhutan." *Journal of Renewable Natural Resources of Bhutan* 10, no. 1 (2012): 13–24.

Ghimiray, Mahesh, and Ronnie Vernooy. "The Importance and Challenges of Crop Germplasm Interdependence: The Case of Bhutan." *Food Security* 9, no. 2 (2017): 301–10.

Gibbs, H. K., and J. Meghan Salmon. "Mapping the World's Degraded Lands." *Applied Geography* 57 (2015): 12–21.

Gibson, R. S. "Strategies for Preventing Multi-Micronutrient Deficiencies: A Review of Experiences with Food-Based Approaches in Developing Countries." In *Combating Micronutrient Deficiencies: Food-Based Approaches*, ed. Brian Thompson and Leslie Amoroso, 7–27. Wallingford, UK: CABI, 2011.

Giller, Ken E. "No Silver Bullets for African Soil Problems." *Nature* 485, 41 (2012). doi.org/10.1038/485041c.

Giller, Ken E., Ernst Witter, Marc Corbeels, and Pablo Tittonell. "Conservation Agriculture and Smallholder Farming in Africa: The Heretics' View." *Field Crops Research* 114, no. 1 (2009): 23–34.

Gilligan, Daniel O. "Safety Nets for Agriculture and Nutrition." In *Agriculture for Improved Nutrition: Seizing the Momentum*, ed. Shenggen Fan, Sivan Yosef, and Rajul Pandya-Lorch, 104–12. Wallingford, UK: International Food Policy Research Institute and CABI, 2019.

Global Agriculture and Food Security Program (GAFSP). "Our Impact." Accessed August 10, 2021. https://www.gafspfund.org/index.php/impact.

Global Alliance for Improved Nutrition. *Food Fortification: The Unfinished Agenda*. Briefing paper. Geneva, Switzerland: Global Alliance for Improved Nutrition, 2018. https://www.gainhealth.org/sites/default/files/publications/documents/food-fortification-unfinished-agenda-2018.pdf.

——. "Standing Together for Nutrition (STfN)." July 28, 2020. https://www.gainhealth.org/media/news/standing-together-nutrition-stfn.

Global Fund to Fight AIDS, Tuberculosis and Malaria. "Global Fund Overview." Accessed August 10, 2021. https://www.theglobalfund.org/en/overview/.

Global Nutrition Report. "Country Nutrition Profiles: Madagascar." Accessed August 14, 2021. https://globalnutritionreport.org/resources/nutrition-profiles/africa/eastern-africa/madagascar/.

——. "Country Nutrition Profiles: United Republic of Tanzania." Accessed August 24, 2021. https://globalnutritionreport.org/resources/nutrition-profiles/africa/eastern-africa/united-republic-tanzania/#profile.

Godfray, H. C. J., and T. Garnett. "Food Security and Sustainable Intensification." *Philosophical Transactions of the Royal Society B* 369 (2014): 20120273.

Goedert, W. J. "Management of the Cerrado Soils of Brazil: A Review." *Journal of Soil Science* 34, no. 3 (1983): 405–28.

Goh, L. "How AgriTech Is Transforming Traditional Agriculture in Emerging Markets." In *Breakthrough: The Promise of Frontier Technologies for Sustainable Development*, ed. Homi Kharas, John W. McArthur, and Izumi Ohno. Washington, D. C.: Brookings Institution Press, 2022.

Gollin, Douglas. *Smallholder Agriculture in Africa: An Overview and Implications for Policy*. International Institute for Environment and Development Working Paper. London: IIED, 2014. http://pubs.iied.org/14640IIED.

Gorsuch, Joyce. *Bhutan–IRRI Project: Local Tradition Meets Modern Know-How*. Manila: International Rice Research Institute, 2001. http://books.irri.org/9712201589_content.pdf.

Govaerts, Bram, Nele Verhulst, A. Castellanos-Navarrete, Ken D. Sayre, John Dixon, and L. Dendooven. "Conservation Agriculture and Soil Carbon Sequestration: Between Myth and Farmer Reality." *Critical Reviews in Plant Sciences* 28, no. 3 (2009): 97–122.

Government of Timor-Leste. "Prime Minister Launched the SDP During the Timor-Leste and Development Partners Meeting." News release, July 13, 2011. http://timor-leste.gov.tl/?p=5337&lang=en.

——. *Report on the Implementation of the Sustainable Development Goals: From Ashes to Reconciliation, Reconstruction and Sustainable Development*. Voluntary National Review of Timor-Leste 2019. Dili: Government of Timor-Leste, 2019. https://sustainabledevelopment.un.org/content/documents/23417TimorLeste_VNR_2019_FINAL.pdf.

Greatrex H., J. W. Hansen, S. Garvin, R. Diro, S. Blakeley, M. Le Guen, K. N. Rao, and D. E. Osgood. *Scaling Up Index Insurance for Smallholder Farmers: Recent Evidence and Insights*. CCAFS Report no. 14 Copenhagen: CGIAR Research Program on Climate Change, Agriculture, and Food Security (CCAFS), 2015.

Greenland, Dennis J., ed. *The Sustainability of Rice Farming.* Wallingford, UK: CABI, 1997.

Greenpeace International. "#Food." Accessed August 15, 2021. https://www.greenpeace.org/international/tag/food/.

——. "Nobel Laureates Sign Letter on Greenpeace 'Golden' Rice Position—Statement." June 30, 2016. https://www.greenpeace.org/international/press-release/6866/nobel-laureates-sign-letter-on-greenpeace-golden-rice-position-statement/.

Gregory, Peter. "Dennis James Greenland. 13 June 1930–23 December 2012." *Biographical Memoirs of Fellows of the Royal Society* 66 (2019): 225–41. https://royalsocietypublishing.org/doi/10.1098/rsbm.2018.0030.

Griffiths, Anthony J. F. "Recombinant DNA, Genetic Engineering." *Britannica.* Accessed August 15, 2021. https://www.britannica.com/science/recombinant-DNA-technology.

Gupta, Raj, and Ashok Seth. "A Review of Resource Conserving Technologies for Sustainable Management of the Rice–Wheat Cropping Systems of the Indo-Gangetic Plains (IGP)." *Crop Protection* 26, no. 3 (2007): 436–47.

Gusmão, Xanana. "Address by His Excellency The Prime Minister Kay Rala Xanana Gusmão at the Timor-Leste and Development Partners Meeting, Dili, July 12, 2011." Democratic Republic of Timor-Leste. http://timor-leste.gov.tl/wp-content/uploads/2011/07/Launch-of-the-SDP_TLDPM_12.07.11.pdf.

Gustavsson, Jenny, Christel Cederberg, Ulf Sonesson, Robert Van Otterdijk, and Alexandre Meybeck. *Global Food Losses and Food Waste: Extent, Causes and Prevention.* Rome: FAO, 2011.

Gwede, Wanga. "Lack of Ventilators in Malawi Will See More Die to Coronavirus." *Nyasa Times,* March 27, 2020. https://www.nyasatimes.com/lack-of-ventilators-in-malawi-will-see-more-die-to-coronavirus/.

Hadley, Craig, David Lindstrom, Fasil Tessema, and Tefara Belachew. "Gender Bias in the Food Insecurity Experience of Ethiopian Adolescents." *Social Science and Medicine* 66, no. 2 (2008): 427–38.

Hager, T. *The Alchemy of Air.* New York: Broadway Books, 2008.

Haggblade, Steven, Peter B. R. Hazell, and Thomas Reardon, eds. *Transforming the Rural Nonfarm Economy: Opportunities and Threats in the Developing World.* International Food Policy Research Institute Issue Brief 58. Baltimore: Johns Hopkins University Press, 2007.

Hahn, F., and V. Nekrasov. "CRISPR/Cas precision: do we need to worry about off-targeting in plants?" *Plant Cell Reports* 38, no. 4 (2019): 437–41.

Haire, Benjamin. "Blockchain: Revolutionising the Agriculture Industry." *Agribusiness Bulletin,* October 2018. https://www2.deloitte.com/au/en/pages/consumer-business/articles/blockchain-revolutionising-agriculture-industry.html.

Hansen, James W., Walter E. Baethgen, Daniel E. Osgood, Pietro N. Ceccato, and Robinson Kinuthia Ngugi. "Innovations in Climate Risk Management: Protecting and Building Rural Livelihoods in a Variable and Changing Climate." *Journal of Semi-Arid Tropical Agricultural Research* 4, no. 1 (2007).

Hanson, C., K. Flanagan, K. Robertson, H. Axmann, H. Bos-Brouwers, J. Broeze, C. Kneller, et al. *Reducing Food Loss and Waste: Ten Interventions to Scale Impact.* Washington, D.C.: World Resources Institute, 2019.

Hargrove, Thomas R. *A Dragon Lives Forever: War and Rice in Vietnam's Mekong Delta, 1969–1991, and Beyond.* New York: Ivy Books, 1994.

——. "I Remember Honda Rice." *Rice Today* 5, no. 4 (2006): 39–44.

Harkup, Kathryn. "Chlorine: The Gas of War Crimes." *Guardian,* September 16, 2016. https://www.theguardian.com/science/blog/2016/sep/16/chlorine-the-gas-of-war-crimes.

Harper, Justin. "Why Orange Juice Prices Are Soaring on Global Markets." *BBC News,* March 26, 2020. https://www.bbc.com/news/technology-52030133.

Harrigan, Jane. "Food Insecurity, Poverty and the Malawian Starter Pack: Fresh Start or False Start?" *Food Policy* 33, no. 3 (2008): 237–49.

Hartz, T. K., and G. J. Hochmuth. "Fertility Management of Drip-Irrigated Vegetables." *HortTechnology* 6, no. 3 (1996): 168–72.

Hawkes, Corinna, Bettina Gerken Brazil, Inês Rugani Ribeiro de Castro, and Patricia Constante Jaime. "How to Engage Across Sectors: Lessons from Agriculture and Nutrition in the Brazilian School Feeding Program." *Revista de Saúde Pública* 50 (2016): 47. doi.org/10.1590/S1518-8787.2016050006506.

Hawkes, Corinna, Marie T. Ruel, Leah Salm, Bryony Sinclair, and Francesco Branca. "Double-Duty Actions: Seizing Programme and Policy Opportunities to Address Malnutrition in All its Forms." *Lancet* 395, no. 10218 (2020): 142–55. Erratum in *Lancet* 395, no. 10221 (February 1, 2020): 338. PMID: 31852603.

Hazell, Peter B. R. "Transforming Agriculture: The Green Revolution in Asia." In *Millions Fed: Proven Successes in Agricultural Development*, ed. David J. Spielman and Rajul Pandya-Lorch, 25–32. Washington, D.C.: International Food Policy Research Institute, 2009.

Headey, Derek, and Shenggen Fan. *Reflections on the Global Food Crisis: How Did It Happen? How Has It Hurt? And How Can We Prevent the Next One.* Research Reports 165. Washington, D.C.: International Food Policy Research Institute, 2010.

Headey, Derek, Rebecca Heidkamp, Saskia Osendarp, Marie Ruel, Nick Scott, Robert Black, and Meera Shekar, et al. "Impacts of COVID-19 on Childhood Malnutrition and Nutrition-Related Mortality." *Lancet* 396, no. 10250 (2020): 519–21.

Hell, Kerstin, and Charity Mutegi. "Aflatoxin Control and Prevention Strategies in Key Crops of Sub-Saharan Africa." *African Journal of Microbiology Research* 5, no. 5 (2011): 459–66.

Heller, J. L. "The Nomenclature of Soils, or What's in a Name?" *Soil Science Society of America Journal* 27, no. 2 (1963): 216–20.

Hellmich, R. L., and K. A. Hellmich. "Use and Impact of Bt Maize." *Nature Education Knowledge* 3, no. 10 (2012): 4.

Herforth, Anna, Mary Arimond, Cristina Álvarez-Sánchez, Jennifer Coates, Karin Christianson, and Ellen Muehlhoff. "A Global Review of Food-Based Dietary Guidelines." *Advances in Nutrition* 10, no. 4 (2019): 590–605.

Herrero, Mario, Benjamin Henderson, Petr Havlík, Philip K. Thornton, Richard T. Conant, Pete Smith, Stefan Wirsenius, et al. "Greenhouse Gas Mitigation Potentials in the Livestock Sector." *Nature Climate Change* 6, no. 5 (2016): 452–61.

Hesser, Leon. *The Man Who Fed the World.* Dallas, Tex.: Durban House, 2006.

Hettel, Gene. *The IRRI Pioneer Interviews.* Los Baños, Philippines: International Rice Research Institute, 2010. http://books.irri.org/Pioneer_Interviews.pdf.

Hewitt, Chris, Simon Mason, and David Walland, "The Global Framework for Climate Services." *Nature Climate Change* 2, no. 12 (2012): 831–32.

Higginbottom, Thomas P., Roshan Adhikari, Ralitza Dimova, Sarah Redicker, and Timothy Foster. "Performance of Large-Scale Irrigation Projects in Sub-Saharan Africa." *Nature Sustainability* 4 (2021): 501–8. doi.org/10.1038/s41893-020-00670-7.

High Level Panel of Experts on Food Security and Nutrition. *Food Losses and Waste in the Context of Sustainable Food Systems: A Report by the High Level Panel of Experts on Food Security and Nutrition of the Committee on World Food Security.* Rome: Committee on World Food Security, 2014.

——. "HLPE Reports." Accessed August 10, 2021. http://www.fao.org/cfs/cfs-hlpe/reports/en/.

——. *Impacts of COVID-19 on Food Security and Nutrition: Developing Effective Policy Responses to Address the Hunger and Malnutrition Pandemic.* Rome: Committee on World Food Security, 2020. http://www.fao.org/3/cb1000en/cb1000en.pdf.

——. *Nutrition and Food Systems. A Report by the High Level Panel of Experts on Food Security and Nutrition.* Rome: Committee on World Food Security, 2017. http://www.fao.org/3/i7846e/i7846e.pdf.

Himmel, Jeffrey. *Khmer Rouge Irrigation Development in Cambodia.* Phnom Penh: Documentation Center of Cambodia (DC-Cam), 2007.

Hirvonen, Kalle, Y. Bai, D. Headey, and W. A. Masters. "Affordability of the EAT–Lancet Reference Diet: A Global Analysis." *Lancet Global Health* 8, no. 1 (2020): e59–e66.

Hirvonen, Kalle, Alan de Brauw, and Gashaw T. Abate. "Food Consumption and Food Security During the COVID-19 Pandemic in Addis Ababa." *American Journal of Agricultural Economics* 103, no. 3 (2021): 772–89.

Hirvonen, Kalle, and John F. Hoddinott. *Payment Modality Preferences: Evidence from Ethiopia's Productive Safety Net Programme.* ESSP Working Paper 125. Washington, D.C.: International Food Policy Research Institute and Addis Ababa: Ethiopian Development Research Institute, 2018.

Hoekstra, Arjen Y. "Water Footprint Assessment: Evolvement of a New Research Field." *Water Resources Management* 31, no. 10 (2017): 3061–81.

Hofste, Rutger W., Samantha Kuzma, Sara Walker, Edwin H. Sutanudjaja, Marc F. P. Bierkens, Marijn J. M. Kuijper, Marta Faneca Sanchez, et al. *Aqueduct 3.0: Updated Decision-Relevant Global Water Risk Indicators.* Technical Note. Washington, D.C.: World Resources Institute, 2019. https://www.wri.org/publication/aqueduct-30.

Hopkins, Caroline. "Covid-19 Patients Are Flooding New York Hospitals, and the Peak May Be 3 Weeks Away." *Vox*, March 27, 2020. https://www.vox.com/2020/3/27/21197400/new-york-covid-19-hospitals-coronavirus.

Horton, Susan, and Richard H. Steckel. "Malnutrition: Global Economic Losses Attributable to Malnutrition 1900–2000 and Projections to 2050." In *How Much Have Global Problems Cost the Earth? A Scorecard from 1900 to 2050*, ed. B. Lomborg, 247–72. Cambridge: Cambridge University Press, 2013.

Houthakker, Hendrik Samuel. "An International Comparison of Household Expenditure Patterns, Commemorating the Centenary of Engel's Law." *Econometrica* 25 (1957): 532–51.

Huang, Jikun, Martín Piñeiro, Valeria Piñeiro, Kym Anderson, Nelson Illescas, David Laborde Debucquet, and Laura Wellesley. *Global Food Security and Market Stability: The Role and Concerns of Large Net Food Importers and Exporters.* T20 Argentina: Food Security and Sustainable Development Task Force Brief. https://www.g20-insights.org/policy_briefs/global-food-security-and-market-stability-the-role-and-concerns-of-large-net-food-importers-and-exporters/.

Huang, Yiping. "Understanding China's Belt and Road Initiative: Motivation, Framework and Assessment." *China Economic Review* 40 (2016): 314–21.

Hunger Task Force. *Report to the Government of Ireland.* Limerick: Irish Aid, 2008. https://www.irishaid.ie/media/irishaid/allwebsitemedia/20newsandpublications/publicationpdfsenglish/hunger-task-force.pdf.

Hunt, Pascale. "'Great Reboot' or Short-Term Saviour? Bali's Seaweed Farming Revival." *The Fish Site*, April 30, 2021. https://thefishsite.com/articles/great-reboot-or-short-term-saviour-balis-seaweed-farming-revival.

Hunter, Mitchell C., Richard G. Smith, Meagan E. Schipanski, Lesley W. Atwood, and David A. Mortensen. "Agriculture in 2050: Recalibrating Targets for Sustainable Intensification." *BioScience* 67, no. 4 (2017): 386–91.

ICT Update 88 (September 2018). https://ictupdate.cta.int/en/issues/88-blockchain.

Ingram, John, Raquel Ajates, Alex Arnall, Lauren Blake, Rosina Borrelli, Rosemary Collier, Annabel de Frece, et al. "A Future Workforce of Food-System Analysts." *Nature Food* 1, no. 1 (2020): 9–10.

Innovation Development Centre, Republic of Tajikistan. *Country Strategic Review: Food Security and Nutrition.* Dushanbe, Tajikistan: Innovation Development Centre, 2018. https://www.wfp.org/publications/tajikistan-zero-hunger-strategic-review-2018.

InterAcademy Council. *Realizing the Promise and Potential of African Agriculture: Science and technology Strategies for Improving Agricultural Productivity and Food Security in Africa.* N.p.: InterAcademy Council, 2004.

Intergovernmental Panel on Climate Change. "Annex I: Glossary," ed. J. B. R. Matthews. In *Global Warming of 1.5°C. An IPCC Special Report on the Impacts of Global Warming of 1.5°C Above Pre-Industrial Levels ond Related Global Greenhouse Gas Emission Pathways, in the Context of Strengthening the Global Response to the Threat of Climate Change, Sustainable Development, and Efforts to Eradicate Poverty*, ed. V. Masson-Delmotte, P. Zhai, H.-O. Pörtner, D. Roberts, J. Skea, P. R. Shukla, A. Piran, et al. Geneva: IPCC, 2018. https://www.ipcc.ch/sr15/chapter/glossary/.

——. *Climate Change and Land: An IPCC Special Report on Climate Change, Desertification, Land Degradation, Sustainable Land Management, Food Security, and Greenhouse Gas Fluxes in Terrestrial Ecosystems*, ed. P. R. Shukla, J. Skea, E. Calvo Buendia, V. Masson-Delmotte, H.-O. Pörtner, D. C. Roberts, P. Zhai, et al. Geneva: IPCC, 2019.

——. "Definition of Terms Used Within the DDC Pages, Glossary L–M." Data Distribution Centre. Accessed August 23, 2021. https://www.ipcc-data.org/guidelines/pages/glossary/glossary_lm.html.

——. "Summary for Policymakers." In *Climate Change 2021: The Physical Science Basis. Contribution of Working Group I to the Sixth Assessment Report of the Intergovernmental Panel on Climate Change*, ed. V. Masson-Delmotte, V., P. Zhai, A. Pirani, S. L. Connors, C. Péan, S. Berger, N. Caud, Y. Chen, L. Goldfarb, M. I. Gomis, M. Huang, K. Leitzell, E. Lonnoy, J. B. R. Matthews, T. K. Maycock, T. Waterfield, O. Yelekçi, R. Yu, and B. Zhou. Cambridge: Cambridge University Press, in press.

International Commission on Education for Sustainable Development Practice. *Final Report October 2008*. New York: Earth Institute at Columbia University, 2008. https://www.macfound.org/media/article_pdfs/develcomm-execsumm.pdf.

International Dietary Data Expansion Project. "Data4Diets: Building Blocks for Diet-Related Food Security Analysis." Accessed August 21, 2021. https://inddex.nutrition.tufts.edu/data4diets.

——. "Food Insecurity Experience Scale (FIES)." Accessed August 24, 2021. https://inddex.nutrition.tufts.edu/data4diets/indicator/food-insecurity-experience-scale-fies?back=/data4diets/indicators.

International Fertilizer Development Center (IFDC). "Slow-Release Nitrogen Fertilizers Measure Up." June 3, 2019. https://ifdc.org/2019/06/03/slow-release-nitrogen-fertilizers-measure-up/.

International Finance Corporation. "How Warehouse Receipts Can Improve Lives." May 13, 2015. https://www.ifc.org/wps/wcm/connect/news_ext_content/ifc_external_corporate_site/news+and+events/news/how+warehouse+receipts+can+improve+lives

International Institute for Climate and Society. "Seasonal Climate Forecasts." https://iri.columbia.edu/our-expertise/climate/forecasts/#Seasonal_Climate_Forecasts.

International Institute of Tropical Agriculture. "Major Breakthrough for Farmers and Scientists as Nigerian Biotech Body Approves Commercial Release of Genetically Modified Cowpea." News release, February 9, 2019. https://www.iita.org/news-item/major-breakthrough-for-farmers-and-scientists-as-nigerian-biotech-body-approves-commercial-release-of-genetically-modified-cowpea/.

International Maize and Wheat Improvement Center (CIMMYT). "Water Efficient Maize for Africa (WEMA)." Accessed August 15, 2021. https://www.cimmyt.org/projects/water-efficient-maize-for-africa-wema/.

International Rice Research Institute. *Annual Report for 1975*. Manila: IRRI, 1976.

——. *Cambodia and IRRI*. Manila: IRRI, October 2020. http://books.irri.org/Cambodia_IRRI_brochure.pdf.

——. "Golden Rice." Accessed August 15, 2021. https://www.irri.org/golden-rice.

——. "Hybrid Rice." Accessed August 15, 2021. https://www.irri.org/hybrid-rice.

——. "International Rice Genebank." Accessed August 15, 2021. https://www.irri.org/international-rice-genebank.

——. *IR36, the World's Most Popular Rice*. Los Baños, Philippines: IRRI, n.d. http://books.irri.org/IR36.pdf.

——. "Philippines Approves Golden Rice for Direct Use as Food and Feed, or for Processing." December 18, 2019. https://www.irri.org/news-and-events/news/philippines-approves-golden-rice-direct-use-food-and-feed-or-processing.

——. "Philippines Becomes First Country to Approve Nutrient-Enriched 'Golden Rice' for Planting." July 23, 2021. https://www.irri.org/news-and-events/news/philippines-becomes-first-country-approve-nutrient-enriched-golden-rice.

——. "Redux: U.S. President Lyndon Johnson Visits IRRI, 26 October 1966." YouTube, February 24, 2011. https://www.youtube.com/watch?v=WEwdkz7XK2s.

International Service for the Acquisition of Agri-biotech Applications. "About ISAAA." Accessed August 15, 2021. https://www.isaaa.org/.

——. *Global Status of Commercialized Biotech/GM Crops: 2018*, ISAAA Brief No. 54. Ithaca, N.Y.: ISAAA, 2018.

International Union of Soil Sciences (IUSS) Working Group WRB. *World Reference Base for Soil Resources 2014, Update 2015 International Soil Classification System for Naming Soils and Creating Legends for Soil Maps*. World Soil Resources Reports no. 106. Rome: FAO, 2015. http://www.fao.org/3/i3794en/I3794en.pdf.

Internet Classics Archive. "*Critias* by Plato. Written 360 BCE. Translated by Benjamin Jowett." Accessed August 14, 2021. http://classics.mit.edu/Plato/critias.html.

Investing.com. "Orange Juice Futures." Accessed August 29, 2021. https://www.investing.com/commodities/orange-juice-historical-data.

Investopedia. "A Look at Coca-Cola's Advertising Expenses." Updated August 10, 2021. https://www.investopedia.com/articles/markets/081315/look-cocacolas-advertising-expenses.asp.

IPC Global Partners. *Integrated Food Security Phase Classification Technical Manual Version 3.0. Evidence and Standards for Better Food Security and Nutrition Decisions.* Rome: FAO, 2019. http://www.ipcinfo.org/fileadmin/user_upload/ipcinfo/manual/IPC_Technical_Manual_3_Final.pdf.

Jayagoda, Dimithri Devinda. "Community-Based Mangrove Forest Management in Association with Sustainable Tourism in Puerto Princesa City of the Philippines." *International Journal of Sustainable Future for Human Security* 3, no. 2 (2015): 23–30.

Jayne, Thomas S., and Pedro A. Sanchez. "Agricultural Productivity Must Improve in Sub-Saharan Africa." *Science* 372, no. 6546 (2021): 1045–47.

Jensen, Robert, T., and Nolan H. Miller. "Do Consumer Price Subsidies Really Improve Nutrition?" *Review of Economics and Statistics* 93, no. 4 (2011): 1205–23. doi.org/10.1162/REST_a_00118.

Jinek, Martin, Krzysztof Chylinski, Ines Fonfara, Michael Hauer, Jennifer A. Doudna, and Emmanuelle Charpentier. "A Programmable Dual-RNA–Guided DNA Endonuclease in Adaptive Bacterial Immunity." *Science* 337, no. 6096 (2012): 816–21.

Johnston, Bruce F., and John W. Mellor. "The Role of Agriculture in Economic Development." *American Economic Review* 51, no. 4 (1961): 566–93.

Ju, Xiao-Tang, Guang-Xi Xing, Xin-Ping Chen, Shao-Lin Zhang, Li-Juan Zhang, Xue-Jun Liu, Zhen-Ling Cui, et al. "Reducing Environmental Risk by Improving N Management in Intensive Chinese Agricultural Systems." *Proceedings of the National Academy of Sciences* 106, no. 9 (2009): 3041–46.

Kathage, Jonas, and Matin Qaim. "Economic Impacts and Impact Dynamics of Bt (Bacillus thuringiensis) Cotton in India." *Proceedings of the National Academy of Sciences* 109, no. 29 (2012): 11652–56.

Kaufman, Mark. "Grim Space Photo Shows the Horror of Demolishing Forests." *Mashable*, July 13, 2018. https://mashable.com/article/nasa-space-photo-madagascar/.

Kawabata, Mariko, Andrea Berardo, Paolo Mattei, and Saskia de Pee. "Food Security and Nutrition Challenges in Tajikistan: Opportunities for a Systems Approach." *Food Policy* 96 (2020): 101872.

Kelly, T., W. Yang, C. S Chen., K. Reynolds, and J. He. "Global Burden of Obesity in 2005 and Projections to 2030." *International Journal of Obesity* 32, no. 9 (September 2008): 1431–37.

Kennedy, Emma. "Can Kitchen Tech Reduce Excessive Food Waste?" *CNN Business*, July 11, 2018. https://money.cnn.com/2018/07/11/technology/startups/smarterware-fridgecam-food-waste/index.html.

Kennedy, G. L., M. R. Pedro, C. Seghieri, G. Nantel, and I. Brouwer. "Dietary Diversity Score Is a Useful Indicator of Micronutrient Intake in Non-Breast-Feeding Filipino Children." *Journal of Nutrition* 137, no. 2 (2007): 472–77.

Khush, G. S., D. W. Puckridge, and G. L. Denning. *Trip Report to People's Republic of Kampuchea, 23–30 January 1986.* Manila: IRRI, 1986.

Kiernan, Ben. *Cambodia's Twisted Path to Justice.* The History Place: Points of View, 1999. https://www.historyplace.com/pointsofview/kiernan.htm.

——. *The Pol Pot Regime: Race, Power, and Genocide in Cambodia Under the Khmer Rouge 1975–79.* New Haven, Conn.: Yale University Press, 2014.

Kitaoka, Kaori. "The National School Meal Program in Brazil: A Literature Review." *Japanese Journal of Nutrition and Dietetics* 76, suppl. (2018): S115–S125.

Kulish, Nicholas. "'Never Seen Anything Like It:' Cars Line Up for Miles at Food Banks." *New York Times*, April 8, 2020. https://www.nytimes.com/2020/04/08/business/economy/coronavirus-food-banks.html.

Kupferschmidt, Kai. "Activists Destroy 'Golden Rice' Field Trial." *Science*, August 9 (corrected August 13), 2013. https://www.sciencemag.org/news/2013/08/activists-destroy-golden-rice-field-trial.

Lal, Rattan. "Soil Erosion and Land Degradation: The Global Risks." In *Advances in Soil Science*, ed. R. Lal and B. A. Stewart, 129–72. New York: Springer, 1990.

Lando, Richard P., and Solieng Mak. *Rainfed Lowland Rice in Cambodia: A Baseline Survey.* International Rice Research Institute Research Paper Series, no. 152. Los Baños, Philippines: IRRI, 1994.

L'Aquila Food Security Initiative. "L'Aquila Joint Statement on Global Food Security." L'Aquila Summit, July 9, 2009. http://www.g8.utoronto.ca/summit/2009laquila/2009-food.html.

Laurance, William F., Gopalasamy Reuben Clements, Sean Sloan, Christine S. O'Connell, Nathan D. Mueller, Miriam Goosem, Oscar Venter, et al. "A Global Strategy for Road Building." *Nature* 513, no. 7517 (2014): 229–32.

Laurance, William F., Jeffrey Sayer, and Kenneth G. Cassman. "Agricultural Expansion and its Impacts on Tropical Nature." *Trends in Ecology and Evolution* 29, no. 2 (2014): 107–16.

Lebergott, S. "Labor Force and Employment, 1800–1960." National Bureau of Economic Research. https://www.nber.org/system/files/chapters/c1567/c1567.pdfNBER. 1966.

Lele, Uma, William A. Masters, Joyce Kinabo, J. V. Meenakshi, Bharat Ramaswami, Julia Tagwireyi, Winnie F. L. Bell, et al. *Measuring Food and Nutrition Security: An Independent Technical Assessment and User's Guide for Existing Indicators, Measuring Food and Nutrition Security Technical Working Group.* Rome: Food Security Information Network, 2016. https://sites.tufts.edu/willmasters/files/2016/06/FSIN-TWG_UsersGuide_12June2016.pdf.

Leroy, Jef L., Paola Gadsden, Teresa González de Cossío, and Paul Gertler. "Cash and In-Kind Transfers Lead to Excess Weight Gain in a Population of Women with a High Prevalence of Overweight in Rural Mexico." *Journal of Nutrition*, 143, no. 3 (March 2013): 378–83. doi.org/10.3945/jn.112.167627.

Leroy, Jef L., Marie Ruel, Jean-Pierre Habicht, and Edward A. Frongillo. "Linear Growth Deficit Continues to Accumulate Beyond the First 1000 Days in Low- and Middle-Income Countries: Global Evidence from 51 National Surveys." *Journal of Nutrition* 144, no. 9 (2014): 1460–66.

Levy, Sarah, ed. *Starter Packs: A Strategy to Fight Hunger in Developing Countries? Lessons from the Malawi Experience 1998–2003.* Wallingford, UK: CABI, 2005.

Liebig, Justus. *Die Organische Chemie in Ihrer Anwendung auf Agricultur und Physiologie.* Braunschweig, Germany: F. Vieweg und Sohn, 1840.

Lipper, Leslie, Philip Thornton, Bruce M. Campbell, Tobias Baedeker, Ademola Braimoh, Martin Bwalya, Patrick Caron, et al. "Climate-Smart Agriculture for Food Security." *Nature Climate Change* 4, no. 12 (2014): 1068–72.

Loc, Ho Huu, Mindy Low Lixian, Edward Park, Tran Duc Dung, Sangam Shrestha, and Yong-Jin Yoon. "How the Saline Water Intrusion Has Reshaped the Agricultural Landscape of the Vietnamese Mekong Delta, a Review." *Science of the Total Environment* 794 (2021): 148651.

Locker, Melissa. "Tom Hanks and Rita Wilson Had Very Different Reactions to COVID-19." *Time*, July 6, 2020. https://time.com/5863326/tom-hanks-coronavirus-reaction/.

Luo, Tianyi, Robert Young, and Paul Reig. *Aqueduct Projected Water Stress Country Rankings*, Technical Note. Washington, D.C.: World Resources Institute, 2015. www.wri.org/publication/aqueduct-projected-water-stress-country-rankings.

Lyall, Sarah. "Past as Prologue: Blair Faults Britain in Irish Potato Blight." *New York Times*, June 3, 1997. https://www.nytimes.com/1997/06/03/world/past-as-prologue-blair-faults-britain-in-irish-potato-blight.html.

MacDonald, A. M., H. C. Bonsor, K. M. Ahmed, W. G. Burgess, M. Basharat, R. C. Calow, A. Dixit, S. S. D. Foster, et al. "Groundwater Quality and Depletion in the Indo-Gangetic Basin Mapped from In Situ Observations." *Nature Geoscience* 9, no. 10 (2016): 762–66.

MacDonald, A. M., H. C. Bonsor, R. Taylor, M. Shamsudduha, W. G. Burgess, K. M. Ahmed, A. Mukherjee, et al. "Groundwater Resources in the Indo-Gangetic Basin: Resilience to Climate Change and Abstraction,' data made live September 29, 2015. Nottingham, UK: British Geological Survey (OR/15/047), unpublished. http://nora.nerc.ac.uk/id/eprint/511898.

Mackill, David J., and Gurdev S. Khush. "IR64: A High-Quality and High-Yielding Mega Variety." *Rice* 11, no. 1 (2018): 18. doi.org/10.1186/s12284-018-0208-3.

Mahon, N., I. Crute, M. Di Bonito, E. A. Simmons, and Md Mofakkarul Islam. "Towards a Broad-Based and Holistic Framework of Sustainable Intensification Indicators." *Land Use Policy* 77 (2018): 576–97.

Malabo Montpellier Panel. *Water-Wise: Smart Irrigation Strategies for Africa*. Dakar, Senegal: International Food Policy Research Institute (IFPRI) and Malabo Montpellier Panel, 2018.

Malthus, Thomas Robert. *An Essay on the Principle of Population as It Affects the Future Improvement of Society, with Remarks on the Speculations of Mr Godwin, M. Condorcet, and Other Writers*. London: J. Johnson, 1798.

Mann, Charles C. "The Book That Incited a Worldwide Fear of Overpopulation." *Smithsonian Magazine*, January/February 2018. https://www.smithsonianmag.com/innovation/book-incited-worldwide-fear-overpopulation-180967499/.

——. "Our Good Earth: The Future Rests on the Soil Beneath Our Feet. Can We Save It?" *National Geographic* 214, no. 3 (2008): 80–106.

Mann, Jim, and A. Stewart Truswell, eds. *Essentials of Human Nutrition*. 5th ed. Oxford: Oxford University Press, 2017.

Masina, Lameck. "WHO, US Name Malawi a High Risk COVID-19 Country as Cases Spike." *Voice of America*, July 10, 2021. https://www.voanews.com/covid-19-pandemic/who-us-name-malawi-high-risk-covid-19-country-cases-spike.

Maxwell, Daniel. "Famine Early Warning and Information Systems in Conflict Settings: Challenges for Humanitarian Metrics and Response." November 2019. LSE and UKaid. http://eprints.lse.ac.uk/102836/1/Maxwell_famine_early_warning_and_information_systems_published.pdf.

Mbow, C., C. Rosenzweig, L.G. Barioni, T.G. Benton, M. Herrero, M. Krishnapillai, E. Liwenga, et al. "Food Security." In *Climate Change and Land: An IPCC Special Report on Climate Change, Desertification, Land Degradation, Sustainable Land Management, Food Security, and Greenhouse Gas Fluxes in Terrestrial Ecosystems*, ed. International Panel on Climate Change [P. R. Shukla, J. Skea, E. Calvo Buendia, V. Masson-Delmotte, H.-O. Pörtner, D.C. Roberts, P. Zhai, et al.]. Geneva: IPCC, 2019.

McCulloch, J. S. G. "Sir Herbert Charles Pereira. 12 May 1913–19 December 2004." *Biographical Memoirs of Fellows of the Royal Society* 54 (2008): 245–56. doi.org/10.1098/rsbm.2008.0001.

McDonald, Christine M., Ibironke Olofin, Seth Flaxman, Wafaie W Fawzi, Donna Spiegelman, Laura E Caulfield, Robert E Black, et al. for the Nutrition Impact Model Study. "The Effect of Multiple Anthropometric Deficits on Child Mortality: Meta-Analysis of Individual Data in 10 Prospective Studies from Developing Countries." *American Journal of Clinical Nutrition* 97, no. 4 (2013): 896–901. doi.org/10.3945/ajcn.112.047639.

McMichael, Anthony. *Climate Change and the Health of Nations: Famines, Fevers, and the Fate of Populations*. Oxford: Oxford University Press, 2017.

Meadows, Donella. "What Would the World Be If There Were No Hunger?" Donella Meadows Archives, December 17, 1987. https://donellameadows.org/archives/what-would-the-world-be-if-there-were-no-hunger/.

Meadows, Donella H., Dennis L. Meadows, Jorgen Randers, and William W. Behrens. *The Limits to Growth*. New York: Universe Books, 1972.

Meehl, Gerald A., Warren M. Washington, William D. Collins, Julie M. Arblaster, Aixue Hu, Lawrence E. Buja, Warren G. Strand, and Haiyan Teng. "How Much More Global Warming and Sea Level Rise?" *Science* 307, no. 5716 (2005): 1769–72.

Mertz, Ole, Christine Padoch, Jefferson Fox, R. A. Cramb, Stephen J. Leisz, Nguyen Thanh Lam, and Tran Duc Vien. "Swidden Change in Southeast Asia: Understanding Causes and Consequences." *Human Ecology* 37, no. 3 (2009): 259–64.

Meybeck, Alexandre, and Vincent Gitz. *"Climate-Smart" Agriculture: Policies, Practices and Financing for Food Security, Adaptation and Mitigation*. Rome: FAO, 2010.

Micha, Renata, Venkatesh Mannar, Ashkan Afshin, Lorena Allemandi, Phillip Baker, Jane Battersby, Zulfiqar Bhutta, et al. *2020 Global Nutrition Report: Action on Equity to End Malnutrition*. Bristol, UK: Development Initiatives, 2020. https://globalnutritionreport.org/reports/2020-global-nutrition-report/.

Michigan State University. "RUSLE On-Line Soil Erosion Assessment Tool." Accessed August 14, 2021. http://www.iwr.msu.edu/rusle/.

Micklin, Philip. "The Aral Sea Disaster." *Annual Review of Earth and Planetary Sciences* 35 (2007): 47–72.

Miko, Ilona. "Gregor Mendel and the Principles of Inheritance." *Nature Education* 1, no. 1 (2008): 134.

Milhorance, Carolina, Marcel Bursztyn, and Eric Sabourin. "The Politics of the Internationalisation of Brazil's 'Zero Hunger' Instruments." *Food Security* 11, no. 2 (2019): 447–60.

Millennium Development Goals Centre. *Africa's Green Revolution: A Call to Action. Proceedings of the July 5, 2004 High-Level Seminar, Addis Ababa, Ethiopia.* Nairobi: MDG Centre, 2004.

Ministry of Health, Malawi. "COVID-19 National Information Dashboard." Accessed October 27, 2020. https://covid19.health.gov.mw/.

Minten, Bart, and Christopher B. Barrett. "Agricultural Technology, Productivity, and Poverty in Madagascar." *World Development* 36, no. 5 (2008): 797–822.

Mitchell, Shira, Andrew Gelman, Rebecca Ross, Joyce Chen, Sehrish Bari, Uyen Kim Huynh, Matthew W. Harris, et al. "The Millennium Villages Project: A Retrospective, Observational, Endline Evaluation." *Lancet Global Health* 6, no. 5 (2018): e500–e513. doi.org/10.1016/s2214-109x(18)30065-2.

Molden, David, ed. *Water for Food Water for Life: A Comprehensive Assessment of Water Management in Agriculture.* London: Routledge, 2007.

Montgomery, David R. "Soil Erosion and Agricultural Sustainability." *Proceedings of the National Academy of Sciences* 104, no. 33 (2007): 13268–72.

Morris, M., V. A. Kelly, R. J. Kopicki, and D. Byerlee. *Fertilizer Use in African Agriculture: Lessons Learned and Good Practice Guidelines.* Washington, D.C.: World Bank, 2007.

Mosher, Arthur T. *Getting Agriculture Moving: Essentials for Development and Modernization.* New York: Frederick A. Praeger, 1966.

Mozaffarian, Dariush, Sonia Y. Angell, Tim Lang, and Juan A. Rivera. "Role of Government Policy in Nutrition—Barriers to and Opportunities for Healthier Eating." *BMJ* 361 (2018).

Murdock, Larry L., Venu Margam, Ibrahim Baoua, Susan Balfe, and Richard E. Shade. "Death by Desiccation: Effects of Hermetic Storage on Cowpea Bruchids." *Journal of Stored Products Research* 49 (2012): 166–70.

Mutharika, Bingu wa. "Capitalizing on Opportunity." World Economic Forum on Africa, 4 June 2008, Cape Town, South Africa. http://www.youtube.com/watch?v=rmUjw1J35V4.

Myers, N., R. A. Mittermeier, C.G. Mittermeier, G. A. B. De Fonseca, and J. Kent. "Biodiversity Hotspots for Conservation Priorities." *Nature* 403 (2000): 853–58.

Naja, Farah, and Rena Hamadeh. "Nutrition Amid the COVID-19 Pandemic: A Multi-Level Framework for Action." *European Journal of Clinical Nutrition* 74 (2020): 1117–21. doi.org/10.1038/s41430-020-0634-3.

Nash, J. Madeleine. "This Rice Could Save a Million Kids a Year." *Time*, July 31, 2000. http://content.time.com/time/magazine/article/0,9171,997586,00.html.

National Human Genome Research Institute. "How Does Genome Editing Work?" Accessed August 15, 2021. https://www.genome.gov/about-genomics/policy-issues/Genome-Editing/How-genome-editing-works.

National Statistics Directorate, Timor-Leste, Ministry of Finance, Timor-Leste, and ICF Macro. *Timor-Leste Demographic and Health Survey 2009–10.* Dili, Timor-Leste: NSD, Timor-Leste, and ICF Macro, 2010. http://dhsprogram.com/pubs/pdf/FR235/FR235.pdf.

Nature. "Does the Fight against Hunger Need Its Own IPCC?" *Nature* 595, 332 (2021). doi.org/10.1038/d41586-021-01904-0.

Nature. "Super-Tomato Shows What Plant Scientists Can Do." *Nature* 562, 8 (2018).

NCD Risk Factor Collaboration (NCD-RisC). "Rising Rural Body-Mass Index Is the Main Driver of the Global Obesity Epidemic in Adults." *Nature* 569 (2019): 260–64. doi.org/10.1038/s41586-019-1171-x.

Nesbitt, H. J. "Harnessing People Power for Technology Uptake." In *Food for the Future: Opportunities for a Crowded Planet*, ed. A. G. Brown, 55–62. Fyshwick, Australia: Crawford Fund, 2002.

——, ed. *Rice Production in Cambodia.* Manila: IRRI, 1997.

New South Wales Department of Primary Industries. *Determining Soil Texture Using the Ribboning Technique*, Primefact 1363. NSW, Australia: DPI, 2014. https://www.dpi.nsw.gov.au/__data/assets/pdf_file/0005/164615/determining_soil_texture_using_-ribboning_technique.pdf.

Newton, Isaac to Robert Hooke. February 5, 1675. Simon Gratz collection, Digital Library. https://digitallibrary.hsp.org/index.php/Detail/objects/9792.

Nielsen, R. L. "Historical Corn Grain Yields in the U.S." Updated August 2021. https://www.agry.purdue.edu/ext/corn/news/timeless/yieldtrends.html.

New York Times. "Tracking Coronavirus in New York: Latest Map and Case Count." Accessed October 27, 2020. https://www.nytimes.com/interactive/2021/us/new-york-covid-cases.html.

——. "Yuan Longping, Plant Scientist Who Helped Curb Famine, Dies at 90." May 23, 2021. https://www.nytimes.com/2021/05/23/world/asia/yuan-longping-dead.html.

NIH National Heart, Lung, and Blood Institute. "Assessing Your Weight and Health Risk." Accessed August 24, 2021. https://www.nhlbi.nih.gov/health/educational/lose_wt/risk.htm.

Njuki, J. "Women are the nucleus of our global food system—but their value is undermined by the patriarchy." *Independent,* July 12, 2021. https://www.independent.co.uk/voices/food-systems-sustainable-poverty-hunger-patriarchal-system-b1879840.html.

NOAA National Ocean Service. "What Are El Niño and La Niña?" Accessed August 23, 2021. https://oceanservice.noaa.gov/facts/ninonina.html.

Nobel Prize. "The Nobel Prize in Chemistry 2020." News release, October 7, 2020. https://www.nobelprize.org/prizes/chemistry/2020/press-release/.

Noble, I. R., S. Huq, Y. A. Anokhin, J. Carmin, D. Goudou, F. P. Lansigan, B. Osman-Elasha, and A. Villamizar. "Adaptation Needs and Options." In *Climate Change 2014: Impacts, Adaptation, and Vulnerability. Part A: Global and Sectoral Aspects. Contribution of Working Group II to the Fifth Assessment Report of the Intergovernmental Panel on Climate Change,* ed. C. B. Field, V. R. Barros, D. J. Dokken, K. J. Mach, M. D. Mastrandrea, T. E. Bilir, M. Chatterjee, et al. Cambridge: Cambridge University Press, 2014.

Nord, Roger, Yuri V. Sobolev, David G. Dunn, Alejandro Hajdenberg, Niko A. Hobdari, Samar Maziad, and Stéphane Roudet. *Tanzania: The Story of an African Transition.* No. 09/02. Washington, D.C.: International Monetary Fund, 2009.

Normile, Dennis. "Crossing Rice Strains to Keep Asia's Rice Bowls Brimming." *Science* 283, no. 5400 (1999): 313.

NPR. "Green Revolutionary Norman Borlaug Dies." September 13, 2009. https://www.npr.org/templates/story/story.php?storyId=112791886.

Nye, P. H., and D. J. Greenland. "The Soil Under Shifting Cultivation." *Soil Science* 92, no. 5 (1961): 354.

Nyerere, Julius K. "Ujamaa: The Basis of African Socialism." *Journal of Pan African Studies* 1, no. 1 (1987): 4–11.

Nziguheba, G., C. A. Palm, T. Berhe, G. Denning, A. Dicko, O. Diouf, W. Diru, et al. "The African Green Revolution: Results from the Millennium Villages Project." *Advances in Agronomy* 109 (2010): 75–115.

Office for the Coordination of Humanitarian Affairs Services. "Malawi 2005." Accessed August 11, 2021. https://fts.unocha.org/appeals/186/flows.

——. "UN Humanitarian Chief: After COVID-19, It's in Everyone's Interest to Help the World's Poorest Countries." May 4, 2020. https://www.unocha.org/story/un-humanitarian-chief-after-covid-19-it%E2%80%99s-everyone%E2%80%99s-interest-help-worlds-poorest-countries.

Oikeh, Sylvester, Dianah Ngonyamo-Majee, Stephen N. Mugo, Kingstone Mashingaidze, Vanessa Cook, and Michael Stephens. "The Water Efficient Maize for Africa Project as an Example of a Public–Private Partnership." In *Convergence of Food Security, Energy Security and Sustainable Agriculture,* ed. David D. Songstad, Jerry L. Hatfield, and Dwight T. Tomes, 317–29. Berlin: Springer, 2014.

Oliveira, Victor, Laura Tiehen, Mark Prell, and David Smallwood. "Evolution and Implementation of the Supplemental Nutrition Assistance Program in the United States." In *The 1.5 Billion People Question,* ed. Harold Alderman, Ugo Gentilini, and Ruslan Yemtsov, 209–63. Washington, D.C.: World Bank, 2018.

Osborn, Fairfield. *Our Plundered Planet.* Boston: Little, Brown, 1948.

Otsuka, Keijiro, and Rie Muraoka. "Green Revolution for Sub-Saharan Africa: Past Failures and Future Prospects." *Journal of African Economies* 26, suppl. 1 (August 2017): i73–i98.

Oweis, Theib, Ahmed Hachum, and Jacob Kijne. *Water Harvesting and Supplemental Irrigation for Improved Water Use Efficiency in Dry Areas.* SWIM Paper 7. Colombo, Sri Lanka: International Water Management Institute, 1999.

Owens, B. "Golden Rice Is Safe to Eat, Says FDA." *Nature Biotechnology* 36 (2018): 559–60. doi.org/10.1038/nbt0718-559a.

Paddock, William, and Paul Paddock. *Famine—1975! America's Decision: Who Will Survive?* Boston: Little, Brown, 1967.

Paine, Jacqueline A., Catherine A. Shipton, Sunandha Chaggar, Rhian M. Howells, Mike J. Kennedy, Gareth Vernon, Susan Y. Wright, et al. "Improving the Nutritional Value of Golden Rice Through Increased Pro-Vitamin A Content." *Nature Biotechnology* 23, no. 4 (2005): 482.

Palm, C. A., H. K. Murwira, and S. E. Carter. "Organic matter management: From science to practice." In *Soil Fertility Research for Maize-based Farming Systems in Malawi and Zimbabwe*, edited by S. R. Waddington, H. K. Murwira, J. D. T. Kumwenda, D. Hikwa, and F. Tagwira, 21–27. Harare: CIMMYT, 1998.

Palm, Cheryl, Stephen A. Vosti, Pedro A. Sanchez, and Polly J. Ericksen, eds. *Slash-and-Burn Agriculture: The Search for Alternatives.* New York: Columbia University Press, 2005.

Parfitt, Julian, Mark Barthel, and Sarah Macnaughton. "Food Waste Within Food Supply Chains: Quantification and Potential for Change to 2050." *Philosophical Transactions of the Royal Society B: Biological Sciences* 365, no. 1554 (2010): 3065–81.

Peng, S., K. T. Ingram, H. U. Neue, and L. H. Ziska, eds. *Climate Change and Rice.* Berlin: Springer-Verlag and Manila: IRRI, 1995.

Perez, Marvin G., Michael Hirtzer, and Megan Durisin. "Avocados Are In, Pork Bellies Out in Era of Pandemic Eating." *Bloomberg*, April 26, 2020. https://www.bloomberg.com/news/articles/2020-04-26/avocados-are-in-pork-bellies-out-in-the-era-of-pandemic-eating.

Phiri, N. A., and G. Otieno. *Managing Pests of Stored Maize in Kenya, Malawi and Tanzania.* Nairobi, Kenya: CABI Africa, 2008.

Pingali, Prabhu L. "Green Revolution: Impacts, Limits, and the Path Ahead." *Proceedings of the National Academy of Sciences* 109, no. 31 (2012): 12302–8.

Pinstrup-Andersen, Per, ed. *Food Subsidies in Developing Countries: Costs, Benefits, and Policy Options.* Baltimore: Johns Hopkins University Press, 1988.

Ploetz, Randy C. "Panama Disease: A Classic and Destructive Disease of Banana." *Plant Health Progress* 1, no. 1 (2000): 10. doi.org/10.1094/PHP-2000-1204-01-HM.

Pollan, Michael. "The Way We Live Now: The Great Yellow Hype." *New York Times*, March 4, 2001. https://michaelpollan.com/articles-archive/the-way-we-live-now-the-great-yellow-hype/.

Popkin, Barry M. "Global Nutrition Dynamics: The World Is Shifting Rapidly Toward a Diet Linked with Noncommunicable Diseases." *American Journal of Clinical Nutrition* 84, no. 2 (2006): 289–98.

——. "The Nutrition Transition in Low-Income Countries: An Emerging Crisis." *Nutrition Reviews* 52, no. 9 (1994): 285–98.

Popkin, Barry M., Linda S. Adair, and Shu Wen Ng. "Global Nutrition Transition and the Pandemic of Obesity in Developing Countries." *Nutrition Reviews* 70, no. 1 (2012): 3–21. doi.org/10.1111/j.1753-4887.2011.00456.x.

Popkin, Barry M., Shufa Du, William D. Green, Melinda A. Beck, Taghred Algaith, Christopher H. Herbst, Reem F. Alsukait, et al. "Individuals with Obesity and COVID-19: A Global Perspective on the Epidemiology and Biological Relationships." *Obesity Reviews* 21, no. 11 (2020): e13128.

Porter-Bolland, Luciana, Edward A. Ellis, Manuel R. Guariguata, Isabel Ruiz-Mallén, Simoneta Negrete-Yankelevich, and Victoria Reyes-García. "Community Managed Forests and Forest Protected Areas: An Assessment of Their Conservation Effectiveness Across the Tropics." *Forest Ecology and Management* 268 (2012): 6–17.

Presley, DeAnn, and John Tatarko. *Principles of Wind Erosion and its Control.* Manhattan: Kansas State University, September 2009. https://www.ars.usda.gov/ARSUserFiles/30200525/419PrinciplesofWinderosionanditscontrol.pdf.

Pretty, Jules N. "The Sustainable Intensification of Agriculture." *Natural Resources Forum* 21, no. 4 (1997): 247–56.

Pretty, Jules N., and Zareen Pervez Bharucha. "Sustainable Intensification in Agricultural Systems." *Annals of Botany* 114 (2014): 1571–96.

Puckridge, Don. *The Burning of the Rice: A Cambodian Success Story*. Hartwell, Australia: Temple House, 2004.

Purugganan, Michael D., and Dorian Q. Fuller. "The Nature of Selection During Plant Domestication." *Nature* 457, no. 7231 (2009): 843–48.

Quinn, Kenneth M. "Extended Biography." World Food Prize Foundation. Accessed August 29, 2021. https://www.worldfoodprize.org/en/dr_norman_e_borlaug/about_norman_borlaug/.

——. "Roads and Rice: How Innovation and Infrastructure Can Feed the World." *Guardian, Poverty Matters* blog, June 21, 2011. https://www.theguardian.com/global-development/poverty-matters/2011/jun/21/roads-rice-to-feed-world.

Ramalingam, Ben, David Nabarro, Arkebe Oqubuy, D. Carnall, and Leni Wild. "Principles to Guide Adaptive Leadership." *Harvard Business Review*, September 11, 2020. https://hbr.org/2020/09/5-principles-to-guide-adaptive-leadership.

Ramankutty, Navin, Amato T. Evan, Chad Monfreda, and Jonathan A. Foley. "Farming the Planet: 1. Geographic Distribution of Global Agricultural Lands in the Year 2000." *Global Biogeochemical Cycles* 22, no. 1 (2008): GB1003.

Randrianarijaona, P. "The Erosion of Madagascar." *Ambio* 12 (1983): 308–11.

Rao, Nitya. *M. S. Swaminathan in Conversation with Nitya Rao*. Chennai, India: M. S. Swaminathan Research Foundation, 2013.

Reardon, Thomas, and Julio A. Berdegué. "The Rapid Rise of Supermarkets in Latin America: Challenges and Opportunities for Development." *Development Policy Review* 20, no. 4 (2002): 371–88.

Reardon, Thomas, C. Peter Timmer, Christopher B. Barrett, and Julio Berdegué. "The Rise of Supermarkets in Africa, Asia, and Latin America." *American Journal of Agricultural Economics* 85, no. 5 (2003): 1140–46.

Reardon, Thomas, C. Peter Timmer, and Bart Minten. "Supermarket Revolution in Asia and Emerging Development Strategies to Include Small Farmers." *Proceedings of the National Academy of Sciences* 109, no. 31 (2012): 12332–37.

Reardon, Thomas, David Tschirley, Lenis Saweda O. Liverpool-Tasie, Titus Awokuse, Jessica Fanzo, Bart Minten, Rob Vos, et al. "The Processed Food Revolution in African Food Systems and the Double Burden of Malnutrition." *Global Food Security* 28 (2021): 100466.

Reeves, Timothy. "Sustainable Intensification of Agriculture—for Food and Nutritional Security." Farrer Memorial Oration, August 2019. https://www.dpi.nsw.gov.au/__data/assets/pdf_file/0008/1147967/Farrer-memorial-oration-2019-reeves.pdf.

Republic of Tajikistan. *National Development Strategy of the Republic of Tajikistan for the Period up to 2030*. Dushanbe: Republic of Tajikistan, 2016.

Resilience of the UK Food System in a Global Context. "What Is Food System Resilience?" Accessed August 23, 2021. https://www.foodsystemresilienceuk.org/what-is-food-system-resilience/.

Rice Today. "Coping with Climate Change." *Rice Today* 6, no. 3, 10–15.

Richardson, David P., and Julie A. Lovegrove. "Nutritional Status of Micronutrients as a Possible and Modifiable Risk Factor for COVID-19: A UK Perspective." *British Journal of Nutrition* 125, no. 6 (2021): 678–84.

Robins, Lisa, Steven Crimp, Monica van Wensveen, Robyn G. Alders, R. Michael Bourke, James Butler, Michaela Cosijn, et al. *COVID-19 and Food Systems in the Indo-Pacific: An Assessment of Vulnerabilities, Impacts and Opportunities for Action*. Australian Centre for International Agricultural Research Technical Report 96. Canberra: ACIAR, 2020.

Rockefeller Foundation. "A Digital History: Mexico." Accessed August 10, 2021. https://rockfound.rockarch.org/mexico.

Rockström, Johan, and Malin Falkenmark. "Agriculture: Increase Water Harvesting in Africa." *Nature* 519, no. 7543 (2015): 283–85.

Rockström, Johan, Mats Lannerstad, and Malin Falkenmark. "Assessing the Water Challenge of a New Green Revolution in Developing Countries." *Proceedings of the National Academy of Sciences* 104, no. 15 (2007): 6253–60.

Rockström, Johan, Will Steffen, Kevin Noone, Åsa Persson, F. Stuart Chapin III, Eric Lambin, Timothy M. Lenton, et al. "Planetary Boundaries: Exploring the Safe Operating Space for Humanity." *Ecology and Society* 14, no. 2 (2009): 32.

Rockström, Johan, John Williams, Gretchen Daily, Andrew Noble, Nathanial Matthews, Line Gordon, Hanna Wetterstrand, et al. "Sustainable Intensification of Agriculture for Human Prosperity and Global Sustainability." *Ambio* 46, no. 1 (2017): 4–17.

Roe, Brian E., Kathryn Bender, and Danyi Qi. "The Impact of COVID-19 on Consumer Food Waste." *Applied Economic Perspectives and Policy* 43, no. 1, (2020): 401–11.

Roe, Stephanie, Charlotte Streck, Michael Obersteiner, Stefan Frank, Bronson Griscom, Laurent Drouet, Oliver Fricko, et al. "Contribution of the Land Sector to a 1.5°C World." *Nature Climate Change* 9, no. 11 (2019): 817–28.

Ronald, Pamela. "Moving Beyond Pro/Con Debates Over Genetically Engineered Crops." *The Conversation*, June 2, 2016. https://theconversation.com/moving-beyond-pro-con-debates-over-genetically-engineered-crops-59564.

Roosevelt, Franklin D., *Letter to all State Governors on a Uniform Soil Conservation Law*, February 26, 1937. Online by Gerhard Peters and John T. Woolley, American Presidency Project. https://www.presidency.ucsb.edu/node/209378.

Roser, Max, Hannah Ritchie, and Esteban Ortiz-Ospina. "World Population Growth." 2013, most recently revised May 2019. https://ourworldindata.org/world-population-growth.

Rothamsted Research. "The History of Rothamsted Research." Accessed August 14, 2021. https://www.rothamsted.ac.uk/history-and-heritage.

Ruel, M. T. "Operationalizing Dietary Diversity: A Review of Measurement Issues and Research Priorities." *Journal of Nutrition*, 133, no. 11 Suppl 2 (2003): 3911S–3926S.

Ruel, M. T., and C. E. Levin. "Discussion Paper 92. Assessing the Potential for Food-Based Strategies to Reduce Vitamin A and Iron Deficiencies: A Review of Recent Evidence." *Food and Nutrition Bulletin* 22 no. 1 (2001): 94–95.

Rusike, Joseph, and Carl K. Eicher. "Institutional Innovations in the Maize Seed Industry." In *Africa's Emerging Maize Revolution*, ed. Derek Byerlee and Carl K. Eicher, 173–91. London: Lynne Rienner, 1977.

Russell, E. John. "Rothamsted and Its Experiment Station." *Agricultural History* 16, no. 4 (1942): 161–83.

Russell, R. Ross, ed. *Cambodia: A Country Study*. Washington, D.C.: GPO for the Library of Congress, 1987. http://countrystudies.us/cambodia/40.htm.

Ruttan, Vernon W. "Integrated Rural Development Programmes: A Historical Perspective." *World Development* 12, no. 4 (1984): 393–401.

Sachs, Jeffrey D., Guido Schmidt-Traub, Mariana Mazzucato, Dirk Messner, Nebojsa Nakicenovic, and Johan Rockström. "Six Transformations to Achieve the Sustainable Development Goals." *Nature Sustainability* 2, no. 9 (2019): 805–14.

Salem, B. Ben. "Prevention and Control of Wind Erosion in Arid Regions." *Unasylva No. 164 Watershed Management* 42, no. 1 (1991). http://www.fao.org/3/u1510e/u1510e07.htm.

Sanchez, Pedro A. "En Route to Plentiful Food Production in Africa." *Nature Plants* 1 (2015): 14014.

——. *Properties and Management of Soils in the Tropics*. 2nd ed. Cambridge: Cambridge University Press, 2019.

——. "Science in agroforestry." *Agroforestry Systems* 30, no. 1–2 (1995): 5–55.

——. "Soil Fertility and Hunger in Africa." *Science* 295, no. 5562 (2002): 2019–20.

——. "Tripling Crop Yields in Tropical Africa." *Nature Geoscience* 3, no. 5 (2010): 299–300.

Sanchez, Pedro A., Glenn L. Denning, and Generose Nziguheba. "The African Green Revolution Moves Forward." *Food Security* 1, no. 1 (2009): 37–44.

Sanchez, Pedro, Cheryl Palm, Jeffrey Sachs, Glenn Denning, Rafael Flor, Rebbie Harawa, Bashir Jama, et al. "The African Millennium Villages." *Proceedings of the National Academy of Sciences* 104, no. 43 (2007): 16775–80.

Sánchez, Pedro A., and José G. Salinas. "Low-Input Technology for Managing Oxisols and Ultisols in Tropical America." *Advances in Agronomy* 34 (1981): 279–406.

Sanchez, Pedro A., and M. S. Swaminathan. "Cutting World Hunger in Half." *Science* 307, no. 5708 (2005): 357–59.

Sánchez-Bayo, Francisco, Luis de Almeida, Robert Williams, Graeme Wright, Ivan R. Kennedy, and Angus Crossan. "The Aflatoxin QuicktestTM—a Practical Tool for Ensuring Safety in Agricultural Produce." In *Poisoning—from Specific Toxic Agents to Novel Rapid and Simplified Techniques for Analysis*, ed. Ntambwe Malangu, 193–208. Rijeka, Croatia: Intech, 2017.

Sasakawa Africa Association. *Take It to the Farmer: The Sasakawa Experience in Africa*. Tokyo: Sasakawa Africa Association, 2015.

Schanes, Karin, Karin Dobernig, and Burcu Gözet. "Food Waste Matters—a Systematic Review of Household Food Waste Practices and Their Policy Implications." *Journal of Cleaner Production* 182 (2018): 978–91.

Schleussner, Carl-Friedrich, Delphine Deryng, Christoph Müller, Joshua Elliott, Fahad Saeed, Christian Folberth, Wenfeng Liu, et al. "Crop Productivity Changes in 1.5°C and 2°C Worlds Under Climate Sensitivity Uncertainty." *Environmental Research Letters* 13, no. 6 (2018): 064007.

Schleussner, Carl-Friedrich, Tabea K. Lissner, Erich M. Fischer, Jan Wohland, Mahé Perrette, Antonius Golly, Joeri Rogelj, et al. "Differential Climate Impacts for Policy-Relevant Limits to Global Warming: The Case of 1.5°C and 2°C." *Earth System Dynamics* 7, no. 2 (2016): 327–51.

School Nutrition Association. "School Meal Trends & Stats." Accessed August 11, 2021. https://schoolnutrition.org/aboutschoolmeals/schoolmealtrendsstats/.

Schouten, H. J., F. A. Krens, and E. Jacobsen. "Cisgenic Plants Are Similar to Traditionally Bred Plants: International Regulations for Genetically Modified Organisms Should Be Altered to Exempt Cisgenesis." *EMBO Reports* 7, no. 8 (2016): 750–53. doi.org/10.1038/sj.embor.7400769.

Schroth, Götz, Celia A. Harvey, and Grégoire Vincent. "Complex Agroforests: Their Structure, Diversity, and Potential Role in Landscape Conservation." In *Agroforestry and Biodiversity Conservation in Tropical Landscapes*, ed. Götz Schroth, Gustavo A. B. da Fonseca, Celia A. Harvey, Claude Gascon, Heraldo L. Vasconcelos, and Anne-Marie N. Izac, 227–60. Washington, D.C.: Island Press, 2004.

Searchinger, Tim, Richard Waite, Craig Hanson, Janet Ranganathan, Patrice Dumas, and Emily Matthews. *Creating a Sustainable Food Future: A Menu of Solutions to Feed Nearly 10 Billion People by 2050*. Final Report, July 2019. Washington, D.C.: World Resources Institute, 2019. https://research.wri.org/sites/default/files/2019-07/WRR_Food_Full_Report_0.pdf.

Security Council Report. "Negotiations on Conflict and Food Security Draft Presidential Statement." What's in Blue, March 18, 2021. https://www.securitycouncilreport.org/whatsinblue/2021/03/negotiations-on-conflict-and-food-security-draft-presidential-statement.php.

Seufert, Verena, Navin Ramankutty, and Jonathan A. Foley. "Comparing the Yields of Organic and Conventional Agriculture." *Nature* 485, no. 7397 (2012): 229–32.

SharathKumar, Malleshaiah, Ep Heuvelink, and Leo F. M. Marcelis. "Vertical Farming: Moving from Genetic to Environmental Modification." *Trends in Plant Science* 25, no. 8 (2020): 724–27.

Shawcross, William. *Sideshow: Kissinger, Nixon and the Destruction of Cambodia*. New York: Simon and Schuster, 1979.

Sheahan, Megan, and Christopher B. Barrett. "Food Loss and Waste in Sub-Saharan Africa." *Food Policy* 70 (2017): 1–12.

Shekar, Meera, and Barry Popkin, eds. *Obesity: Health and Economic Consequences of an Impending Global Challenge*. Human Development Perspectives series. Washington, D.C.: World Bank, 2020. https://openknowledge.worldbank.org/handle/10986/32383.

Shelton, Anthony M., Sayed H. Sarwer, Md J. Hossain, Graham Brookes, and Vijay Paranjape. "Impact of Bt Brinjal Cultivation in the Market Value Chain in Five Districts of Bangladesh." *Frontiers in Bioengineering and Biotechnology* 8 (2020): 498.

Sicat, Gerardo P. *The Economic Legacy of Marcos*. Discussion Paper no. 2011–11. Manila: University of the Philippines School of Economics, November 2011. https://econ.upd.edu.ph/dp/index.php/dp/article/viewFile/679/144.

Simunek, Michal, Uwe Hoßfeld, and Volker Wissemann. "'Rediscovery' Revised—the Cooperation of Erich and Armin von Tschermak-Seysenegg in the Context of the 'Rediscovery' of Mendel's Laws in 1899–1901." *Plant Biology* 13, no. 6 (2011): 835–41.

Smale, Melinda, and T. S. Jayne, *Maize Breeding in East and Southern Africa, 1900–2000*, 2020 Vision Focus 12, no. 4. Washington, D.C.: International Food Policy Research Institute, 2004.

Smil, Vaclav. "Global Population and the Nitrogen Cycle." *Scientific American* 277, no. 1 (1997): 76–81.

Smyth, T. J., and P. A. Sanchez. "Effects of Lime, Silicate, and Phosphorus Applications to an Oxisol on Phosphorus Sorption and Ion Retention." *Soil Science Society of America Journal* 44, no. 3 (1980): 500–505.

Smythe, William J. "The Story of the Great Irish Famine 1845–52: A Geographical Perspective." In *Atlas of the Great Irish Famine, 1845–52*, ed. John Crowley, William J. Smyth, and Michael Murphy, 4–12. Cork, Ireland: Cork University Press, 2012.

Social Protection Interagency Cooperation Board. *A Joint Statement on the Role of Social Protection in Responding to the COVID-19 Pandemic*. New York: SPIAC-B, 2020. https://socialprotection.org/sites/default/files/publications_files/Joint%20SPIAC-B%20COVID-19%20statement.pdf.

Soil Science Society of America. "Glossary of Soil Science Terms." Accessed August 14, 2021. https://www.soils.org/publications/soils-glossary#.

Sonenthal, Paul D., Jones Masiye, Noel Kasomekera, Regan H Marsh, Emily B Wroe, Kirstin W Scott, Ruoran Li, et al. "COVID-19 Preparedness in Malawi: A National Facility-Based Critical Care Assessment." *Lancet Global Health*, 8, no. 7 (2020): E890–92. doi.org/10.1016/S2214-109X(20)30250-3.

Soni, Pooja, Sunil S. Gangurde, Alejandro Ortega-Beltran, Rakesh Kumar, Sejal Parmar, Hari K. Sudini, Yong Lei, et al. "Functional Biology and Molecular Mechanisms of Host-Pathogen Interactions for Aflatoxin Contamination in Groundnut (Arachis hypogaea L.) and Maize (Zea mays L.)." *Frontiers in Microbiology* 11 (2020): 227.

Statistical Agency under the President of the Republic of Tajikistan (SA), Tajikistan Ministry of Health (MOH), and ICF International. *Tajikistan Demographic and Health Survey 2012*. Dushanbe, Tajikistan: SA, MOH and Calverton, Md.: ICF International, 2013. http://dhsprogram.com/pubs/pdf/FR279/FR279.pdf.

——. *Tajikistan Demographic and Health Survey 2017*. Dushanbe, Tajikistan: SA, MOH and Rockville, Md.: ICF International, 2018. http://dhsprogram.com/pubs/pdf/FR341/FR341.pdf.

Steffen, Will, Katherine Richardson, Johan Rockström, Sarah E. Cornell, Ingo Fetzer, Elena M. Bennett, Reinette Biggs, et al. "Planetary Boundaries: Guiding Human Development on a Changing Planet." *Science* 347, no. 6223 (2015): 1259855.

Stevenson, James R., Nelson Villoria, Derek Byerlee, Timothy Kelley, and Mywish Maredia. "Green Revolution Research Saved an Estimated 18 to 27 Million Hectares from Being Brought Into Agricultural Production." *Proceedings of the National Academy of Sciences* 110, no. 21 (2013): 8363–68.

Strohmaier, R., J. Rioux, A. Seggel, A. Meybeck, M. Bernoux, M. Salvatore, J. Miranda, and A. Agostini, *The Agriculture Sectors in the Intended Nationally Determined Contributions: Analysis*. Environment and Natural Resources Management Working Paper no. 62. Rome: FAO, 2016.

Support Precision Agriculture. "Laureates Letter Supporting Precision Agriculture (GMOs)." June 29, 2016. https://www.supportprecisionagriculture.org/nobel-laureate-gmo-letter_rjr.html.

Swaminathan, M. S. "An Evergreen Revolution." *Crop Science* 46, no. 5 (2006): 2293–303.

——. "Combating Hunger." *Science* 338, no. 6110 (2012): 1009.

——. "Obituary: Norman E. Borlaug (1914–2009)." *Nature* 461, no. 7266 (2009): 894.

Tadesse, Addis Getachew. "600 Million Africans Go Without Electricity: IEA Chief." *Anadolu Agency*, June 12, 2019. https://www.aa.com.tr/en/africa/600-million-africans-go-without-electricity-iea-chief/1502097.

Taylor, N. J., M. Halsey, E. Gaitán-Solís, P. Anderson, S. Gichuki S, D. Miano, A. Bua, T. Alicai, and C. M. Fauquet. "The VIRCA Project: Virus Resistant Cassava for Africa." *GM Crops Food* 3, no. 2 (2012): 93–103.

TechnoServe. "About Us." Accessed August 28, 2021. https://www.technoserve.org/about-us/.

Tefera, T. "Post-Harvest Losses in African Maize in the Face of Increasing Food Shortage." *Food Security* 4, (2012): 267–77. doi.org/10.1007/s12571-012-0182-3.

Telem, R. S., H. Wani, N. B. Singh, R. Nandini, R. Sadhukhan, S. Bhattacharya, and N. Mandal. "Cisgenics—a Sustainable Approach for Crop Improvement." *Current Genomics* 14, no. 7 (2013): 468–76.

Thurow, Roger. *The First Thousand Days: A Crucial Time for Mothers and Children—and the World*. New York: PublicAffairs, 2016.

Tian, Hanqin, Rongting Xu, Josep G. Canadell, Rona L. Thompson, Wilfried Winiwarter, Parvadha Suntharalingam, Eric A. Davidson, et al. "A Comprehensive Quantification of Global Nitrous Oxide Sources and Sinks." *Nature* 586, no. 7828 (2020): 248–56.

Tibke, Gary. "5. Basic Principles of Wind Erosion Control." *Agriculture, Ecosystems and Environment* 22–23 (1988): 103–22.

Tiehen, Laura. "Taking a Closer Look at Supplemental Nutrition Assistance Program (SNAP) Participation and Expenditures." Amber Waves, U.S. Department of Agriculture, Economic Research Service, August 3, 2020. https://www.ers.usda.gov/amber-waves/2020/august/taking-a-closer-look-at-supplemental-nutrition-assistance-program-snap-participation-and-expenditures/.

Tilakaratna, Ganga, and Chinthani Sooriyamudali. "Food-Based Social Assistance Programs in Sri Lanka: Evolution and Transition to Cash Transfers." In *The 1.5 Billion People Question*, ed. Harold Alderman, Ugo Gentilini, and Ruslan Yemtsov, 151–71. Washington, D.C.: World Bank, 2018.

Tilman, David, Christian Balzer, Jason Hill, and Belinda L. Befort. "Global Food Demand and the Sustainable Intensification of Agriculture." *Proceedings of the National Academy of Sciences* 108, no. 50 (2011): 20260–64.

Times (UK). "Sir Charles Pereira." January 13, 2005. https://www.thetimes.co.uk/article/sir-charles-pereira-vkth68j796r.

Timmer, C. Peter. "Agriculture and Economic Development." In *Handbook of Agricultural Economics*, ed. B. L. Gardner and G. C. Rausser, 2:14871546. Amsterdam: Elsevier/North Holland, 2002.

——. *Food Security and Scarcity: Why Ending Hunger Is So Hard*. Philadelphia: University of Pennsylvania Press, 2015.

——. "Reflections on Food Crises Past." *Food Policy* 35, no. 1 (2010): 1–11.

——. "The Role of Agriculture in 'Catching Up': A Gerschenkronian Perspective." In *Diverse Development Paths and Structural Transformation in the Escape from Poverty*, ed. Martin Andersson and Tobias Axelsson, 69–80. Oxford: Oxford University Press, 2016. https://oxford.universitypressscholarship.com/view/10.1093/acprof:oso/9780198737407.001.0001/acprof-9780198737407.

——. "The Structural Transformation and the Changing Role of Agriculture in Economic Development." Wendt Lecture. American Enterprise Institute, Washington, D.C. , October 30, 2007.

Timmer, Peter, Hastuti, and Sudarno Sumarto. "Evolution and Implementation of the RASTRA Program in Indonesia." In *The 1.5 Billion People Question*, ed. Harold Alderman, Ugo Gentilini, and Ruslan Yemtsov, 265–310. Washington, D.C.: World Bank, 2018.

Timmer, C. Peter, Walter P. Falcon, and Scott R. Pearson. *Food Policy Analysis*. Baltimore: Johns Hopkins University Press for the World Bank, 1983.

Tirivayi, Nyasha, Marco Knowles, and Benjamin Davis. "The Interaction Between Social Protection and Agriculture: A Review of Evidence." *Global Food Security* 10 (2016): 52–62.

Tomich, Thomas P., Peter Kilby, and Bruce F. Johnston. *Transforming Agrarian Economies: Opportunities Seized, Opportunities Missed*. Ithaca, N.Y.: Cornell University Press, 1995. http://www.jstor.org/stable/10.7591/j.ctv1nhms5.

Tropical Agriculture Platform. "Alliance for a Green Revolution in Africa." Accessed August 14, 2021. https://tapipedia.org/content/alliance-green-revolution-africa.

Tubiello, Francesco N., Cynthia Rosenzweig, Giulia Conchedda, Kevin Karl, Johannes Gütschow, Pan Xueyao, Griffiths Obli-Laryea, et al. "Greenhouse Gas Emissions from Food Systems: Building the Evidence Base." *Environmental Research Letters* 16, no. 6 (2021): 065007.

Turner, C., A. Aggarwal, H. Walls, A. Herforth, A. Drewnowski, J. Coates, S. Kalamantianou, et al. "Concepts and Critical Perspectives for Food Environment Research: A Global Framework with Implications for Action in Low- and Middle-Income Countries." *Global Food Security* 18, (2018): 93–101.

Tyagi, Swati, Robin Choudhary, Aparup Das, So Youn Won, and Pratyoosh Shukla. "CRISPR-Cas9 System: A Genome-Editing Tool with Endless Possibilities." *Journal of Biotechnology* 319, no. 1 (2020): 36–53.

UN News. "'If You Don't Feed People, You Feed Conflict,' UN Chief Tells Security Council." *UN News*, March 11, 2021. https://news.un.org/en/story/2021/03/1087032.

United Nations. *The Future We Want: Outcome Document of the United Nations Conference on Sustainable Development, Rio de Janeiro, Brazil, 20–22 June 2012*. New York: United Nations, 2012.

——. "Goal 12: Ensure sustainable consumption and production patterns." Accessed August 28, 2021. https://www.un.org/sustainabledevelopment/sustainable-consumption-production/.

——. *Malawi 2005 Flash Appeal September 2005—March 2006*. New York: Office for the Coordination of Humanitarian Affairs, 2005. https://www.humanitarianresponse.info/sites/www.humanitarianresponse.info/files/documents/files/flash_2005_malawi.pdf.

——. *The Millennium Development Goals Report 2015*. New York: United Nations, 2015. https://www.un.org/millenniumgoals/2015_MDG_Report/pdf/MDG%202015%20rev%20(July%201).pdf.

——. "Secretary-General Calls for 'Uniquely African Green Revolution' in 21st Century, to End Continent's Plague of Hunger, in Addis Ababa Remarks." News release SG/SM/9405-AFR/988, July 6, 2004. https://www.un.org/press/en/2004/sgsm9405.doc.htm.

——. "Secretary-General's Chair Summary and Statement of Action on the UN Food Systems Summit." September 23, 2021. https://www.un.org/en/food-systems-summit/news/making-food-systems-work-people-planet-and-prosperity

——. *Transforming Our World: The 2030 Agenda for Sustainable Development*. New York: UN Division for Sustainable Development Goals, 2015. https://sustainabledevelopment.un.org/post2015/transformingourworld/publication.

——. "Universal Declaration of Human Rights." Accessed August 11, 2021. https://www.un.org/en/about-us/universal-declaration-of-human-rights.

United Nations Children's Fund. "Goal 2: Zero Hunger." Accessed August 24, 2021. https://data.unicef.org/sdgs/goal-2-zero-hunger/.

——. "In Malawi, Health Workers Bring Hope and Experience to COVID-19 Fight." July 27, 2021. https://www.unicef.org/coronavirus/malawi-health-workers-bring-hope-and-experience-covid-19-fight.

——. "Nutrition and Care for Children with Wasting." Accessed August 24, 2021. https://www.unicef.org/nutrition/index_sam.html.

——. *Strategy for Improved Nutrition of Children and Women in Developing Countries*. UNICEF Policy Review. New York: UNICEF, 1990.

——. "Water, Sanitation and Hygiene (WASH)." Accessed August 24, 2021. https://www.unicef.org/wash.

United Nations Children's Fund, World Health Organization, International Bank for Reconstruction and Development/The World Bank. *Levels and Trends in Child Malnutrition: Key Findings of the 2021 Edition of the Joint Child Malnutrition Estimates*. Geneva: World Health Organization, 2021.

United Nations Children's Fund (UNICEF) Office of Research–Innocenti. *Food Systems for Children and Adolescents: Working Together to Secure Nutritious Diets*. New York: UNICEF, Global Alliance for Improved Nutrition, 2018. https://www.unicef.org/media/94086/file/Food-systems-brochure.pdf.

United Nations, Department of Economic and Social Affairs, Population Division. *World Population Prospects 2019, Volume I: Comprehensive Tables* (ST/ESA/SER.A/426). New York: United Nations, 2019.

United Nations Development Programme. *Human Development Report 2005: International Cooperation at a Crossroads—Aid, Trade and Security in an Unequal World*. New York: UNDP, 2005. http://hdr.undp.org/sites/default/files/reports/266/hdr05_complete.pdf.

United Nations Environment Programme. *Becoming #GenerationRestoration: Ecosystem Restoration for People, Nature and Climate*. Nairobi: UNEP, 2021.

——. *Food Waste Index Report 2021*. Nairobi: UNEP, 2021.

United Nations Food Systems Summit 2021. "About the Summit." Accessed August 10, 2021. https://www.un.org/en/food-systems-summit/about.

United Nations Framework Convention on Climate Change. *Nationally Determined Contributions under the Paris Agreement: Synthesis Report by the Secretariat.* Bonn, Germany: UNFCC, February 26, 2021. https://unfccc.int/process-and-meetings/the-paris-agreement/nationally-determined-contributions-ndcs/nationally-determined-contributions-ndcs/ndc-synthesis-report.

United Nations Human Rights Office of the High Commissioner. "International Covenant on Economic, Social and Cultural Rights." Accessed August 11, 2021. https://www.ohchr.org/en/professionalinterest/pages/cescr.aspx.

United Nations Millennium Project Task Force on Hunger. *Halving Hunger: It Can Be Done.* London: Earthscan, 2005.

United Nations Statistics Division. "Official List of MDG Indicators." January 15, 2008. http://mdgs.un.org/unsd/mdg/Host.aspx?Content=Indicators/OfficialList.htm.

United Nations Sustainable Development Goals Knowledge Platform. "Tajikistan Voluntary National Review 2017." Accessed August 29, 2021. https://sustainabledevelopment.un.org/memberstates/tajikistan.

United Nations Water. "Water, Food and Energy." Accessed August 15, 2021. https://www.unwater.org/water-facts/water-food-and-energy/.

United Nations World Food Systems Summit 2021. "About the Summit." Accessed August 29, 2021. https://www.un.org/en/food-systems-summit/about.

United States Agency for International Development (USAID). "Community-Based Management of Acute Malnutrition: Technical Guidance Brief." Accessed April 2, 2022. https://www.usaid.gov/sites/default/files/documents/1864/CMAM-technical-guidance-brief-508-revFeb2017.pdf

——. *US International Food Assistance Report 2008.* Washington, D.C.: USAID, 2009.

United States Department of Agriculture. *Food Distribution Program on Indian Reservations.* Washington, D.C.: USDA, January 2020. https://www.fns.usda.gov/fdpir/fdpir-fact-sheet

——. "Farming and Farm Income." Accessed August 24, 2021. https://www.ers.usda.gov/data-products/ag-and-food-statistics-charting-the-essentials/farming-and-farm-income/.

——. *Soil Taxonomy, a Basic System of Soil Classification for Making and Interpreting Soil Surveys.* 2nd ed. Washington, D.C.: USDA, 1999. https://www.nrcs.usda.gov/Internet/FSE_DOCUMENTS/nrcs142p2_051232.pdf.

United States Department of Agriculture. Economic Research Service. "Ag and Food Sectors and the Economy." Accessed August 24, 2021. https://www.ers.usda.gov/data-products/ag-and-food-statistics-charting-the-essentials/ag-and-food-sectors-and-the-economy/.

United States Department of Agriculture. National Agricultural Statistics Service. *Crop Production 2019 Summary.* Washington, D.C.: USDA, January 2020.

——. *Crop Production Historical Track Records.* Washington, D.C.: USDA, April 2018.

United States Department of Agriculture. Natural Resources Conservation Service. "Erosion." Accessed August 14, 2021. https://www.nrcs.usda.gov/wps/portal/nrcs/main/national/landuse/crops/erosion/.

——. "Soil Biology." Accessed August 14, 2021. https://www.nrcs.usda.gov/wps/portal/nrcs/main/soils/health/biology/.

United States Environmental Protection Agency. *Advancing Sustainable Materials Management: 2018 Fact Sheet. Assessing Trends in Materials Generation and Management in the United States, December 2020.* Washington, D.C.: EPA, 2020. https://www.epa.gov/sites/default/files/2021-01/documents/2018_ff_fact_sheet_dec_2020_fnl_508.pdf.

——. "Understanding Global Warming Potentials." Accessed August 23, 2021. https://www.epa.gov/ghgemissions/understanding-global-warming-potentials.

United States Geological Survey. Mineral Commodity Summaries, *Phosphate Rock.* Washington, D.C.: U.S. Geological Survey, January 2020. https://pubs.usgs.gov/periodicals/mcs2020/mcs2020-phosphate.pdf.

United States Treaties and Other International Agreements. Paris Agreement, December 12, 2015, U. S. T. I-54113. https://unfccc.int/sites/default/files/english_paris_agreement.pdf.

University of Minnesota. "The Beginning of the Green Revolution." Last modified June 12, 2004. https://web.archive.org/web/20041227090100/http://www.coafes.umn.edu/The_Beginning_of_the_Green_Revolution.html.

van der Ploeg, Rienk R., Wolfgang Böhm, and Mary Beth Kirkham. "On the Origin of the Theory of Mineral Nutrition of Plants and the Law of the Minimum." *Soil Science Society of America Journal* 63, no. 5 (1999): 1055–62.

Van Es, Harold. "A New Definition of Soil." *CSA News* 62, no. 10 (2017): 20–21.

Vanlauwe, B., A. Bationo, J. Chianu, K. E. Giller, R. Merckx, U. Mokwunye, O. Ohiokpehai et al. "Integrated soil fertility management: operational definition and consequences for implementation and dissemination." *Outlook on Agriculture* 39, no. 1 (2010): 17–24.

Vanlauwe, Bernard, and Ken E. Giller. "Popular Myths Around Soil Fertility Management in Sub-Saharan Africa." *Agriculture, Ecosystems and Environment* 116, no. 1–2 (2006): 34–46.

Vanlauwe, Bernard, J. Wendt, Ken E. Giller, Marc Corbeels, B. Gerard, and Christian Nolte. "A Fourth Principle Is Required to Define Conservation Agriculture in Sub-Saharan Africa: The Appropriate Use of Fertilizer to Enhance Crop Productivity." *Field Crops Research* 155 (2014): 10–13.

Vaughan, Duncan A., Bao-Rong Lu, and Norihiko Tomooka. "The Evolving Story of Rice Evolution." *Plant Science* 174, no. 4 (2008): 394–408.

Vermeulen, S. J., J. R. Porter, and E. Bennetzen. "Climate-Smart Food Systems." Plenary paper presented at 3rd Climate Smart Agriculture Global Science Conference, Montpellier, France, March 16–18, 2015.

Vidal, John. "'If It Comes, It Will Overwhelm Us': Malawi Braces for Coronavirus." *Guardian*, April 3, 2020. https://www.theguardian.com/global-development/2020/apr/03/if-it-comes-it-will-overwhelm-us-malawi-braces-for-coronavirus.

Vieilledent, Ghislain, Clovis Grinand, Fety A. Rakotomalala, Rija Ranaivosoa, Jean-Roger Rakotoarijaona, Thomas F. Allnutt, and Frédéric Achard. "Combining Global Tree Cover Loss Data with Historical National Forest Cover Maps to Look at Six Decades of Deforestation and Forest Fragmentation in Madagascar." *Biological Conservation* 222 (2018): 189–97. doi.org/10.1016/j.biocon.2018.04.008.

Vireak, Thou. "2020 Rice Exports Nearly 3.6M Tonnes." *Phnom Penh Post*, January 4, 2021. https://www.phnompenhpost.com/business/2020-rice-exports-nearly-36m-tonnes.

Vitousek, Peter M., John D. Aber, Robert W. Howarth, Gene E. Likens, Pamela A. Matson, David W. Schindler, William H. Schlesinger, and David G. Tilman. "Human Alteration of the Global Nitrogen Cycle: Sources and Consequences." *Ecological Applications* 7, no. 3 (1997): 737–50.

Vogt, William. *Road to Survival*. New York: Sloane Associates, 1948.

Waltz, Emily. "Gene-Edited CRISPR Mushroom Escapes US Regulation." *Nature* 532, no. 7599 (2016): 293.

——. "With CRISPR and Machine Learning, Startups Fast-Track Crops to Consume Less, Produce More." *Nature Biotechnology* 37, no. 11 (2019): 1251.

Wang, Bin, Xiao Luo, Young-Min Yang, Weiyi Sun, Mark A. Cane, Wenju Cai, Sang-Wook Yeh, and Jian Liu. "Historical Change of El Niño Properties Sheds Light on Future Changes of Extreme El Niño." *Proceedings of the National Academy of Sciences* 116, no. 45 (2019): 22512–17.

Wang, Maria. *Grow Your Own? Rice Self-sufficiency in Timor-Leste*. SIPA-12-001.0. New York: Columbia University School of International and Public Affairs, 2012. http://ccnmtl.columbia.edu/projects/caseconsortium/casestudies/84/casestudy/www/layout/case_id_84.html.

Wassmann, R., H-U. Neue, R. S. Lantin, L. V. Buendia, and H. Rennenberg. "Characterization of Methane Emissions from Rice Fields in Asia. I. Comparison Among Field Sites in Five Countries." *Nutrient Cycling in Agroecosystems* 58, no. 1 (2000): 1–12.

Weaver, Warren, and J. George Harrar. "Research on Rice." October 21, 1954. Appendix 1 in Rockefeller Foundation Board Meeting, November 30—December 1, 1954. https://iiif.rockarch.org/pdfs/7CGpU9AhiMfZGuQsr5L64K.

Whitfield, Stephen, Andrew Juan Challinor, and Robert M. Rees. "Frontiers in Climate Smart Food Systems: Outlining the Research Space." *Frontiers in Sustainable Food Systems* 2 (2018): 2.

Whitman, Elizabeth. "A Land Without Water: The Scramble to Stop Jordan from Running Dry." *Nature* 573 (2019): 20–23.

Willett, Walter, Johan Rockström, Brent Loken, Marco Springmann, Tim Lang, Sonja Vermeulen, Tara Garnett, et al. "Food in the Anthropocene: The EAT–Lancet Commission on Healthy Diets from Sustainable Food Systems." *Lancet* 393, no. 10170 (2019): 447–92.

Willett, Walter, and Patrick J. Skerrett. *Eat, Drink, and Be Healthy: The Harvard Medical School Guide to Healthy Eating*. New York: Simon and Schuster, 2017.

Williams, Liana, and Rob Cramb. "Adapting the Green Revolution for Laos." In *White Gold: The Commercialisation of Rice Farming in the Lower Mekong Basin*, ed. Rob Cramb, 121–49. Singapore: Palgrave Macmillan, 2020.

Winterflood, Julia. "Post-Pandemic, Will Bali Rethink Tourism?" *Diplomat*, June 10, 2020. https://thediplomat.com/2020/06/post-pandemic-will-bali-rethink-tourism/.

World Bank. "Climate-Smart Agriculture." Accessed August 23, 2021. https://www.worldbank.org/en/topic/climate-smart-agriculture.

——. *Enabling the Business of Agriculture 2019*. Washington, D.C.: World Bank, 2019.

——. "Food Security and COVID-19, Brief." Accessed October 15, 2020. https://www.worldbank.org/en/topic/agriculture/brief/food-security-and-covid-19.

——. "GNI per Capita, Atlas Method (Current US$)—Malawi." Accessed August 24, 2021. https://data.worldbank.org/indicator/NY.GNP.PCAP.CD?locations=MW.

——. "Joint Statement on COVID-19 Impacts on Food Security and Nutrition," April 21, 2020. https://www.worldbank.org/en/news/statement/2020/04/21/joint-statement-on-covid-19-impacts-on-food-security-and-nutrition.

——. "Livestock and Dairy Development Project (Bangladesh)." Accessed August 23, 2021. https://projects.worldbank.org/en/projects-operations/project-detail/P161246.

——. "Madagascar Overview." Last updated July 31, 2020. https://www.worldbank.org/en/country/madagascar/overview.

——. *Malawi—Poverty and Vulnerability Assessment: Investing in Our Future*. Washington, D.C.: World Bank, December 2007.

——. "Prevalence of Stunting, Height for Age (% of Children Under 5)—Afghanistan, Yemen, Rep., Burundi, Timor-Leste." Accessed August 28, 2021. https://data.worldbank.org/indicator/SH.STA.STNT.ZS?end=2017&locations=AF-YE-BI-TL&start=1997.

——. "Prevalence of Stunting, Height for Age (% of Children Under 5)—Malawi." Accessed August 24, 2021. https://data.worldbank.org/indicator/SH.STA.STNT.ZS?locations=MW.

——. *World Development Report 2008: Agriculture for Development*. Washington, D.C.: World Bank, 2007. https://openknowledge.worldbank.org/handle/10986/5990.

——. *World Development Report 2016: Digital Dividends*. Washington, D.C.: World Bank, 2016. https://www.worldbank.org/en/publication/wdr2016.

——. *World Development Report 2020: Trading for Development in the Age of Global Value Chains*. Washington, D.C.: World Bank, 2020. https://www.worldbank.org/en/publication/wdr2020.

World Bank, International Finance Corporation and Multilateral Investment Guarantee Agency. *World Bank Group Climate Change Action Plan 2016–2020*. Washington, D.C.: World Bank, 2016.

World Commission on Environment and Development. *Our Common Future*. New York: United Nations, 1987. https://sustainabledevelopment.un.org/content/documents/5987our-common-future.pdf.

World Food Prize Foundation. "1996: Beachell and Khush." Accessed August 15, 2021. https://www.worldfoodprize.org/en/laureates/19871999_laureates/1996_beachell_and_khush.

——. "2004: Jones and Yuan." Accessed August 15, 2021. https://www.worldfoodprize.org/en/laureates/20002009_laureates/2004_jones_and_yuan/.

World Food Programme. "Cash Assistance." *WFP Year in Review: The Last Resort, the First Defence*, 2018. Accessed August 28, 2021. https://publications.wfp.org/2018/en/annual-report/section2.html#Cash_assistance.

——. "Futures of 370 Million Children in Jeopardy as School Closures Deprive Them of School Meals—UNICEF and WFP." News release, April 29, 2020. https://www.wfp.org/news/futures-370-million-children-jeopardy-school-closures-deprive-them-school-meals-unicef-and-wfp.

——. "Global Monitoring of School Meals During COVID-19 School Closures." https://cdn.wfp.org/2020/school-feeding-map/?_ga=2.72314643.1844293323.1604985491-67437893.1603343218.

——. *The Impact of School Feeding Programmes*. Rome: WFP, 2019. https://docs.wfp.org/api/documents/WFP-0000102338/download/.

——. *Making the Mid-day Meals Functional Following School Reopening, COVID-19 Response*. New Dehli: WFP, 2020. https://www.wfp.org/publications/wfp-india-making-mid-day-meals-functional-following-school-re-opening-covid-19.

——. "Smallholder Support." *WFP Year in Review 2018: The Last Resort, the First Defence*. Accessed August 28, 2021. https://publications.wfp.org/2018/en/annual-report/section2.html#Smallholder_support.

——. "WFP and Australia Provide Funds to Help 80,000 School Children Receive Take-Home Food Rations Amidst School Closures." News release, May 14, 2020. https://www.wfp.org/news/wfp-and-australia-provide-funds-help-80000-school-children-receive-take-home-food-rations.

——. "The WFP Food Basket." Accessed August 11, 2021. https://www.wfp.org/wfp-food-basket.

——. *WFP School Feeding Programmes in 2018*. Rome: WFP, 2019. https://docs.wfp.org/api/documents/WFP-0000107178/download/.

World Food Summit. "Rome Declaration on World Food Security." November 13–17, 1996. http://www.fao.org/3/w3613e/w3613e00.htm.

World Health Organization. *The Double Burden of Malnutrition*. Policy brief. Geneva: World Health Organization, 2017.

——. *Mycotoxins*. May 9, 2018. https://www.who.int/news-room/fact-sheets/detail/mycotoxins.

——. "Obesity and Overweight." June 9, 2021. https://www.who.int/news-room/fact-sheets/detail/obesity-and-overweight.

——. "Vitamin A Deficiency." Accessed August 24, 2021. https://www.who.int/data/nutrition/nlis/info/vitamin-a-deficiency.

——. *WHA Global Nutrition Targets 2025: Wasting Policy Brief*. Washington, D.C.: WHO, 2014. https://apps.who.int/iris/bitstream/handle/10665/149023/WHO_NMH_NHD_14.8_eng.pdf.

——. "WHO Director-General's Opening Remarks at the Media Briefing on COVID-19–11 March 2020." March 11, 2020. https://www.who.int/director-general/speeches/detail/who-director-general-s-opening-remarks-at-the-media-briefing-on-covid-19---11-march-2020.

World Resources Institute. "Country Rankings." Aqueduct Beta. Accessed August 14, 2021. https://www.wri.org/applications/aqueduct/country-rankings/.

World Water Assessment Programme. *The United Nations World Water Development Report 3: Water in a Changing World*. Paris: UNESCO and London: Earthscan, 2009.

Worldometer. "Malawi." Accessed July 31, 2021. https://www.worldometers.info/coronavirus/country/malawi/.

WRAP. "The Courtauld Commitment 2030." Accessed August 28, 2021. https://wrap.org.uk/taking-action/food-drink/initiatives/courtauld-commitment.

——. *The Courtauld Commitment Annual Report*. Banbury, UK: WRAP, 2021. https://wrap.org.uk/sites/default/files/2021-07/WRAP-Courtauld-Commitment-Annual-Report-2021.pdf.

——. *Labelling Guidance—Best Practice on Food Date Labelling and Storage Advice*. Banbury, UK: WRAP, November 2019. https://wrap.org.uk/sites/default/files/2020-07/WRAP-Food-labelling-guidance.pdf.

Wu, Shellen X. "Yuan Longping (1930–2021)." *Nature* 595, no. 7865 (2021): 26.

WWF. "Cerrado-Threats." Critical Ecosystem Partnership Fund. Accessed September 5, 2021. https://www.cepf.net/our-work/biodiversity-hotspots/cerrado/threats.

Xinhuanet. "Xi Focus: Xi Confers Highest State Honors on Individuals Ahead of National Day." September 29, 2019. http://www.xinhuanet.com/english/2019-09/29/c_138433793.htm.

Xu, Kenong, Xia Xu, Takeshi Fukao, Patrick Canlas, Reycel Maghirang-Rodriguez, Sigrid Heuer, Abdelbagi M. Ismail, Julia Bailey-Serres, Pamela C. Ronald, and David J. Mackill. "Sub1A Is an Ethylene-Response-Factor-Like Gene That Confers Submergence Tolerance to Rice." *Nature* 442, no. 7103 (2006): 705–8.

Yaffe-Bellany, David, and Michael Corker. "Dumped Milk, Smashed Eggs, Plowed Vegetables: Food Waste of the Pandemic." *New York Times*, April 11, 2020. https://www.nytimes.com/2020/04/11/business/coronavirus-destroying-food.html.

Yale University. "Cambodia Genocide Program." Accessed August 29, 2021. https://gsp.yale.edu/case-studies/cambodian-genocide-program.

Yoon, Jim, Christian Klassert, Philip Selby, Thibaut Lachaut, Stephen Knox, Nicolas Avisse, Julien Harou, et al. "A Coupled Human–Natural System Analysis of Freshwater Security under Climate and Population Change." *Proceedings of the National Academy of Sciences* 118, no. 14 (2021): e2020431118. doi.org/10.1073/pnas.2020431118.

Yudelman, Montague. "Agriculture in Integrated Rural Development: The Experience of the World Bank." *Food Policy* 1, no. 5 (1976): 367–81.

Zaidi, Syed Shan-e-Ali, Hervé Vanderschuren, Matin Qaim, Magdy M. Mahfouz, Ajay Kohli, Shahid Mansoor, and Mark Tester. "New Plant Breeding Technologies for Food Security." *Science* 363, no. 6434 (2019): 1390–91.

Zandstra, Hubert, Kenneth Swanberg, Carlos Zulberti, and Barry Nestel. *Cáqueza: Living Rural Development*. Ottawa, Canada: International Development Research Centre, 1979.

Zsögön, Agustin, Tomáš Èermák, Emmanuel Rezende Naves, Marcela Morato Notini, Kai H. Edel, Stefan Weinl, Luciano Freschi, Daniel F. Voytas, Jörg Kudla, and Lázaro Eustáquio Pereira Peres. "De Novo Domestication of Wild Tomato Using Genome Editing." *Nature Biotechnology* 36 (2018): 1211–16. doi.org/10.1038/nbt.4272.

Zuidema, Pieter A., and René G. A. Boot. "Demography of the Brazil Nut Tree (*Bertholletia excelsa*) in the Bolivian Amazon: Impact of Seed Extraction on Recruitment and Population Dynamics." *Journal of Tropical Ecology* 18, no. 1 (2002): 1–31.

INDEX

Printed and bound by CPI Group (UK) Ltd, Croydon, CR0 4YY

11/05/2023

03218084-0001